LONDON MATHEMATICAL SOCIETY LECTURE NOTE SERIES

Managing Editor: Professor J.W.S. Cassels, Department of Pure Mathematics and Mathematical Statistics, University of Cambridge, 16 Mill Lane, Cambridge CB2 1SB, England

The books in the series listed below are available from booksellers, or, in case of difficulty, from Cambridge University Press.

4 Algebraic topology, J.F. ADAMS
5 Commutative algebra, J.T. KNIGHT
8 Integration and harmonic analysis on compact groups, R.E. EDWARDS
11 New developments in topology, G. SEGAL (ed)
12 Symposium on complex analysis, J. CLUNIE & W.K. HAYMAN (eds)
13 Combinatorics, T.P. McDONOUGH & V.C. MAVRON (eds)
16 Topics in finite groups, T.M. GAGEN
17 Differential germs and catastrophes, Th. BROCKER & L. LANDER
18 A geometric approach to homology theory, S. BUONCRISTIANO, C.P. ROURKE & B.J. SANDERSON
20 Sheaf theory, B.R. TENNISON
21 Automatic continuity of linear operators, A.M. SINCLAIR
23 Parallelisms of complete designs, P.J. CAMERON
24 The topology of Stiefel manifolds, I.M. JAMES
25 Lie groups and compact groups, J.F. PRICE
26 Transformation Groups, C. KOSNIOWSKI (ed)
27 Skew field constructions, P.M. COHN
29 Pontryagin duality and the structure of LCA groups, S.A. MORRIS
30 Interaction models, N.L. BIGGS
31 Continuous crossed products and type III von Neumann algebras, A. VAN DAELE
32 Uniform algebras and Jensen measures, T.W. GAMELIN
34 Representation theory of Lie groups, M.F. ATIYAH *et al*
35 Trace ideals and their applications, B. SIMON
36 Homological group theory, C.T.C. WALL (ed)
37 Partially ordered rings and semi-algebraic geometry, G.W. BRUMFIEL
38 Surveys in combinatorics, B. BOLLOBAS (ed)
39 Affine sets and affine groups, D.G. NORTHCOTT
40 Introduction to H_p spaces, P.J. KOOSIS
41 Theory and applications of Hopf bifurcation, B.D. HASSARD, N.D. KAZARINOFF & Y-H. WAN
42 Topics in the theory of group presentations, D.L. JOHNSON
43 Graphs, codes and designs, P.J. CAMERON & J.H. VAN LINT
44 Z/2-homotopy theory, M.C. CRABB
45 Recursion theory: its generalisations and applications, F.R. DRAKE & S.S. WAINER (eds)
46 p-adic analysis: a short course on recent work, N. KOBLITZ
47 Coding the universe, A. BELLER, R. JENSEN & P. WELCH
48 Low-dimensional topology, R. BROWN & T.L. THICKSTUN (eds)
49 Finite geometries and designs, P. CAMERON, J.W.P. HIRSCHFELD & D.R. HUGHES (eds)
50 Commutator calculus and groups of homotopy classes, H.J. BAUES
51 Synthetic differential geometry, A. KOCK
52 Combinatorics, H.N.V. TEMPERLEY (ed)
54 Markov processes and related problems of analysis, E.B. DYNKIN
55 Ordered permutation groups, A.M.W. GLASS
56 Journees arithmetiques, J.V. ARMITAGE (ed)
57 Techniques of geometric topology, R.A. FENN
58 Singularities of smooth functions and maps, J.A. MARTINET
59 Applicable differential geometry, M. CRAMPIN & F.A.E. PIRANI
60 Integrable systems, S.P. NOVIKOV *et al*
61 The core model, A. DODD

62 Economics for mathematicians, J.W.S. CASSELS
63 Continuous semigroups in Banach algebras, A.M. SINCLAIR
64 Basic concepts of enriched category theory, G.M. KELLY
65 Several complex variables and complex manifolds I, M.J. FIELD
66 Several complex variables and complex manifolds II, M.J. FIELD
67 Classification problems in ergodic theory, W. PARRY & S. TUNCEL
68 Complex algebraic surfaces, A. BEAUVILLE
69 Representation theory, I.M. GELFAND *et al*
70 Stochastic differential equations on manifolds, K.D. ELWORTHY
71 Groups - St Andrews 1981, C.M. CAMPBELL & E.F. ROBERTSON (eds)
72 Commutative algebra: Durham 1981, R.Y. SHARP (ed)
73 Riemann surfaces: a view towards several complex variables, A.T. HUCKLEBERRY
74 Symmetric designs: an algebraic approach, E.S. LANDER
75 New geometric splittings of classical knots, L. SIEBENMANN & F. BONAHON
76 Spectral theory of linear differential operators and comparison algebras, H.O. CORDES
77 Isolated singular points on complete intersections, E.J.N. LOOIJENGA
78 A primer on Riemann surfaces, A.F. BEARDON
79 Probability, statistics and analysis, J.F.C. KINGMAN & G.E.H. REUTER (eds)
80 Introduction to the representation theory of compact and locally compact groups, A. ROBERT
81 Skew fields, P.K. DRAXL
82 Surveys in combinatorics, E.K. LLOYD (ed)
83 Homogeneous structures on Riemannian manifolds, F. TRICERRI & L. VANHECKE
84 Finite group algebras and their modules, P. LANDROCK
85 Solitons, P.G. DRAZIN
86 Topological topics, I.M. JAMES (ed)
87 Surveys in set theory, A.R.D. MATHIAS (ed)
88 FPF ring theory, C. FAITH & S. PAGE
89 An F-space sampler, N.J. KALTON, N.T. PECK & J.W. ROBERTS
90 Polytopes and symmetry, S.A. ROBERTSON
91 Classgroups of group rings, M.J. TAYLOR
92 Representation of rings over skew fields, A.H. SCHOFIELD
93 Aspects of topology, I.M. JAMES & E.H. KRONHEIMER (eds)
94 Representations of general linear groups, G.D. JAMES
95 Low-dimensional topology 1982, R.A. FENN (ed)
96 Diophantine equations over function fields, R.C. MASON
97 Varieties of constructive mathematics, D.S BRIDGES & F. RICHMAN
98 Localization in Noetherian rings, A.V. JATEGAONKAR
99 Methods of differential geometry in algebraic topology, M. KAROUBI & C. LERUSTE
100 Stopping time techniques for analysts and probabilists, L. EGGHE
101 Groups and geometry, ROGER C. LYNDON
102 Topology of the automorphism group of a free group, S.M. GERSTEN
103 Surveys in combinatorics 1985, I. ANDERSON (ed)
104 Elliptic structures on 3-manifolds, C.B. THOMAS
105 A local spectral theory for closed operators, I. ERDELYI & WANG SHENGWANG
106 Syzygies, E.G. EVANS & P. GRIFFITH
107 Compactification of Siegel moduli schemes, C-L. CHAI
108 Some topics in graph theory, H.P. YAP
109 Diophantine Analysis, J. LOXTON & A. VAN DER POORTEN (eds)
110 An introduction to surreal numbers, H. GONSHOR
111 Analytical and geometric aspects of hyperbolic space, D.B.A.EPSTEIN (ed)
112 Low-dimensional topology and Kleinian groups, D.B.A. EPSTEIN (ed)
113 Lectures on the asymptotic theory of ideals, D. REES
114 Lectures on Bochner-Riesz means, K.M. DAVIS & Y-C. CHANG
115 An introduction to independence for analysts, H.G. DALES & W.H. WOODIN
116 Representations of algebras, P.J. WEBB (ed)
117 Homotopy theory, E. REES & J.D.S. JONES (eds)
118 Skew linear groups, M. SHIRVANI & B. WEHRFRITZ

London Mathematical Society Lecture Note Series. 124

Lie groupoids and Lie algebroids in Differential Geometry

K. MACKENZIE
Department of Mathematics, University of Melbourne

The right of the University of Cambridge to print and sell all manner of books was granted by Henry VIII in 1534. The University has printed and published continuously since 1584.

CAMBRIDGE UNIVERSITY PRESS
Cambridge
New York New Rochelle Melbourne Sydney

CAMBRIDGE UNIVERSITY PRESS
Cambridge, New York, Melbourne, Madrid, Cape Town, Singapore, São Paulo

Cambridge University Press
The Edinburgh Building, Cambridge CB2 8RU, UK

Published in the United States of America by Cambridge University Press, New York

www.cambridge.org
Information on this title: www.cambridge.org/9780521348829

First published 1987
Re-issued in this digitally printed version 2007

A catalogue record for this publication is available from the British Library

ISBN 978-0-521-34882-9 paperback

CONTENTS

INTRODUCTION

The concept of groupoid is one of the means by which the twentieth century reclaims the original domain of application of the group concept. The modern, rigorous concept of group is far too restrictive for the range of geometrical applications envisaged in the work of Lie. There have thus arisen the concepts of Lie pseudogroup, of differentiable and of Lie groupoid, and of principal bundle - as well as various related infinitesimal concepts such as Lie equation, graded Lie algebra and Lie algebroid - by which mathematics seeks to acquire a precise and rigorous language in which to study the symmetry phenomenae associated with geometrical transformations which are only locally defined.

This book is both an exposition of the basic theory of differentiable and Lie groupoids and their Lie algebroids, with an emphasis on connection theory, and an account of the author's work, not previously published, on the abstract theory of transitive Lie algebroids, their cohomology theory, and the integrability problem and its relationship to connection theory.

The concept of groupoid was introduced into differential geometry by Ehresmann in the 1950's, following his work on the concept of principal bundle. Indeed the concept of Lie groupoid - a differentiable groupoid with a local triviality condition - is, modulo some details, equivalent to that of principal bundle. Since the appearance of Kobayashi and Nomizu (1963), the concept of principal bundle has been recognized as a natural setting for the formulation and study of general geometric problems; both the theory of G-structures and the theory of general connections are set in the context of principal bundles, and so too is much work on gauge theory. As an analytical tool in differential geometry, the importance of the principal bundle concept undoubtedly goes back to the fact that it abstracts the moving frame technique of Cartan. An important secondary aim of these notes is to establish that the theory of principal bundles and general connection theory is illuminated and clarified by its groupoid formulation; it will be shown in Chapter III that the Lie theory of Lie groupoids with a given base is coextensive with the standard theory of connections.

To summarize very briefly the work done on groupoids within differential geometry since Ehresmann, there are the following two main areas.

(1) Work on groupoid theory itself. The construction by Pradines (1966, 1967, 1968a,b) of a first-order infinitesimal invariant of a differential groupoid, the Lie algebroid, and his announcement of a full Lie theory for differentiable

groupoids, paralleling the Lie theory of Lie groups and Lie algebras.

Proofs of many of the Lie theoretic results announced by Pradines were given by Almeida (1980); the construction of counterexamples to the integrability of Lie algebroids was announced by Almeida and Molino (1985).

The general theory of differentiable and microdifferentiable groupoids is a generalization of foliation theory, and the techniques used are largely foliation - theoretic in character.

A very recent article on general differentiable groupoids, expanding considerably on Pradines (1966), is Pradines (1986).

(2) Work in which Lie groupoids have been used as a tool or language. Here there is firstly a range of work which may be somewhat loosely described as the theory of Lie equations and Spencer cohomology - see, for example, Ngô Van Quê (1967, 1968, 1969), Kumpera and Spencer (1972) and Kumpera (1975). Secondly, much of the theory of higher-order connections is in terms of Lie groupoids - see Virsik (1969, 1971), Bowshell (1971), and ver Eecke (1981), for example.

Much of this work has also contributed to the theory of differentiable and Lie groupoids per se.

Outside of differential geometry, there are the following major areas.

(3) The work of Brown and a number of co-authors on the theory of general topological groupoids. See Brown and Hardy (1976) and Brown et al (1976).

For references to the considerable body of work by Brown, Higgins and others on multiple groupoid structures and homotopy theory, see the survey by Brown ("Some non-abelian methods in homotopy theory and homological algebra", in Categorical Topology: Proc. Conf. Toledo, Ohio, 1983. Ed. H.L. Bentley et al, Helderman-Verlag, Berlin (1984), 108-146).

(4) Work on the algebraic theory of groupoids, and their application to problems in group theory. See Higgins (1971).

(5) Work on the cohomology of classifying spaces associated with groupoids, usually having non-Hausdorff, sheaf-like topologies. See the survey by Stasheff (1978).

(6) A rapidly growing body of work on the C*-algebras associated with a topological or measured groupoid. See Renault (1980) and Connes ("A survey of

foliations and operator algebras." Proc. Symp. Pure Mathematics, 38 (1), 1982, 521-628. American Mathematical Society, Providence, R.I.).

For the measure theory of groupoids and its use in functional analytic questions see also Seda (1980) and references given there.

A bibliography on all aspects of groupoid theory up to 1976 is given in Brown and Hardy (1976) and Brown et al (1976). The list of references to the present work is not a bibliography.

The primary aim of this book is to present certain new results in the theory of transitive Lie algebroids, and in their connection and cohomology theory; we intend that these results establish a significant theory of abstract Lie algebroids independent of groupoid theory. As a necessary preliminary, we give the first full account of the basic theory of differentiable groupoids and Lie algebroids, with emphasis on the case of Lie groupoids and transitive Lie algebroids. One important secondary aim has already been mentioned - to integrate the standard theory of connections in principal bundles with the Lie theory of Lie groupoids on a given base, to the benefit of both theories. As a matter of exposition, we describe the principal bundle versions of groupoid concepts and constructions whenever this appears to clarify the groupoid theory.

The concept of Lie algebroid was introduced by Pradines (1967), as the first-order invariant attached to a differentiable groupoid, generalizing the construction of the Lie algebra of a Lie group. In the case of Lie groupoids, the Lie algebroid is the Atiyah sequence of the corresponding principal bundle, as introduced by Atiyah (1957). For a differentiable groupoid arising from a foliation, the Lie algebroid is the corresponding involutive distribution. The closely related concept of Lie pseudo-algebra has also been introduced by a number of authors, under a variety of names - see III§2 for references.

In Chapter IV, and in Chapter III§§2,5,7 we undertake the first development of the abstract theory of transitive Lie algebroids and of their connection and cohomology theory. The condition of transitivity for Lie algebroids is related to that of local triviality for groupoids - for example, the Lie algebroid of a differentiable groupoid on a connected base is transitive iff the groupoid is locally trivial. (However that the transitivity condition implies a true local triviality condition for the Lie algebroid is non-trivial - see IV§4.) A transitive Lie algebroid is naturally written as an exact sequence $L \rightarrowtail A \twoheadrightarrow TB$, where TB is the tangent bundle of the base manifold and L is, a priori, a vector bundle whose fibres are Lie algebras; it is, in fact, a Lie algebra bundle.

Exact sequences are generally classified by cohomology in the second degree. Using this point of view, we develop two separate cohomological classifications of transitive Lie algebroids. Firstly, there is a "global" classification in terms of curvature forms and what we propose to call adjoint connections. A transitive Lie algebroid $L \rightarrowtail A \twoheadrightarrow TB$ is characterized by the curvature 2-form $\overline{R}_\gamma : TB \oplus TB \to L$ of any connection $\gamma: TB \to A$ in it, together with the connection ∇^γ in the Lie algebra bundle L induced by γ. Thus, for example, we obtain simple algebraic criteria for a 2-form, with values in a Lie algebra bundle, to be the curvature of a connection in a Lie algebroid. The criteria are a Bianchi identity and a compatibility condition between the given form and the curvature properties of the Lie algebra bundle. At the simplest level, this generalizes the observation that the curvature form of a connection in a principal bundle with abelian structure group must be closed. In cohomological terms this classification is a specialization of the classification of non-abelian extensions of Lie algebroids.

Secondly we give a "local" classification of transitive Lie algebroids by what we propose to call transition forms. These are Lie algebra valued Maurer-Cartan forms. The classification is analogous to that of principal bundles by transition functions, and indeed for a Lie algebroid which is given as the Atiyah sequence of a principal bundle, the transition forms may be obtained as the right-derivatives of transition functions for the bundle. This classification establishes that transitive Lie algebroids are locally trivial in a sense precisely analogous to that true of Lie groupoids. The author obtained this result in 1979 at a time when it was generally believed that all transitive Lie algebroids were the Lie algebroids of Lie groupoids; it is now known that this is not so, and this classification is the more interesting. The key to this result is that a transitive Lie algebroid on a contractible base admits a flat connection, and we obtain this from the cohomological classification of extensions. In turn, the classification of transitive Lie algebroids by systems of transition forms may be regarded as an element in a non-abelian cohomology theory for manifolds with values in Lie algebra bundles, in the same way that the classification of principal bundles by transition functions may be regarded as a cohomological classification.

In §5 of Chapter IV we show that there is a spectral sequence associated in a natural algebraic manner with a transitive Lie algebroid, which generalizes the Leray-Serre spectral sequence for de Rham cohomology of a principal bundle and, in particular, allows coefficients in general vector bundles to be introduced. This algebraization allows the transfer to principal bundle theory of techniques developed for the cohomology of discrete groups and Lie algebras, and we believe it will also provide the correct setting for the study of the cohomology structure of principal bundles with noncompact structure group. Here we only make a beginning on these questions.

In Chapter V we present a cohomological obstruction to the integrability of a transitive Lie algebroid on a simply-connected base. In this case, this obstruction gives a complete resolution of the problem of when a transitive Lie algebroid is the Lie algebroid of a Lie groupoid.

Combining the obstruction to integrability with the global classification of transitive Lie algebroids by curvature forms, we obtain necessary and sufficient conditions for a Lie algebra bundle valued 2-form to be the curvature of a connection in a principal bundle, providing that the base manifold is simply-connected. These conditions generalize and reformulate the integrality lemma of Weil (1958); see also Kostant (1970).

The methods developed in Chapter IV and in Chapter V represent a rather intricate combination of cohomological and connection-theoretic techniques. We believe we have in fact shown that these two subjects are even more inextricably linked, in a nontrivial fashion, than has been realized.

Indeed it should perhaps be emphasized that this is a book about the general theory of connections, since this may not be fully evident from a glance at the table of contents. General connection theory has traditionally taken place on principal bundles, but we argue here that the proper setting for much of connection theory is on a Lie algebroid, and that the relationship between principal bundles and Lie algebroids is best understood by replacing principal bundles by Lie groupoids.

A reader who is interested in the abstract theory of Lie algebroids and/or the integrability obstruction, and who is familiar with principal bundle theory, but does not wish to acquire the Lie groupoid language, could read Chapter III§§2, 5, Appendix A and Chapters IV and V, though they will miss much explanatory material by so doing.

In Chapters I, II and III we give a detailed account of the basic theory of differentiable groupoids and Lie algebroids, with emphasis on the locally trivial case. The presentation is intended to resemble, as far as is possible, the standard treatment of the theory of Lie groups and Lie algebras. Chapter I is an introduction to the algebra of groupoids. In Chapter II we treat topological groupoids, not so much for their own interest - which is considerable - but as a device for setting down the formal content of certain later constructions without the need to address questions of differentiability. Thus - with a few brief exceptions - we address only those matters which have meaning in the differentiable case.

The main business of the book starts in Chapter III and the resemblance between this part of the subject and the standard treatment of Lie groups and Lie algebras will be evident. We first treat questions of differentiability for the constructions of Chapter II and then, in §§2-3, introduce the Lie algebroid of a differentiable groupoid. In §4 we construct the exponential map and use it to compute the Lie algebroids of several important Lie groupoids, central to connection theory. In §6 we establish two of the main results of the Lie theory of Lie groupoids with a given base. In §5 and §7 we present an account of the connection theory of Lie groupoids and transitive Lie algebroids; §5 giving the infinitesimal theory and §7 those aspects which depend on path-lifting or holonomy. In §5 we also begin the classification of transitive Lie algebroids by transition forms.

Much of Chapters I, II and III is the work of other minds. I have given references to the original literature in the text itself, but I have not attempted to write a comparative history. The following features of these chapters are, I believe, new and significant.

——— The construction in II§6 of the monodromy groupoid of a locally trivial topological groupoid, and the proof in III§6 that there is a bijective correspondence between α-connected Lie subgroupoids of a given Lie groupoid, and transitive Lie subalgebroids of its Lie algebroid, and between base-preserving local morphisms of Lie groupoids and base-preserving morphisms of their Lie algebroids.

These results were announced, for general differentiable groupoids and general morphisms, by Pradines (1966, 1967) and proofs in that generality were given by Almeida (1980) and Almeida and Kumpera (1981). The proofs given here make essential use of local triviality to bypass questions of holonomy, and are new and considerably simpler.

——— The circle of ideas concerning frame groupoids of a geometric structure on a vector bundle: The proof of Ngô's theorem III 1.20 by use of III 1.9 - and thus, ultimately, by Pradines' theorem III 1.4; the calculation III 4.7 of the Lie algebroids of isotropy subgroupoids and of the induced representations III 4.8; and the derivation of III 7.11 from these results.

——— The separation of standard connection theory into the infinitesimal connection theory of abstract transitive Lie algebroids (III§5, IV§1) and the path connection theory of locally trivial topological or Lie groupoids (II§7, III§7). The deduction of the Ambrose-Singer theorem (III 7.27) from the correspondence

III 6.1 between α-connected Lie subgroupoids and Lie subalgebroids.

The concept of transition form in III§5, and the results and techniques of Chapters IV and V, have already been referred to above.

Three appendices follow the main text. Appendices B and C are brief summaries of relevant formulas for Lie groups and vector bundles, respectively, and also serve to fix some matters of notation. Appendix A, however, is substantial, and gives a detailed translation of the elementary theory of connections in principal bundles (as given, for example, by Kobayashi and Nomizu (1963) or Greub et al (1973)) into the language of Atiyah sequences. This Appendix is entirely in terms of principal bundles, and makes no use of groupoid concepts. The Atiyah sequence formulation of connection theory has been mentioned in passing by many writers on gauge theory but - to the knowledge of the author - this is the first full account of its equivalence with the usual formulation. Care has been taken with matters of signs, especially since it is necessary to use the right-hand bracket on the Lie algebra of the structure group.

Two major topics have been omitted from these notes. Firstly there is the theory of jet prolongations of differentiable groupoids and Lie algebroids. This is thoroughly treated in existing accounts - see, for example, Kumpera and Spencer (1972), Kumpera (1975) and ver Eecke (1981).

Secondly there is the important body of work revolving around the concept of microdifferentiable groupoid. This is a generalization of the theory of foliations, both in results and techniques. For the construction of the holonomy groupoid of a microdifferentiable groupoid, announced by Pradines (1966), and its applications, see Almeida (1980). Some very brief indications of the results of this theory are included here. The author hopes that this book will also facilitate a wider appreciation of the importance and depth of the general theory of microdifferentiable groupoids. See Pradines (1986) and references given there.

Some nonstandard terminology deserves comment. In I 2.18 and III 2.1 we have used the word "anchor" where Pradines uses "flèche". It seems to us that the English word, "arrow", is overused and colourless. A possible alternative, "transitivity projection", is cumbersome. The anchor ties - or fails to tie - the structure of the groupoid or algebroid to the topology of the base. Secondly, in II 2.22 we use the word "produced" to describe what in principal bundle terms is the bundle $\frac{P \times H}{G}(B,H)$ constructed from a given P(B, G) and a morphism $G \longrightarrow H$. The usual terms "prolongation" and "extension" have other uses in this subject, and "produced" has the virtue of being clearly antonymous to the word "reduced", used to describe the dual concept.

The background needed for this book is slight. A knowledge of the elementary theory of Lie groups and Lie algebras (not including any structure theory), of vector bundles (not including the homotopy classification), and of de Rham cohomology, is essential. Some acquaintance with the theory of connections in principal bundles is desirable, but only so that the purpose of the constructions given here will be clear. For Chapter IV a familiarity with the cohomology theory of either discrete groups of Lie algebras will help, but - as with connection theory - proofs of almost all results are given in full.

This book is designed primarily for those interested in differential geometry. The methods given here are essentially algebraic and since much recent differential geometry is very firmly rooted in analysis, we have given the algebraic constructions in some detail. We feel that the use of algebraic methods to produce cohomological invariants has a substantial history in differential geometry and is capable of much further development.

We use the words 'category' and 'functor' when it is convenient, but we make no actual use of category theory.

In conclusion, there is a point to be made about the need in differential geometry for the general connection theory of principal bundles, as distinct from that merely of vector bundles. So long as one is interested only in geometries with a matrix structure group (that is, in G-structures), the two approaches are, of course, perfectly equivalent. However one of the points of global Lie group theory is that not all Lie groups are realizable as matrix Lie groups (unlike Lie algebras, which always admit faithful finite-dimensional representations), and to work in this generality it is essential to use principal bundles - or Lie groupoids.

Throughout this book we have given most proofs and constructions in considerable detail. In the case of the first three chapters, we have found that even quite simple details can be difficult to supply quickly, on account of the eclectic nature of groupoid theory. In the case of Chapter IV, we have not wanted to presuppose a knowledge of homological algebra. In any case, we believe that there is enough good mathematics to go around, and there seems no reason why anyone should have to do for themselves what the author has done in preparing this book.

This book has been some time in the making. Some of the work recorded here comes from the author's Ph.D. thesis of 1979 written at Monash University and supervised by Dr. Juraj Virsik. I acknowledge with gratitude my debt to Dr. Virsik for proposing the notes of Pradines (1966, 1967, 1968a,b), work well worthy of study. I am also grateful to Professor Jean Pradines, who provided the information which has become III 1.4, and Professor James Stasheff, for detailed stylistic criticism of the material in that thesis in 1979. I am of course myself solely responsible for the form and content of the present book.

Much of the detailed writing of this book was done in 1981-83 as a Queen Elizabeth II Fellow in the Research School of Physical Sciences, Australian National University, and I gratefully acknowledge the benefit of that support.

It is a pleasure to at last thank Professor Antônio Kumpera, for arranging a visit of six months to the IMECC, Universidade Estadual de Campinas, state of São Paulo, Brazil, in 1982, which provided an invaluable opportunity to lecture on much of this material, and for his open hospitality and courtesy there.

I am also grateful to Professor Ronald Brown for much correspondence over the years concerning topological groupoids, and for references to the work of Almeida and Molino.

I would like to acknowledge here the work of three authors which has had a profound effect on the overall orientation of the work presented here, but which has seldom had the specific influence which one can acknowledge in text: the work of van Est (1953, 1955a,b) on the cohomology of Lie groups determined the cohomological approach taken in Chapter IV and elsewhere in the author's work; the paper of Kostant (1970) had a seminal influence on the concept of transition form and the construction of the elements e_{ijk} of Chapter V, as well as on IV§3; and the notes of Koszul (1960) first made the author aware of the power of algebraic methods in differential geometry

For typing and re-typing the whole text with great meticulousness and care I am most grateful to Ms M. Funston of the University of Melbourne. I also wish to record my appreciation of the considerable work done on an earlier version by Mrs H.Daish of the ANU; and of the myriad corrections made to the final copy by Mrs J. Gibson of the University of Durham.

To Lew Luton, who encouraged and fostered this work at several critical stages in its development over many years, and to Margaret Lazner, who gave me her complete support over the final year of its writing, I am profoundly grateful.

The typescript of this work was completed in July 1985; minor changes were made up to February 1987.

To Lew Luton

CHAPTER I THE ALGEBRA OF GROUPOIDS

In this chapter we define groupoids and their morphisms and give the basic algebraic definitions and constructions of subgroupoids, quotient groupoids, kernels of morphisms, products of groupoids and other standard concepts. We do not address the algebraic theory of groupoids for its own sake, and we do not prove any of the deeper results from the algebraic theory.

An interesting algebraic theory of groupoids exists, and was begun by Brandt and by Baer in the 1920's, well before Ehresmann made the concept of groupoid central to his vision of differential geometry. However the algebraic theory is primarily concerned with problems which are largely trivial for categories of transitive groupoids and there is therefore no reason for us to treat it here. See Higgins (1971) for a full account and further references, and Brown (1968) for an account which is more accessible to the non-algebraist, though less comprehensive than Higgins'. Much material on the algebraic theory of groupoids, from a different point of view to that of the work cited above, can be extracted from Ehresmann (1965). See also Clifford and Preston (1961, §3.3).

The examples given in this chapter are examples of topological or differentiable groupoids, presented without their topological, or smooth, structures. We have managed to avoid giving examples which can arise only in the purely algebraic setting.

The development of the algebraic theory of groupoids has been succinctly chronicled by Higgins (1971, pp. 171-172). The examples, as has been said, belong to the topological and differentiable theories, and will be sourced when they reappear in full in later chapters.

§1. Groupoids

A groupoid is a complicated structure and we will spend a little time in giving a full definition and, in the process, introduce the notation to be used in these notes.

Groups commonly arise as the structures natural to sets of automorphisms of mathematical objects. In differential geometry one frequently encounters families of mutually isomorphic objects, the basic example being the set of tangent spaces to a manifold, and the way in which the members of such a family relate to each other is captured by taking as the set of 'automorphisms', not merely the automorphisms of each individual object, but all isomorphisms between each pair of objects in the family. The resulting system of isomorphisms has the structure of a groupoid. Of course, like groups, groupoids also often arise in other ways, not related to automorphisms.

To illustrate the concept of groupoid, we take as example the set, denoted $\Pi(TB)$, of all linear isomorphisms between the various tangent spaces to a manifold B. Each such isomorphism $\xi: T(B)_x \to T(B)_y$ has associated with it two points of B, namely the points x and y which label the tangent spaces which are its domain and range; we denote x by $\alpha(\xi)$ and y by $\beta(\xi)$ and call $\alpha,\beta: \Pi(TB) \to B$ the source and target projections of $\Pi(TB)$; the isomorphism ξ can be composed with an isomorphism $\eta: T(B)_{y'} \to T(B)_z$ iff $y' = y$, that is, iff $\alpha(\eta) = \beta(\xi)$. Thus composition is a partial multiplication on $\Pi(TB)$ with domain the set $\Pi(TB)*\Pi(TB) = \{(\eta,\xi) \in \Pi(TB) \times \Pi(TB) \mid \alpha(\eta) = \beta(\xi)\}$. Note that when the composition $\eta\xi$ is defined, we have $\alpha(\eta\xi) = \alpha(\xi)$ and $\beta(\eta\xi) = \beta(\eta)$. This partial multiplication has properties which resemble the properties of a group multiplication as closely as is possible: each point $x \in B$ has associated with it the identity isomorphism $id_{T(B)_x}$, here denoted $\tilde{x}$, and the elements $\tilde{x}, x \in B$, act as unities for every multiplication in which they can take part; each isomorphism $\xi: T(B)_x \to T(B)_y$ has an inverse isomorphism $\xi^{-1}: T(B)_y \to T(B)_x$ and $\xi\xi^{-1}$ and $\xi^{-1}\xi$ are the unities $\widetilde{\beta(\xi)}$ and $\widetilde{\alpha(\xi)}$ respectively. These properties are abstracted into the following definition.

<u>Definition 1.1.</u> A <u>groupoid</u> consists of two sets Ω and B, called respectively the <u>groupoid</u> and the <u>base</u>, together with two maps α and β from Ω to B, called respectively the <u>source</u> and <u>target projections</u>, a map $\varepsilon: x \mapsto \tilde{x}$, $B \to \Omega$ called the <u>object inclusion map</u>, and a partial multiplication $(\eta,\xi) \mapsto \eta\xi$ in Ω defined on the set $\Omega*\Omega = \{(\eta,\varepsilon) \in \Omega \times \Omega \mid \alpha(\eta) = \beta(\xi)\}$, all subject to the following conditions:

(i) $\alpha(\eta\xi) = \alpha(\xi)$ and $\beta(\eta\xi) = \beta(\eta)$ for all $(\eta,\xi) \in \Omega*\Omega$;

(ii) $\zeta(\eta\xi) = (\zeta\eta)\xi$ for all $\zeta,\eta,\xi \in \Omega$ such that $\alpha(\zeta) = \beta(\eta)$ and $\alpha(\eta) = \beta(\xi)$;

(iii) $\alpha(\tilde{x}) = \beta(\tilde{x}) = x$ for all $x \in B$;

(iv) $\xi\widetilde{\alpha(\xi)} = \xi$ and $\widetilde{\beta(\xi)}\xi = \xi$ for all $\xi \in \Omega$;

(v) each $\xi \in \Omega$ has a (two-sided) inverse ξ^{-1} such that $\alpha(\xi^{-1}) = \beta(\xi)$, $\beta(\xi^{-1}) = \alpha(\xi)$ and $\xi^{-1}\xi = \widetilde{\alpha(\xi)}$, $\xi\xi^{-1} = \widetilde{\beta(\xi)}$. //

Elements of B may be called <u>objects</u> of the groupoid Ω and elements of Ω may be called <u>arrows</u>. The arrow $\tilde{x}$ corresponding to an object $x \in B$ may also be called the <u>unity</u> or <u>identity</u> corresponding to x. To justify this terminology and to prove that the inverse in (v) is unique, we have the following proposition.

Proposition 1.2. Let Ω be a groupoid with base B, and consider $\xi \in \Omega$ with $\alpha(\xi) = x$ and $\beta(\xi) = y$.

(i) If $\eta \in \Omega$ has $\alpha(\eta) = y$ and $\eta\xi = \xi$, then $\eta = \tilde{y}$.
If $\zeta \in \Omega$ has $\beta(\zeta) = x$ and $\xi\zeta = \xi$, then $\zeta = \tilde{x}$.

(ii) If $\eta \in \Omega$ has $\alpha(\eta) = y$ and $\eta\xi = \tilde{x}$, then $\eta = \xi^{-1}$.
If $\zeta \in \Omega$ has $\beta(\zeta) = x$ and $\xi\zeta = \tilde{y}$, then $\zeta = \xi^{-1}$.

Proof. Exercise. //

In place of the phrase "a groupoid with base B", we will often write "a groupoid on B". For a groupoid Ω on B and $x, y \in B$ we will write Ω_x for $\alpha^{-1}(x)$, Ω^y for $\beta^{-1}(y)$ and Ω_x^y for $\Omega_x \cap \Omega^y$. To avoid cumbersome suffices we will sometimes denote "$\xi \in \Omega_x^y$" by "$\xi: x \to y$". The set Ω_x is the α-fibre over x and Ω^y is the β-fibre over y. The set Ω_x^x, obviously a group under the restriction of the partial multiplication in Ω, is called the vertex group at x. Some writers call Ω_x^x the isotropy group at x. For any subsets $U, V \subseteq B$ we likewise write Ω_U, Ω^V and Ω_U^V for $\alpha^{-1}(U)$, $\beta^{-1}(V)$ and $\Omega_U \cap \Omega^V$, respectively.

Many authors denote Ω_x^y by $\Omega(x,y)$, call Ω_x the star of Ω at x and denote it by $St_\Omega x$, and call Ω^y the co-star of Ω at y and denote it by $Cost_\Omega y$.

The following examples are of basic importance.

Example 1.3. Any set B may be regarded as a groupoid on itself with $\alpha = \beta = id_B$ and every element a unity. Groupoids in which every element is a unity have been given a variety of names; we will call them base groupoids. //

Example 1.4. Let B be a set and G a group. We give $B \times G \times B$ the structure of a groupoid on B in the following way: α is the projection onto the third factor of $B \times G \times B$ and β is the projection onto the first factor; the object inclusion map is $x \mapsto \tilde{x} = (x,1,x)$ and the partial multiplication is $(z,h,y')(y,g,x) = (z,hg,x)$, defined iff $y' = y$. The inverse of (y,g,x) is (x,g^{-1},y). This is called the trivial groupoid on B with group G.

In particular, any group may be considered to be a groupoid on any singleton set, and any cartesian square $B \times B$ is a groupoid on B. //

Example 1.5. Let X be an equivalence relation on a set B. Then $X \subseteq B \times B$ is a groupoid on B with respect to the restriction of the structure defined in 1.4. Each α-fibre X_x, $x \in B$, may be naturally identified with the equivalence class containing x.

Groupoids Ω such as this, in which each Ω_x^y is either empty or singleton, are sometimes called principal groupoids (see Renault (1980)). We use this term with a different meaning (see II 2.9). //

Example 1.6. Let $G \times B \to B$ be an action of a group G on a set B. Give $G \times B$ the structure of a groupoid on B in the following way: α is the projection onto the second factor of $G \times B$ and β is the action $G \times B \to B$ itself; the object inclusion map is $x \mapsto \tilde{x} = (x,1)$ and the partial multiplication is $(g_2,y)(g_1,x) = (g_2g_1,x)$, defined iff $y = g_1x$. The inverse of (g,x) is (g^{-1},gx). We propose to call $G \times B$ the action groupoid of $G \times B \to B$.

The α-fibre $(G \times B)_x$ is $G \times \{x\}$, and the β-fibre can also be identified with the group G. The vertex group $(G \times B)_x^x$ is naturally isomorphic to the isotropy group G_x.

This construction can be generalized. See II 4.20. //

Example 1.7. Applying the construction of 1.6 to the action $\mathbf{R} \times S^1 \to S^1$, $(t,z) \mapsto e^{2\pi it}z$ gives a groupoid structure on the cylinder $\mathbf{R} \times S^1$. The base may be identified with the circle $t = 0$, the α-fibres are straight lines orthogonal to $t = 0$, the β-fibres are the helices which make an angle of 45° with the circles t = constant, and the vertex groups are the $\mathbf{Z} \times \{z\}$ for $z \in S^1$.

The reader may construct similarly visualizable examples on the torus, using the actions $S^1 \times S^1 \to S^1$, $(w,z) \mapsto w^n z$, for given $n \in \mathbf{Z}$. However no truly typical example of a groupoid of the type with which we shall be concerned in later chapters can be visualized by means of an embedding in $\mathbf{R}^3$. //

Example 1.8. Let B be a topological space. Then the set $\pi(B)$ of homotopy classes $\langle c\rangle$ rel endpoints of paths $c: [0,1] \to B$ is a groupoid on B with respect to the following structure: the source and target projections are $\alpha(\langle c\rangle) = c(0)$ and $\beta(\langle c\rangle) = c(1)$, the object inclusion map is $x \mapsto \tilde{x} = \langle\kappa_x\rangle$, where κ_x is the path constant at x, and the partial multiplication is $\langle c'\rangle\langle c\rangle = \langle c'c\rangle$ where $c'c$ is the standard concatenation of c followed by c', namely $(c'c)(t) = c(2t)$ for $0 \leq t \leq \frac{1}{2}$, $(c'c)(t) = c'(2t-1)$ for $\frac{1}{2} \leq t \leq 1$. The inverse of $\langle c\rangle$ is $\langle c^{\leftarrow}\rangle$ where $c^{\leftarrow}$ is the reverse of the path c, namely $c^{\leftarrow}(t) = c(1-t)$.

Note that many authors take $c'c$ to be c' followed by c, defined iff $c'(1) = c(0)$. The groupoid $\mathcal{T}(B)$ may also be defined using paths of variable length; for this see, for example, Brown (1968).

$\mathcal{T}(B)$ is the fundamental groupoid of B; its vertex groups are the fundamental groups $\pi_1(B,x)$, $x \in B$, and if B is path-connected, locally path-connected and semi-locally simply connected, then its α-fibres are the sets underlying the universal covering spaces of B.

There are now a number of beginning texts on algebraic topology which introduce the concept of fundamental group via that of the fundamental groupoid, but most make little use of the groupoid structure. The first account of elementary homotopy theory to make effective use of the algebraic structure of $\mathcal{T}(B)$ was Brown (1968). //

Example 1.9. Let $p: M \to B$ be a surjective map. Let $\Pi(M)$ denote the set of all bijections $\xi: M_x \to M_y$ for $x,y \in B$, where $M_x = p^{-1}(x)$, $x \in B$. Then $\Pi(M)$ is a groupoid on B with respect to the following structure: for $\xi: M_x \to M_y$, $\alpha(\xi)$ is x and $\beta(\xi)$ is y; the object inclusion map is $x \mapsto \tilde{x} = id_{M_x}$, and the partial multiplication is the composition of maps. The inverse of $\xi \in \Pi(M)$ is its inverse as a map. $\Pi(M)$ is called the frame groupoid of (M,p,B).

Many variants of this fundamental example will be given in later chapters. //

Example 1.10. Let $P(B,G,\pi)$ be a principal bundle. Let G act on $P \times P$ to the right by $(u_2,u_1)g = (u_2g,u_1g)$; denote the orbit of (u_2,u_1) by $\langle u_2,u_1\rangle$ and the set of orbits by $\frac{P \times P}{G}$. Then $\frac{P \times P}{G}$ is a groupoid on B with respect to the following structure: the source and target projections are $\alpha(\langle u_2,u_1\rangle) = \pi(u_1)$, $\beta(\langle u_2,u_1\rangle) = \pi(u_2)$; the object inclusion map is $x \mapsto \tilde{x} = \langle u,u\rangle$, where u is any element of $\pi^{-1}(x)$; and the partial multiplication is defined by

$$\langle u_3,u_2'\rangle\langle u_2,u_1\rangle = \langle u_3,u_1\delta(u_2',u_2)\rangle.$$

Here $\delta: P \underset{\pi}{\times} P \to G$ is the map $(ug,u) \mapsto g$ (see A§1). The condition $\alpha(\langle u_3,u_2'\rangle) = \beta(\langle u_2,u_1\rangle)$ ensures that $(u_2',u_2) \in P \underset{\pi}{\times} P$. Note that one can always choose representatives so that $u_2' = u_2$ and the multiplication is then simply

$$\langle u_3,u_2\rangle\langle u_2,u_1\rangle = \langle u_3,u_1\rangle.$$

The inverse of $\langle u_2,u_1\rangle$ is $\langle u_1,u_2\rangle$. $\frac{P \times P}{G}$ is called the groupoid associated to $P(B,G,\pi)$. //

Example 1.11. Applying 1.10 to the principal bundle $SU(2)(SO(3),\mathbf{Z}_2,\pi)$ yields a groupoid which, though it has dimension 6, is perhaps somewhat visually accessible. Here the action of $\mathbf{Z}_2 = \{I,-I\} \subseteq SU(2)$ on $SU(2)$ is by matrix multiplication and π is essentially the adjoint representation (see, for example, Miller (1972, p. 224)). The groupoid $\frac{SU(2) \times SU(2)}{\mathbf{Z}_2}$ can be naturally identified with $SO(4)$: identifying $SU(2)$ with the unit sphere in the space of quaternions $\mathbf{H}$, each pair $(p,q) \in SU(2) \times SU(2)$ defines a map $\mathbf{H} \to \mathbf{H}$, $x \mapsto pxq^{-1}$ which, as a map $\mathbf{R}^4 \to \mathbf{R}^4$, is a proper rotation. It is well-known that this map $SU(2) \times SU(2) \to SO(4)$ is an epimorphism of Lie groups with kernel $\{(I,I), (-I,-I)\}$ (see, for example, Greub (1967, p. 329)).

Thus we obtain a groupoid structure on $SO(4)$ with base $\mathbf{RP}^3$, α- and β-fibres which are 3-spheres, and vertex groups which are $\mathbf{Z}_2$'s. However it seems that the groupoid multiplication has no clear geometrical significance. //

We shall return to examples 1.4 to 1.10 in chapters II and III.

§2. Morphisms, subgroupoids and quotient groupoids

We treat the concepts listed in the title, and the related concepts of kernel, normal subgroupoid, etc., and consider the factorization of a morphism into a quotient projection, an isomorphism, and an inclusion. This factorization, fundamental in the category of groups, is valid only for certain classes of groupoid morphism, for example, those morphisms which are both piecewise-surjective and base-surjective, and those which are base-injective. In particular, base-preserving morphisms can be so factored. We mention in passing two factorizations of an arbitrary groupoid morphism into a base-preserving morphism and a morphism of another specified type.

The examples given here are tailored to those of later chapters but otherwise the material of this section comes from Higgins (1971) and Brown (1968).

Definition 2.1. Let Ω and Ω' be groupoids on B and B' respectively. A morphism $\Omega \to \Omega'$ is a pair of maps $\phi: \Omega \to \Omega'$, $\phi_o: B \to B'$ such that $\alpha' \circ \phi = \phi_o \circ \alpha$, $\beta' \circ \phi = \phi_o \circ \beta$ and $\phi(\eta\xi) = \phi(\eta)\phi(\xi)$, $\forall(\eta,\xi) \in \Omega*\Omega$. We also say that ϕ is a morphism over ϕ_o. If $B = B'$ and $\phi_o = id_B$ we say that ϕ is a morphism over B, or that ϕ is a base-preserving morphism. //

Note that the conditions $\alpha' \circ \phi = \phi_o \circ \alpha$, $\beta' \circ \phi = \phi_o \circ \beta$ ensure that $\phi(\eta)\phi(\xi)$ is defined whenever $\eta\xi$ is. Morphisms preserve unities and inverses:

Proposition 2.2. Let $\phi: \Omega \to \Omega'$, $\phi_o: B \to B'$ be a groupoid morphism. Then

(i) $\phi(\tilde{x}) = \widetilde{\phi_o(x)}$ $\quad \forall\, x \in B$,

(ii) $\phi(\xi^{-1}) = \phi(\xi)^{-1}$ $\quad \forall\, \xi \in \Omega$.

Proof. Exercise. //

For $x,y \in B$ we denote the restrictions of ϕ to $\Omega_x \to \Omega'_{\phi_o(x)}$, $\Omega^y \to \Omega'^{\phi_o(y)}$ and $\Omega^y_x \to \Omega'^{\phi_o(y)}_{\phi_o(x)}$ by ϕ_x, ϕ^y and ϕ^y_x, respectively.

Definition 2.3. A groupoid morphism $\phi: \Omega \to \Omega'$ over $\phi_o: B \to B'$ is piecewise-surjective (respectively, piecewise-injective, piecewise-bijective) if $\phi^y_x: \Omega^y_x \to \Omega'^{\phi_o(y)}_{\phi_o(x)}$ is surjective (respectively, injective, bijective) $\forall\, x,y \in B$.

ϕ is base-surjective (respectively, base-injective, base-bijective) if $\phi_o: B \to B'$ is surjective (respectively, injective, bijective).

ϕ is an isomorphism if $\phi: \Omega \to \Omega'$ (and hence $\phi_o: B \to B'$) is bijective. //

We will not use the words 'epimorphism' or 'monomorphism' in the algebraic context.

It is trivial to prove that a surjective (injective) morphism is base-surjective (base-injective); further, a morphism is injective iff it is base-injective and piecewise-injective, and a morphism which is base-surjective and piecewise-surjective is itself surjective. All these results are easy to prove. A surjective morphism need not be piecewise-surjective; see example 2.8 below.

Definition 2.4. Let Ω be a groupoid on B. A subgroupoid of Ω is a pair of subsets $\Omega' \subseteq \Omega$, $B' \subseteq B$, such that $\alpha(\Omega') \subseteq B'$, $\beta(\Omega') \subseteq B'$, $\tilde{x} \in \Omega'$ $\forall\, x \in B'$, and Ω' is closed under the partial multiplication and inversion in Ω. A subgroupoid Ω', B' of Ω, B is wide if $B' = B$ and is full if $\Omega'^{y}_{x} = \Omega^{y}_{x}$ $\forall\, x,y \in B'$.

The base subgroupoid or identity subgroupoid of Ω is the subgroupoid $\tilde{B} = \{\tilde{x} \mid x \in B\}$. The inner subgroupoid of Ω is the subgroupoid $G\Omega = \bigcup_{x \in B} \Omega^{x}_{x}$. //

A morphism of groups may be factored into a surjective morphism (the projection of the domain group onto its quotient over the kernel of the given morphism), followed by an isomorphism, followed by an injective morphism (the inclusion of the image of the given morphism into its range). For groupoid morphisms the situation is more complicated. Firstly, the image of a groupoid morphism need not be a subgroupoid; it may happen that a product $\phi(\eta)\phi(\xi)$ is defined but the product $\eta\xi$ is not and that another pair η_1, ξ_1 with $\phi(\eta_1) = \phi(\eta)$, $\phi(\xi_1) = \phi(\xi)$ and $\eta_1\xi_1$ defined cannot be found. This can occur even for morphisms of trivial groupoids: Let B be an interval on the real line, bounded away from infinity and zero, and G' the multiplicative group of positive reals and consider $B \times B \to G'$, $(y,x) \mapsto yx^{-1}$. (It is easy to prove, however, that the image of a base-injective morphism is a subgroupoid.)

Secondly, the concept of kernel for groupoid morphisms does not adequately measure injectivity. To demonstrate this failure and its consequences and describe what factorizations are possible will occupy us until 2.13.

Definition 2.5. Let Ω be a groupoid on B. A normal subgroupoid of Ω is a wide subgroupoid Φ such that for any $\lambda \in G\Phi$ and any $\xi \in \Omega$ with $\alpha\xi = \alpha\lambda = \beta\lambda$, we have $\xi\lambda\xi^{-1} \in \Phi$. //

Note that whether or not a subgroupoid is normal depends only on those of its elements which lie in its inner subgroupoid.

Definition 2.6. Let $\phi\colon \Omega \to \Omega'$, $\phi_o\colon B \to B'$ be a morphism of groupoids. Then the kernel of ϕ is the set $\{\xi \in \Omega \mid \phi(\xi) = \tilde{x},\ \exists\, x \in B'\}$. //

Clearly the kernel of a morphism is a normal subgroupoid. The following construction of quotient groupoids shows that every normal subgroupoid is the kernel of a morphism.

Proposition 2.7. Let Φ be a normal subgroupoid of a groupoid Ω on B. Define an equivalence relation $\sim$ on B by $x \sim y \iff \exists\, \zeta \in \Phi: \alpha\zeta = x,\ \beta\zeta = y$, and denote the equivalence classes by $[x], x \in B$, and the set of equivalence classes by B/Φ. Define a second equivalence relation, also denoted $\sim$, on Ω by $\xi \sim \eta \iff \exists\, \zeta,\zeta' \in \Phi: \zeta\eta\zeta'$ is defined and equals ξ. Denote the equivalence classes by $[\xi], \xi \in \Omega$, and the set of them by Ω/Φ. (Note that, $\forall\, x,y \in B,\ x \sim y \iff \tilde{x} \sim \tilde{y}$.)

Then the following defines the structure of a groupoid Ω/Φ with base B/Φ: the source and target projections are $\bar{\alpha}([\xi]) = [\alpha(\xi)]$, $\bar{\beta}([\xi]) = [\beta(\xi)]$, the object inclusion map is $[x] \mapsto \widetilde{[x]} = [\tilde{x}]$, and the product $[\eta][\xi]$, where $\alpha(\eta) \sim \beta(\xi)$, is defined as $[\eta\zeta^{-1}\xi]$, where ζ is any element of Φ with $\alpha\zeta = \alpha\eta$ and $\beta\zeta = \beta\xi$. The inverse of $[\xi]$ is $[\xi^{-1}]$.

The projections $\natural: \xi \mapsto [\xi],\ \Omega \to \Omega/\Phi$, $\natural_0: x \mapsto [x],\ B \to B/\Phi$ constitute a groupoid morphism. The kernel of $\natural$ is Φ.

Proof. Exercise for the reader. //

Ω/Φ is the quotient groupoid of Ω over the normal subgroupoid Φ. The notation '$\natural$' should be read as 'natural', for 'natural projection'. Note the extreme cases: $\Omega/\tilde{B}$ is isomorphic to Ω under $\natural$, Ω/Ω is a base groupoid (not necessarily singleton), and $\Omega/G\Omega$ is isomorphic to $\tilde{B}$.

It is easy to see that an injective morphism has the base subgroupoid of its domain as kernel, and that a morphism whose kernel is the base subgroupoid is piecewise-injective. The following example is a surjective morphism whose kernel is the base subgroupoid but which is not (always) an isomorphism.

Example 2.8. Let $P(B,G,\pi)$ be a principal bundle and consider the associated groupoid $\frac{P \times P}{G}$ constructed in 1.10. It is easy to see that the map $P \times P \to \frac{P \times P}{G}$, $(u_2,u_1) \mapsto \langle u_2,u_1\rangle$ is a morphism of groupoids over $\pi: P \to B$, where $P \times P$ has the trivial groupoid structure of 1.4. The kernel is the diagonal Δ_P of P, which is the base subgroupoid of $P \times P$. //

A surjective group morphism can be factored into the projection onto a quotient group followed by an isomorphism. This example shows that the straightforward generalization to groupoids is not valid. Surjective groupoid morphisms are not determined by their kernels: both the morphism in 2.8 and $id_{P \times P}$ are surjective morphisms with kernel Δ_P. Notice that the contrasting notations Ω/Φ and $\frac{P \times P}{G}$ are used to emphasize that there is no subgroupoid 'G' of $P \times P$ that makes $\frac{P \times P}{G}$ a quotient of groupoids.

For morphisms which are not only surjective, but are also piecewise-surjective, such a straightforward factorization is possible:

<u>Proposition 2.9.</u> Let $\phi: \Omega \to \Omega'$ be a groupoid morphism over $\phi_o: B \to B'$ with kernel Φ.

(i) If ϕ is base-surjective and piecewise-surjective then $\bar{\phi}: \Omega/\Phi \to \Omega'$, $[\xi] \mapsto \phi(\xi)$ is an isomorphism of groupoids and $\phi = \bar{\phi}\circ\natural$.

(ii) If there is an isomorphism of groupoids $\psi: \Omega/\Phi \to \Omega'$ such that $\phi = \psi\circ\natural$, then ϕ is base-surjective and piecewise-surjective.

<u>Proof.</u> (i) Clearly $\bar{\phi}$ is surjective, since ϕ is. Suppose $\bar{\phi}([\xi]) = \bar{\phi}([\eta])$, that is, $\phi(\xi) = \phi(\eta)$. Then $\beta\xi$ and $\beta\eta$ have the same image, say z, under ϕ_o so, since $\phi^{\beta\eta}_{\beta\xi}: \Omega^{\beta\eta}_{\beta\xi} \to \Omega'^{z}_{z}$ is surjective, there is an element $\zeta \in \Omega^{\beta\eta}_{\beta\xi}$ such that $\phi(\zeta) = \tilde{z}$; such an element must actually be in $\Phi^{\beta\eta}_{\beta\xi}$. Similarly there is an element $\zeta' \in \Phi^{\alpha\eta}_{\alpha\xi}$. Now $\zeta^{-1}\eta\zeta'\xi^{-1}$ is defined, is an element of $\Omega^{\beta\xi}_{\beta\xi}$, and is mapped by ϕ to $\tilde{z}$, so it is actually an element of $\Phi^{\beta\xi}_{\beta\xi}$; denote it by λ. Then $\xi = (\zeta\lambda)^{-1}\eta\zeta'$, which shows that $\xi \sim \eta$; that is, $[\xi] = [\eta]$.

(ii) $\natural$ is base-surjective by construction. To prove that $\natural^{y}_{x}: \Omega^{y}_{x} \to (\Omega/\Phi)^{[y]}_{[x]}$ is surjective, take $[\xi] \in \Omega/\Phi$ with $\bar{\beta}([\xi]) = [y]$, $\bar{\alpha}([\xi]) = [x]$. Then $\beta\xi \sim y$, $\alpha\xi \sim x$ so $\exists\ \zeta,\zeta' \in \Phi$ such that $\zeta: y \to \beta\xi$ and $\zeta': x \to \alpha\xi$. Now $\zeta^{-1}\xi\zeta' \sim \xi$ and $\zeta^{-1}\xi\zeta' \in \Omega^{y}_{x}$. //

In the rest of these notes we will be mostly concerned with morphisms $\phi: \Omega \to \Omega'$ for which $B = B'$ and $\phi_o = id_B$, or at any rate for which ϕ_o is a bijection. For morphisms ϕ with ϕ_o a bijection, surjectivity is equivalent to piecewise-surjectivity so 2.9 shows that in this case we have a factorization of an arbitrary morphism into a natural projection followed by an isomorphism followed by an inclusion, exactly as for group morphisms. Two other simplifications are possible in this case: (i) the kernel of a base-bijective morphism (indeed of a base-injective one) is the union of its vertex groups, that is, in the terminology of 3.1, it is a totally intransitive groupoid, and so there is no need to consider the equivalence relation on the base when quotienting over such a kernel, (ii) when quotienting a groupoid Ω over a totally intransitive normal subgroupoid Φ, the relation "$\xi \sim \eta \iff \exists\ \zeta,\zeta' \in \Phi$: $\zeta\eta\zeta'$ is defined and equals ξ" may be defined by "$\xi \sim \eta \iff \exists\ \lambda \in \Phi$: $\eta\lambda$ is defined and equals ξ" (this is a simple consequence of the facts that ζ, η, ζ' must now all belong to the same Ω^{x}_{x}, and Φ^{x}_{x} is a normal subgroup of Ω^{x}_{x}). It is also true for base-bijective morphisms that a morphism is injective iff its kernel is the base subgroupoid.

A straightforward factorization like that of a group morphism is also possible for base-injective morphisms (Brown (1968, 8.3.2) or Higgins (1971)).

There are two ways in which an arbitrary groupoid morphism can be factored into a base-preserving morphism, to which the factorization given above can be applied, and a second morphism of a specified type. Firstly, one can use a pullback:

Definition 2.10. Let $\phi: \Omega \to \Omega'$, $\phi_o: B \to B'$ be a groupoid morphism. ϕ is a pullback if every groupoid morphism $\psi: \Phi \to \Omega'$, also over $\psi_o = \phi_o: B \to B'$, can be factored uniquely into $\Phi \xrightarrow{\bar{\psi}} \Omega \xrightarrow{\phi} \Omega'$ where $\bar{\psi}$ is a groupoid morphism over B. //

Definition 2.11. Let Ω be a groupoid on B and f: $B' \to B$ a map. The inverse image groupoid of Ω over f is the set $f^*\Omega = \{(y',\xi,x') \in B'\times\Omega\times B' \mid f(y') = \beta\xi,\ f(x') = \alpha\xi\}$ together with the groupoid structure consisting of projections $\alpha'(y',\xi,x') = x'$, $\beta'(y',\xi,x') = y'$, object inclusion map $x' \mapsto \widetilde{x'} = (x',\widetilde{f(x')},x')$, and composition $(z',\eta,y')(y',\xi,x') = (z',\eta\xi,x')$. The inversion is $(y',\xi,x')^{-1} = (x',\xi^{-1},y')$. The inverse image morphism is the morphism $\tilde{f}: f^*\Omega \to \Omega$, $(y',\xi,x') \mapsto \xi$, over f: $B' \to B$. //

Note that $\tilde{f}$ is piecewise-bijective.

Proposition 2.12. A morphism of groupoids is a pullback iff it is piecewise-bijective.

Proof. (<=) If $\phi: \Omega \to \Omega'$ is piecewise-bijective and $\psi: \Phi \to \Omega'$ has $\psi_o = \phi_o: B \to B'$, define $\bar{\psi}^y_x: \Phi^y_x \to \Omega^y_x$ for each $x,y \in B$ by $\bar{\psi}^y_x = (\phi^y_x)^{-1} \circ \psi^y_x$.

(=>) Let $\phi: \Omega \to \Omega'$ be a pullback. By (<=), $\tilde{\phi}_o: \phi_o^*\Omega' \to \Omega'$ is also a pullback so there is a morphism $\psi_1: \Omega \to \phi_o^*\Omega'$ over B such that $\phi = \tilde{\phi}_o \circ \psi_1$. Because ϕ is a pullback, there is a morphism $\psi_2: \phi_o^*\Omega' \to \Omega$ over B such that $\tilde{\phi}_o = \phi\circ\psi_2$. By the uniqueness requirement in 2.10, both of $\psi_1 \circ \psi_2$ and $\psi_2 \circ \psi_1$ are identity morphisms, so ψ_1 and ψ_2 are isomorphisms. Because $\tilde{\phi}_o$ is piecewise-bijective it now follows that ϕ is. //

Hence an arbitrary groupoid morphism $\phi: \Omega \to \Omega'$ can be factored into the inverse image morphism $\tilde{\phi}_o: \phi_o^*\Omega' \to \Omega'$ and a base-preserving morphism $\Omega \to \phi_o^*\Omega'$ over B.

Secondly, and in some sense dually, one can use the concept of universal morphism of Higgins (1971, Chapter 12). A morphism $\phi: \Omega \to \Omega'$ is universal if for all morphisms $\psi: \Omega \to \Omega''$ such that $\psi_o = f \circ \phi_o$ for some $f: B' \to B''$, there is a unique morphism $\theta: \Omega' \to \Omega''$ such that $\psi = \theta\phi$ and $\theta_o = f$. It is a non-trivial fact (Higgins (1971, Chapter 8) or Brown (1968, Chapter 8)) that given any groupoid Ω on B and map $f: B \to B'$ there is a groupoid $U_f(\Omega)$ on B' and a universal morphism $f^*: \Omega \to U_f(\Omega)$ over $f: B \to B'$. Now, given an arbitrary morphism $\phi: \Omega \to \Omega'$ one can factor ϕ through the universal morphism $\Omega \to U_{\phi_o}(\Omega)$ and obtain a morphism $U_{\phi_o}(\Omega) \to \Omega'$ over B' which can be factored in the straightforward way we described above. Note that in the category of groups a universal morphism is an isomorphism.

We close this section with some basic examples.

Example 2.13. A morphism of trivial groupoids $\phi: B \times G \times B \to B' \times G' \times B'$ can be written in the form

$$\phi(y,g,x) = (\phi_o(y), \theta(y)f(g)\theta(x)^{-1}, \phi_o(x))$$

for a group morphism $f: G \to G'$ and a map $\theta: B \to G'$. The maps θ and f are not unique; for any point $b \in B$, such maps are defined by

$$\phi(x,1,b) = (\phi_o(x), \theta(x), \phi_o(b)) \quad \text{and} \quad \phi(b,g,b) = (\phi_o(b), f(g), \phi_o(b)). \qquad //$$

Example 2.14. (Pradines (1966).) If G is a group the 'division map' $\delta(g_2, g_1) = g_2 g_1^{-1}$ is a groupoid morphism $G \times G \to G$ where the domain is the trivial groupoid on the set G and the range is the group G itself.

More generally, for any groupoid Ω on B, the set $\Omega \times_\alpha \Omega = \{(\eta, \xi) \in \Omega \times \Omega \mid \alpha\eta = \alpha\xi\}$ is a subgroupoid of the cartesian square groupoid $\Omega \times \Omega$ on Ω and $\delta: \Omega \times_\alpha \Omega \to \Omega$, $(\eta, \xi) \mapsto \eta\xi^{-1}$ is a groupoid morphism over $\beta: \Omega \to B$. //

Example 2.15. The subgroupoids of a cartesian square groupoid $B \times B$ may be identified with the equivalence relations on the subsets of B. An equivalence relation X on B itself constitutes a normal subgroupoid of $B \times B$ and $(B \times B)/X$ may be identified with the trivial groupoid $(B/X) \times (B/X)$ under $[(y,x)] \mapsto ([y],[x])$. //

Example 2.16. Let G and G' be groups acting on sets B and B', let $\phi: G \to G'$ be a morphism and let $f: B \to B'$ be a map equivariant with respect to ϕ. Then $\phi \times f: G \times B \to G' \times B'$ is a morphism of the action groupoids. //

Example 2.17. If $F(f,\phi)$: $P(B,G) \to P'(B',G')$ is a morphism of principal bundles, then $\langle u_2,u_1\rangle \mapsto \langle F(u_2),F(u_1)\rangle$ is a morphism of the associated groupoids. //

Example 2.18. For any groupoid Ω on base B, the map $[\beta,\alpha]$: $\Omega \to B \times B$, $\xi \mapsto (\beta(\xi),\alpha(\xi))$ is a morphism over B, which we propose to call the anchor of Ω. Its kernel is the inner subgroupoid $G\Omega$. //

These examples will be treated in more depth, and others introduced, in Chapters II and III.

§3. Transitive and totally intransitive groupoids

A groupoid is transitive if any two points of its base can be joined by an element of the groupoid. While the algebraic theory of transitive groupoids is trivial (see 3.2), the main interest of later chapters will be with topological or differentiable groupoids that are transitive, but not topologically trivializable.

A groupoid is totally intransitive if it is the union of its vertex groups. Totally intransitive groupoids are important because a transitive groupoid can be regarded as an extension of a cartesian square groupoid by a totally intransitive one.

This section also treats the concepts of product over a fixed base and abelianity, concepts which are largely meaningless for groupoids which are neither transitive nor totally intransitive.

Definition 3.1. Let Ω be a groupoid on B. Ω is transitive it its anchor $[\beta,\alpha]$: $\Omega \to B \times B$ is surjective. Ω is totally intransitive if the image of $[\beta,\alpha]$ is the base subgroupoid Δ_B of $B \times B$.

In general, the image of $[\beta,\alpha]$ in $B \times B$ is an equivalence relation on B. The equivalence class containing $x \in B$ is denoted M_x and called the transitivity component of Ω containing x. //

Transitive groupoids are sometimes called connected groupoids and totally intransitive groupoids called totally disconnected groupoids. We shall not use this terminology.

Let Ω be any groupoid on B. For $x \in B$, M_x is clearly $\beta(\Omega_x)$ and $\Omega^{M_x}_{M_x}$ is a transitive, full, subgroupoid of Ω; clearly Ω is the disjoint union of the transitive subgroupoids Ω^M_M, as M runs through the transitivity components of Ω.

Now consider a transitive groupoid Ω on B. The anchor is a surjective morphism over B with kernel the inner subgroupoid $G\Omega$ of Ω so we may regard Ω as an exact sequence

$$G\Omega \rightarrowtail \Omega \xrightarrow{[\beta,\alpha]} B \times B$$

or, in some sense, as an extension of $B \times B$ by $G\Omega$. This extension is trivial (or semi-direct) because any right-inverse $\sigma\colon B \to \Omega_b$ to $\beta_b\colon \Omega_b \to B$, for some chosen $b \in B$, defines a morphism $\theta(y,x) = \sigma(y)\sigma(x)^{-1}$, $B \times B \to \Omega$, which is right-inverse to $[\beta,\alpha]$. Further,

<u>Proposition 3.2.</u> Let Ω be a transitive groupoid on B, let b be a point of B, and let $\sigma\colon B \to \Omega_b$ be a right-inverse to $\beta_b\colon \Omega_b \to B$. Then

$$\Sigma\colon B \times \Omega^b_b \times B \to \Omega \qquad (y,\lambda,x) \mapsto \sigma(y)\lambda\sigma(x)^{-1}$$

is an isomorphism of groupoids over B.

<u>Proof.</u> Exercise. //

So every groupoid is the disjoint sum of transitive subgroupoids and every transitive groupoid is isomorphic, though not usually in any natural way, to a trivial groupoid.

<u>Examples 3.3.</u> Trivial groupoids $B \times G \times B$ are of course transitive. The groupoid $\frac{P \times P}{G}$ associated to a principal bundle $P(B,G,\pi)$ is transitive, since π is surjective. The frame groupoid $\Pi(M)$ of a surjection (M,p,B) is transitive if all the fibres M_x, $x \in B$, have the same cardinality. The transitivity components of a fundamental groupoid $\pi(B)$ are the path-components of B, and the transitivity components of an action groupoid $G \times B$ are the orbits of the action. For an equivalence relation X on B, the transitivity components are the equivalence classes. Any inverse image of a transitive groupoid is transitive. //

Given groupoids Ω and Ω' on B and B' one can define a product groupoid $\Omega \times \Omega'$ on $B \times B'$ in an obvious way. However, in the rest of this book we will be mainly concerned with categories of groupoids over a single base. Given groupoids

Ω and Ω' on B we require a product groupoid $\Omega \underset{B\times B}{\times} \Omega'$ on B, not on $B \times B$, and although this construction can be given for arbitrary groupoids, it is largely meaningless unless one of the factors is transitive.

Define $\Phi = \Omega \underset{B\times B}{\times} \Omega'$ to be the pullback

$$\begin{array}{ccc} \Phi & \xrightarrow{\pi'} & \Omega' \\ {\scriptstyle\pi}\downarrow & & \downarrow{\scriptstyle[\beta',\alpha']} \\ \Omega & \xrightarrow[{[\beta,\alpha]}]{} & B \times B \end{array} \quad ;$$

and its anchor $[\bar{\beta},\bar{\alpha}]\colon \Phi \to B \times B$ to be either composite. Thus Φ^y_x may be regarded as $\Omega^y_x \times \Omega'^y_x$. Define the partial multiplication and inversion in Φ componentwise in the obvious fashion. Then it is easy to see that if $\phi\colon \Psi \to \Omega$ and $\phi'\colon \Psi \to \Omega'$ are morphisms over B, then $\bar{\phi}\colon \Psi \to \Phi$, $\xi \mapsto (\phi(\xi),\phi'(\xi))$ is also a morphism over B, so Φ is indeed the product groupoid of Ω and Ω' in the category of groupoids on B and morphisms over B.

Another standard concept which is rather meaningless for general groupoids but has an importance for transitive groupoids is that of abelianity.

Definition 3.4. Let Ω be a transitive groupoid. Ω is abelian if any one (and hence all) of its vertex groups is abelian. //

Though the definition involves only the inner subgroupoid of Ω, it has an effect on the structure of the whole groupoid: For an arbitrary transitive groupoid Ω, all vertex groups are isomorphic for, given $\xi \in \Omega^y_x$, the "inner automorphism" $I_\xi\colon \Omega^x_x \to \Omega^y_y$, $\lambda \mapsto \xi\lambda\xi^{-1}$, is an isomorphism of Ω^x_x onto Ω^y_y (see II 1.2 for the formal definition). If Ω is abelian, then Ω^x_x and Ω^y_y are naturally isomorphic, for in this case $I_\xi = I_\eta$ for all ξ, η in the same Ω^y_x. (Compare the well-known fact that if a path-connected space has abelian fundamental groups, then they are all naturally isomorphic.) Thus there is a well-defined map $B \times G \to \Omega$, where $G = \Omega^b_b$ for some $b \in B$, and an exact sequence

$$B \times G \rightarrowtail \Omega \twoheadrightarrow B \times B.$$

Here $B \times G$ may be regarded as the action groupoid corresponding to the trivial action of G on B.

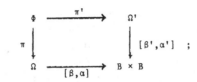

CHAPTER II TOPOLOGICAL GROUPOIDS

It is well-known that despite the body of techniques and results common to the theories of topological groups and Lie groups, the general theory of topological groups scarcely resembles at all the theory of Lie groups. With topological groupoids and differentiable groupoids the divergence is even more marked. This will be particularly clear after III §1, for the central result III 1.4 is proved by a foliation-theoretic method which has no analogue in the general topological case. There is also no analogue of Sard's theorem. At a simpler level, if $f: M \to N$ is a smooth map between manifolds M, N which has the property that a composite $M \xrightarrow{f} N \xrightarrow{g} P$ is smooth iff $g: N \to P$ is smooth (where g is, a priori, not necessarily even continuous), then f is a submersion and, in particular, is open. In the case of topological spaces, the corresponding concept is that of identification map, and such a map need not be open. As a final example, every transitive smooth action of a Lie group on a manifold makes the manifold a homogeneous space; the topological version of this result is false.

This chapter is chiefly concerned with those parts of the theory of topological groupoids which mirror the theory of differentiable and Lie groupoids. Some references to the general theory of topological groupoids are made in §1, §3, §4 and §5 and for further information the reader can consult Brown and Hardy (1976), Brown et al (1976), and Renault (1980).

The point of separating out that part of the theory of differentiable groupoids which is valid in the topological case, is not so much to make a contribution to the theory of topological groupoids, as to demonstrate that these results continue to hold for differentiable groupoids based on more general forms of the manifold concept, such as Banach manifolds or non-paracompact manifolds. Such groupoids may well have applications elsewhere.

There are two natural questions which are still unresolved: Given a morphism $\phi: \Omega \to \Omega'$, where Ω and Ω' are topological groupoids and ϕ is continuous on a neighbourhood of the base in Ω, is it true that ϕ is continuous everywhere? If Ω is a principal topological groupoid, this is easily established (1.21(ii)). Secondly, is it always true that the α-identity component subgroupoid of a suitably locally connected topological groupoid is (α-) open? For differentiable groupoids, this is so (III 1.3), and for principal topological groupoids with connected bases it is so (3.4). If these results turn out to be false in general, it may be that the most general form of the concept of topological groupoid needs re-definition.

We give a brief description of the individual sections. §1 briefly considers the nature of quotient topological groupoids, and then gives the main examples which are used throughout the rest of the text. §2 treats local triviality, its use in reducing global problems to a local problem and a patching problem, and the classification of locally trivial groupoids by cocycles. §3 is concerned with the connected components of the spaces present in a topological groupoid. §4 sets up the apparatus of representations of topological groupoids on fibre bundles and gives the characterization of groupoids of the form $\Omega * M$, where Ω acts on M. §5 is concerned with the concept of left-translation for topological groupoids; the apparatus developed here is needed for the exponential map and adjoint formulas of Chapter III. §6 constructs the monodromy groupoid of a suitably connected principal topological groupoid; this construction generalizes that of the universal covering group of a topological group and of the fundamental groupoid of a topological space. §7 is concerned with path lifting in topological groupoids, and is an introduction to the connection theory of Chapter III.

§1. Basic definitions and examples.

The greater part of this section, from 1.9 to 1.17, consists of examples of topological groupoids. The list is biased towards groupoids which are locally trivial and which admit differentiable structures. It should be noted that the few examples included here which do not meet these criteria are not intended to give a full picture of the range of variation possible in the general theory.

In 1.5 to 1.7 we consider the problems associated with quotient topological groupoids, and in 1.18 to 1.20 we give the equivalence, due to Dakin and Seda (1977), between principal topological groupoids and Cartan principal bundles. This equivalence provides a neat formulation of the correspondence between locally trivial groupoids and principal bundles and, at the same time, shows that a slightly more general class of groupoids shares some of the important properties of locally trivial groupoids.

Definition 1.1. A topological groupoid is a groupoid Ω, B together with topologies on Ω and B such that the five maps which define the groupoid structure are continuous, namely the projections $\alpha,\beta\colon \Omega \to B$, the object inclusion map $\varepsilon\colon x \mapsto \tilde{x}$, $B \to \Omega$, the inversion $\xi \mapsto \xi^{-1}$, $\Omega \to \Omega$, and the partial multiplication $\Omega*\Omega \to \Omega$, where $\Omega*\Omega$ has the subspace topology from $\Omega \times \Omega$.

A morphism of topological groupoids, or, if emphasis is required, a continuous morphism, is a morphism of groupoids $\phi: \Omega \to \Omega'$, $\phi_o: B \to B'$ such that ϕ and ϕ_o are continuous. //

Clearly if one of the projections in a topologized groupoid is continuous, and the inversion is continuous, then the other projection is continuous. In a topological groupoid the inversion is clearly a homeomorphism and the projections are identification maps, since they have the object inclusion map as a right-inverse. Lastly, the object inclusion map is a homeomorphism onto the base subgroupoid $\tilde{B}$, for given $U \subseteq B$ open the set $\tilde{U}$ of unities corresponding to elements of U is the intersection $\Omega_U^U \cap \tilde{B}$, and Ω_U^U is certainly open in Ω. If Ω is Hausdorff then $\tilde{B}$ is closed in Ω, being the image of the map $\xi \mapsto \widetilde{\alpha(\xi)}$, $\Omega \to \Omega$, whose square is itself.

In the definition of a continuous morphism, the requirement that ϕ_o be continuous is superfluous, since ϕ_o is the composite of the object inclusion map of its domain, ϕ, and either projection of the range.

Definition 1.2. Let Ω be a topological groupoid on B, and take $\xi \in \Omega$, $\alpha\xi = x$, $\beta\xi = y$.

The left-translation (right-translation) corresponding to ξ is the map $L_\xi: \Omega^x \to \Omega^y$, $\eta \mapsto \xi\eta$ ($R_\xi: \Omega_y \to \Omega_x$, $\eta \mapsto \eta\xi$). The inner automorphism corresponding to ξ is the map $I_\xi: \Omega_x^x \to \Omega_y^y$, $\lambda \mapsto \xi\lambda\xi^{-1}$. //

If its base B is Hausdorff (or merely T_1), the α-fibres and β-fibres of a topological groupoid Ω are closed subspaces of Ω. Clearly α-fibres (β-fibres) of Ω which are labelled by points of B in a common transitivity component of Ω are homeomorphic under right- (left-) translations. Inner automorphisms (which, of course, are not usually automorphisms at all) are isomorphisms of topological groups.

In many cases the topological properties of the space Ω of a topological groupoid are less important than the properties of its α-fibres. This is the case, for example, with connectedness and simple-connectedness. For a topological property P, therefore, we will say that a topological groupoid is α - P if each of its α-fibres has P. (Each β-fibre is of course homeomorphic to the corresponding α-fibre under the inversion map.) This usage is from Pradines (1966). If P is invariant under continuous maps and finite products then any transitive groupoid Ω which is α - P is itself P, for $\Omega_b \times \Omega_b \to \Omega$, $(\eta,\xi) \mapsto \eta\xi^{-1}$ (any $b \in B$) is a continuous surjection.

It is easy to verify that any subgroupoid Ω', B' of a topological groupoid Ω, B is itself a topological groupoid with respect to the subspace topologies on Ω', B' inherited from Ω, B:

Definition 1.3. Let Ω be a topological groupoid on B. A topological subgroupoid of Ω, B is a subgroupoid Ω', B' of Ω, B equipped with the subspace topologies inherited from Ω, B. //

The problem of giving the factorization results of I§2 validity in the category of topological groupoids is difficult and is not, to the knowledge of the author, completely solved; we will briefly discuss the general situation here, but the only case we will use later is that of base-preserving morphisms of locally trivial groupoids, for which see §2.

Brown and Hardy (1976) show that the universal groupoid construction mentioned in I§2 has a topological validity. Precisely, given a topological groupoid Ω on B and a continuous map $f: B \to B'$ the groupoid $U_f(\Omega)$ has a topology which makes it a topological groupoid on the space B' and makes $f^*: \Omega \to U_f(\Omega)$ continuous; f^* is now universal in the category of topological groupoids in the sense that given any morphism of topological groupoids $\phi: \Omega \to \Omega''$ such that $\phi_o = g \circ f$ for some continuous $g: B' \to B''$, there is a unique morphism of topological groupoids $\psi: U_f(\Omega) \to \Omega''$ such that $\phi = \psi \circ f^*$ and $\psi_o = g$.

Definition 1.4. Let $\phi: \Omega \to \Omega'$ be a morphism of topological groupoids. Then ϕ is a pullback if for every morphism of topological groupoids $\psi: \Phi \to \Omega'$ such that $\psi_o = \phi_o: B \to B'$, there is a unique morphism $\bar{\psi}: \Phi \to \Omega$ over B such that $\psi = \phi \circ \bar{\psi}$. //

If Ω is a topological groupoid on B and $f: B' \to B$ is a continuous map then $f^*\Omega$ with the subspace topology from $B' \times \Omega \times B'$ is a topological groupoid on B' and $\tilde{f}: f^*\Omega \to \Omega$ is a continuous morphism and a pullback (in the sense of 1.4) still called the inverse image of Ω over f. As in the second half of the proof of I 2.12, every pullback $\phi: \Omega \to \Omega'$ is equivalent to the inverse image $\tilde{\phi}_o: \phi_o^*\Omega' \to \Omega'$ and, in particular, is a piecewise homeomorphism. It seems unlikely, however, that every piecewise homeomorphic morphism is a pullback but a counter-example is lacking. See 2.9 for the locally trivial case.

Whether one uses universal morphisms or pullbacks, it is sufficient, from a strictly logical point of view, to restrict the problem of factorization to base-preserving morphisms of topological groupoids. (From a practical point of view neither factorization helps very much to extend results for base-preserving morphisms to general ones.) In particular we may restrict the construction of quotient topological groupoids to the case of totally intransitive normal subgroupoids. The general definition is as follows.

<u>Definition 1.5</u>. Let Ω be a topological groupoid on B and Φ a normal subgroupoid. Then a topological groupoid Ψ and a morphism $\nu\colon \Omega \to \Psi$ are the <u>topological quotient groupoid</u> Ω/Φ and its <u>natural projection</u> $\natural\colon \Omega \to \Omega/\Phi$ if for every morphism of topological groupoids $\phi\colon \Omega \to \Omega'$ such that $\phi(\Phi) \subseteq \widetilde{B'}$ there is a unique morphism $\bar{\phi}\colon \Psi \to \Omega'$ such that $\bar{\phi}\circ\nu = \phi$. //

If G is a topological group and H a normal subgroup it is trivial that $\natural\colon G \to G/H$ is open with respect to the identification topology on G/H and it is thus easy to prove that G/H is the topological quotient group in this topology (see, for example, Higgins (1974)). For groupoids, the natural projection $\natural\colon \Omega \to \Omega/\Phi$ need not be open with respect to the identification topology on Ω/Φ - see example 1.10 below. Even if $\natural$ is open, it is not clear that Ω/Φ, with the identification topology, need be the topological quotient groupoid. However there is the following result.

<u>Theorem 1.6</u>. Let Ω be a topological groupoid on B and let Φ be a normal subgroupoid such that $\natural\colon \Omega \to \Omega/\Phi$ is open, with respect to the identification topology on Ω/Φ, and such that the anchor $[\beta',\alpha']\colon \Phi \to \mathrm{im}[\beta',\alpha']$ is open into its image. Then Ω/Φ, with the identification topology, is the topological quotient groupoid.

<u>Proof</u>. Only the continuity of multiplication in Ω/Φ requires proof. Let D denote the set $(\natural\times\natural)^{-1}(\Omega/\Phi * \Omega/\Phi) = \{(\eta,\xi) \in \Omega\times\Omega \mid \exists\ \zeta \in \Phi\colon \eta\zeta^{-1}\xi \text{ is defined}\}$. Then because D is saturated, the restriction $\natural\times\natural|_D\colon D \to \Omega/\Phi * \Omega/\Phi$ is open.

Denote $\mathrm{im}[\beta',\alpha'] \subseteq B\times B$ by M'. Note that D is also the set $(\alpha\times\beta)^{-1}(M') \subseteq \Omega\times\Omega$. Let $\bar{D}$ denote the set $\{(\eta,\zeta,\xi) \in \Omega\times\Phi\times\Omega \mid \eta\zeta^{-1}\xi \text{ is defined}\}$. Then $\bar{D}$ is the pullback

$$\begin{array}{ccc}
\bar{D} & \longrightarrow & D \\
\big\downarrow & & \big\downarrow{\scriptstyle \alpha\times\beta|_D} \\
\Phi & \xrightarrow{[\beta',\alpha']} & M'
\end{array} \quad .$$

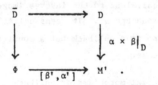

Because $[\beta',\alpha']: \Phi \to M'$ is open, the pullback map $p: \bar{D} \to D$, $(\eta,\zeta,\xi) \mapsto (\eta,\xi)$ is open.

Now the following diagram commutes

$$\begin{array}{ccc} \bar{D} & \xrightarrow{(\eta,\zeta,\xi) \mapsto \eta\zeta^{-1}\xi} & \Omega \\ {\scriptstyle p}\downarrow & & \\ D & & \downarrow{\scriptstyle \natural} \\ {\scriptstyle \natural\times\natural|_D}\downarrow & & \\ \Omega/\Phi * \Omega/\Phi & \xrightarrow{\text{multiplication}} & \Omega/\Phi \end{array}$$

and since $\natural \times \natural|_D$ and p are both open, it follows that the multiplication is continuous. //

This result may be related to a theorem on differentiable quotient groupoids stated by Pradines (1966).

Brown and Hardy (1976) prove the following criterion.

<u>Theorem 1.7.</u> Let Ω be a locally compact, Hausdorff topological groupoid on B and let Φ be a compact normal subgroupoid. (In particular, B must be compact.) Then Ω/Φ, with the identification topology, is the topological quotient groupoid. //

A third criterion is given in 2.18.

It is proved by Brown and Hardy (1976), that topological quotient groupoids always exist, although 1.10 below shows that the topology on Ω/Φ need not be the identification topology. One would like to know, in general, to what extent the topology on a topological quotient groupoid Ω/Φ inherits the topological properties of Ω.

<u>Definition 1.8.</u> Let Ω,Ω' be topological groupoids on B,B'. An <u>isomorphism of topological groupoids</u> $\Omega \to \Omega'$ is a morphism of topological groupoids $\phi: \Omega \to \Omega'$, $\phi_o: B \to B'$ such that ϕ (and hence ϕ_o) is a homeomorphism. //

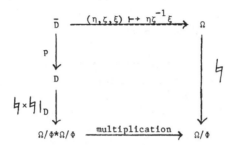

We will not use the terms "epimorphism" or "monomorphism". We now give some basic examples.

Example 1.9. If G is a topological group and B a space, the trivial groupoid $B \times G \times B$ is a topological groupoid in the product topology, called the trivial groupoid, on B with group G. The description of morphisms of trivial groupoids in I 2.13 remains valid for continuous morphisms. //

Example 1.10. Let X be an equivalence relation on a space B, considered as a normal subgroupoid of $B \times B$. In I 2.15 we identified the (algebraic) groupoid $(B \times B)/X$ with the product groupoid $(B/X) \times (B/X)$ and the natural projection $\natural: B \times B \to (B \times B)/X$ with $p \times p: B \times B \to (B/X) \times (B \times X)$.

We now prove that $(B/X) \times (B/X)$, with the cartesian square of the identification topology from p, is the topological quotient groupoid, despite the fact that $p \times p$ need not be an identification. Let $\phi: B \times B \to \Omega'$ be a continuous morphism over $\phi_o: B \to B'$ such that $\phi(X) \subseteq \widetilde{B'}$. Choose $b \in B$ and define $\sigma: B \to \Omega'$ by $\sigma(x) = \phi(x,b)$. Then $\phi(y,x) = \sigma(y)\sigma(x)^{-1}$, $\forall x,y \in B$. Now let $\bar{\sigma}: B/X \to \Omega'$ be $\bar{\sigma}([x]) = \sigma(x)$; then $\bar{\sigma}$ is of course continuous, and so $\bar{\phi}: (B/X) \times (B/X) \to \Omega'$, $([y],[x]) \mapsto \phi(y,x) = \bar{\sigma}([y])\bar{\sigma}([x])^{-1}$ is continuous.

Note that $\natural = p \times p: B \times B \to (B \times B)/X$ may be an identification without being open: let B consist of the two axes in $\mathbb{R}^2$ and let X collapse the y-axis to the origin; p is then the projection onto the x-axis, $B \to \mathbb{R}$. Since B and $\mathbb{R}$ are locally compact, $\natural = p \times p$ is an identification; since p is not open, $\natural$ cannot be.

Any equivalence relation X on any space B is a topological groupoid on B with the subspace topology from $B \times B$. Note that the projections $X \to B$ are not always open maps. //

Example 1.11. Let $G \times B \to B$ be a continuous action of a topological group G on a space B. Then the action groupoid $G \times B$, with the product topology, is a topological groupoid on B. //

Example 1.12. Let $P(B,G,\pi)$ be a Cartan principal bundle (see A 1.1 for definition). Then $\frac{P \times P}{G}$ with the identification topology from $P \times P \to \frac{P \times P}{G}$, $(u_2,u_1) \mapsto \langle u_2,u_1 \rangle$ (see I 1.10) is a topological groupoid on B.

We prove that the groupoid multiplication is continuous. Denote $\frac{P \times P}{G}$ by Ω and $(u_2,u_1) \mapsto \langle u_2,u_1 \rangle$ by p. Now p is open, for if $U \subseteq P \times P$ is any subset, we have $p^{-1}(p(U)) = \bigcup_{g \in G} Ug$ (with respect to the action of G on P × P) and so $p^{-1}(p(U))$ is open if U is. Hence $p \times p\colon P^2 \times P^2 \to \Omega \times \Omega$ is open. Now $(p \times p)^{-1}(\Omega * \Omega)$ $= P \times (P \underset{\pi}{\times} P) \times P$ and

$$\begin{array}{ccc} P \times (P \underset{\pi}{\times} P) \times P & \xrightarrow{(u_4,u_3,u_2,u_1) \mapsto (u_4,u_1\delta(u_3,u_2))} & P \times P \\ {\scriptstyle p \times p}\downarrow & & \downarrow{\scriptstyle p} \\ \Omega * \Omega & \xrightarrow{\text{groupoid multiplication}} & \Omega \end{array}$$

commutes. Since an open map restricted to a saturated set, in this case $(p \times p)^{-1}(\Omega * \Omega)$, is open, it follows that the groupoid multiplication is continuous.

That local triviality of $P(B,G,\pi)$ is not necessary for $\frac{P \times P}{G}$ to be a topological groupoid, was first pointed out by Dakin and Seda (1977).

If $F(f,\phi)\colon P(B,G) \to P'(B',G')$ is a morphism of Cartan principal bundles, then $\langle v,u\rangle \mapsto \langle F(v),F(u)\rangle$ is a continuous morphism of topological groupoids over f.

Consider the special case $G(G/H,H)$, where G is a topological group and H is a subgroup. Then $\frac{G \times G}{H}$ is isomorphic as a topological groupoid to the action groupoid $G \times (G/H)$ (where G acts on G/H to the left in the standard way) under $\langle g_2,g_1\rangle \mapsto (g_2g_1^{-1},g_1H)$.

Returning to general principal bundles, if P(B,G) is locally trivial then the inner subgroupoid $G(\frac{P \times P}{G})$ is a topological group bundle (see A 1.12 for definition) and is naturally isomorphic to $\frac{P \times G}{G}$, the inner group bundle of A§1. The isomorphism is defined by the map

$$\frac{P \times G}{G} \to \frac{P \times P}{G}, \qquad \langle u,g\rangle \mapsto \langle ug,u\rangle .$$

Note that if G is abelian then $\frac{P \times G}{G}$ is naturally isomorphic to B × G under $\langle u,g\rangle \mapsto (\pi(u),g)$. //

<u>Example 1.13.</u> A fibre bundle is a continuous surjection $p\colon M \to B$ with the local triviality property with respect to some fibre type F, where to avoid unnecessary complications we assume that F is locally compact, locally connected and Hausdorff (see A 1.6 for precise definition). By $\Pi(M)$ we from now on understand the groupoid of all homeomorphisms $M_x \to M_y$, $x,y \in B$. We use the local triviality of $p\colon M \to B$ to place a topology on $\Pi(M)$ with respect to which it will be a topological groupoid.

Let $\{\psi_i: U_i \times F \to M_{U_i}\}$ be an atlas for M. Let Homeom(F) denote the group of all homeomorphisms f: F → F with the compact-open topology; it is known that Homeom(F) is a topological group (Arens (1946, §5)), that the evaluation map Homeom(F) × F → F is continuous and that a map X → Homeom(F) from any space X is continuous iff the associated map X × F → F is continuous (for example, Dugundji (1966, Chapter XII)).

For each i and j, define

$$\bar{\psi}_i^j: U_j \times \mathrm{Homeom}(F) \times U_i \to \Pi(M)_{U_i}^{U_j} \quad \text{by} \quad (y,f,x) \mapsto \psi_{j,y} \circ f \circ \psi_{i,x}^{-1} .$$

Clearly each $\bar{\psi}_i^j$ is a bijection and any $(\bar{\psi}_k^{\ell})^{-1} \circ \bar{\psi}_i^j$ which has a nonvoid domain is a homeomorphism. Hence there is a well-defined topology on $\Pi(M)$ for which each $\bar{\psi}_i^j$ is a homeomorphism.

That $\Pi(M)$ is now a topological groupoid is straightforward: one works locally and the details are similar to those for a trivial groupoid.

For a TGB (M,p,B) (see A 1.12 for definition) with locally compact, locally connected and Hausdorff fibres we will always understand by $\Pi(M)$ the topological subgroupoid of topological group isomorphisms; for a vector bundle (E,p,B) of finite rank we will always understand by $\Pi(E)$ the topological subgroupoid of vector space isomorphisms.

For a fibre bundle with fibres which are not locally compact (for example the CVB's of Mackenzie (1978)) or not locally connected, this construction can sometimes be carried through, but we will not need that generality. On the other hand, if p: M → B is merely a continuous surjection, one can presumably adapt the modified compact-open topology of Booth and Brown (1978) to make inversion $\Pi(M) \to \Pi(M)$, $\xi \mapsto \xi^{-1}$ continuous and thus, under some suitable local compactness condition on M, make $\Pi(M)$ into a topological groupoid, even when M → B has no local triviality properties. This is the more interesting of the two generalizations, but we have no specific need for it.

See also Seda (1980, §4). //

<u>Example 1.14</u>. The following example is from Brown and Danesh-Naruie (1975).

Let B be a path-connected, locally path-connected and semi-locally simply connected space. The first condition ensures that the fundamental groupoid $\mathcal{T}(B)$ is transitive; the last two that the topology of B has a basis of open, path-connected sets U_i such that the inclusion $U_i \subseteq B$ maps each fundamental group $\pi_1(U_i,x)$, $x \in U_i$ to the trivial subgroup of $\pi_1(B,x)$. Such sets may be called canonical.

Let N be a normal, totally intransitive subgroupoid of $\mathcal{T}(B)$. We define a topology in $\mathcal{T}(B)/N$ with respect to which it is a topological groupoid on B, and a topological quotient groupoid of $\mathcal{T}(B)$. To reduce the notation, denote $\mathcal{T}(B)/N$ by Ω.

For each canonical U_i and $x \in U_i$ choose a function $\theta_{i,x}$ which to $y \in U_i$ assigns a path in U_i from x to y. By the conditions on U_i, the map $y \mapsto \langle\theta_{i,x}(y)\rangle$, $U_i \to \mathcal{T}(B)_x$ depends only on U_i and x, not on the representative paths chosen. Let $\tilde{U}_{i,x}$ denote the image of U_i under the composition of this map with the projection $\natural: \mathcal{T}(B) \to \Omega$.

It is easy to prove that the sets $\tilde{U}_{j,y}[\langle c\rangle]\tilde{U}_{i,x}^{-1}$, as U_j and U_i range through the basis of canonical sets, y ranges through U_j and x through U_i, and $[\langle c\rangle]$ ranges through Ω_x^y, form a basis for a topology in Ω, and that with this topology Ω is a topological groupoid on B. We verify the continuity of the groupoid multiplication: take $[\langle c'\rangle] \in \Omega_y^z$, $[\langle c\rangle] \in \Omega_x^y$ and write c" for c'c. Consider a basic open neighbourhood $\tilde{U}_{k,z}[\langle c''\rangle]\tilde{U}_{i,x}^{-1}$ of $[\langle c''\rangle]$. Choose any U_j which contains y; it is immediate to verify that

$$\tilde{U}_{k,z}[\langle c'\rangle]\tilde{U}_{j,y}^{-1} \quad \tilde{U}_{j,y}[\langle c\rangle]\tilde{U}_{i,x}^{-1} \subseteq \tilde{U}_{k,z}[\langle c''\rangle]\tilde{U}_{i,x}^{-1},$$

which shows that the multiplication in Ω is continuous.

Clearly $\natural: \mathcal{T}(B) \to \Omega$ maps basic open sets to basic open sets, so $\natural$ is open and Ω is therefore the topological quotient groupoid of $\mathcal{T}(B)$ over N.

The vertex groups $\Omega_x^x = \pi_1(B,x)/N_x$ inherit from Ω the discrete topology, for if $[\langle\lambda\rangle] \in \Omega_x^x$ and $U_i \ni x$, then $\tilde{U}_{i,x}[\langle\lambda\rangle]\tilde{U}_{i,x}^{-1} \cap \Omega_x^x = \{[\langle\lambda\rangle]\}$. It is also clear that the α-fibres Ω_x, $x \in B$, inherit from Ω the standard topology which makes $\beta_x: \Omega_x \to B$ the covering of B determined by $N_x < \pi_1(B,x)$, and that the right action $\Omega_x \times \Omega_x^x \to \Omega_x$ by groupoid multiplication is the deck-transformation action of $\pi_1(B,x)/N_x$ on the covering determined by N_x (see, for example, Hu (1959, III§16)).

It is trivial to check that, given a normal subgroup $H < \pi_1(B,b)$ for some chosen $b \in B$, there is a unique totally intransitive normal subgroupoid N of $\mathcal{T}(B)$

such that $N_b = H$. Thus the groupoid formulation efficiently encapsulates the phenomena of the theory of regular covering spaces. It does not, however, adapt well to the general, non-regular case - any collection whatever of subgroups $H_x < \pi_1(B,x)$, $x \in B$, constitutes a totally intransitive subgroupoid of $\pi(B)$.

The topology on $\pi(B)$ is the identification topology from the compact-open topology on the space of continuous paths in B. This is clear from the constructions in §6.

Returning to the original situation, note that the anchor $[\bar{\beta},\bar{\alpha}]: \pi(B)/N \to B \times B$ is itself a covering: it is easy to see that, given U_j and U_i, the open sets $\tilde{U}_{j,y}[\langle c\rangle]\tilde{U}^{-1}_{i,x}$, as y and x range through U_j and U_i respectively, and $[\langle c\rangle]$ through $(\pi(B)/N)^y_x$, are either disjoint or equal; they are therefore the components of their union, which is $[\bar{\beta},\bar{\alpha}]^{-1}(U_j \times U_i)$, and it is easy to see that each $[\bar{\beta},\bar{\alpha}]: \tilde{U}_{j,y}[\langle c\rangle]\tilde{U}^{-1}_{i,x} \to U_j \times U_i$ is a homeomorphism. Brown and Danesh-Naruie prove the non-trivial result that the fundamental group of $\pi(B)/N$ at a unity $[\langle\kappa_x\rangle]$ is isomorphic to the subgroup $\{(a,b) \in \pi_1(B,x) \times \pi_1(B,x) | a^{-1}b \in N_x\}$; this subgroup, of course, need not be normal.

Lastly, it is easy to see that if B' is a second path-connected, locally path-connected and semi-locally simply-connected space, N' a normal totally intransitive subgroupoid of $\pi(B')$ and $f: B \to B'$ a continuous map such that $f_*: \pi(B) \to \pi(B')$ maps N into N', then the induced morphism $\pi(B)/N \to \pi(B')/N'$ is continuous. //

Example 1.15. Let B be a topological space and Γ a pseudogroup of local homeomorphisms $\phi: U \to V$ of B (that is, Γ contains the identity id_B, and is closed under restriction, inversion and composition). Let $J^\lambda(\Gamma)$ denote the set of germs $g_x\phi$ (or "local jets") of elements of Γ, with the obvious groupoid structure: $\alpha(g_x\phi) = x$, $\beta(g_x\phi) = \phi(x)$, $\tilde{x} = g_x(id_B)$, $(g_x\phi)^{-1} = g_{\phi(x)}(\phi^{-1})$ and $(g_{\phi(x)}\psi)(g_x\phi) = g_x(\psi\circ\phi)$. Then $J^\lambda(\Gamma)$ is called the germ groupoid of Γ. The pseudogroup Γ is transitive if $\forall\, x,y \in B$ there is an element $\phi \in \Gamma$ such that $\phi(x) = y$ (for example, Kobayashi (1972)); clearly $J^\lambda(\Gamma)$ is transitive iff Γ is.

For $\phi \in \Gamma$, define $N_\phi = \{g_x\phi \mid x \in \text{dom}\,\phi\}$. The sets N_ϕ, $\phi \in \Gamma$, form a basis for a topology in $J^\lambda(\Gamma)$, called the sheaf topology, and it is easy to see that $J^\lambda(\Gamma)$ with this topology is a topological groupoid on B, which we will denote by $J^\lambda_{sh}(\Gamma)$. The importance of this topology is well-established (see, for example, Lawson (1977) or the survey by Stasheff (1978)). However for our purposes it is mainly of interest in providing a naturally occurring topological groupoid which is somewhat pathological: it is easy to see, for example, that each $J^{\lambda}_{sh}(\Gamma)_x$ is discrete and that $\tilde{B}$ is open in $J^{\lambda}_{sh}(\Gamma)$. In particular, there are sets $U \subseteq B$ for which each $J^{\lambda}_{sh}(\Gamma)^U_x$, $x \in B$, is open, but for which neither $J^{\lambda}_{sh}(\Gamma)^U$ nor U itself is open. Note also that α and β are étale.

If B is locally compact and Hausdorff one can define a compact-open topology in $J^{\lambda}(\Gamma)$ as follows (Abd-Allah and Brown (1980)): in Γ itself define subsets by

$$N(K,U) = \{\phi \in \Gamma \mid K \subseteq \mathrm{dom}(\phi),\ \phi(K) \subseteq U\}$$

$$N'(U,K) = \{\phi \in \Gamma \mid U \subseteq \mathrm{dom}(\phi),\ \phi(U) \supseteq K\}$$

for $K \subseteq B$ compact and $U \subseteq B$ open. Take the topology in Γ generated by these sets as subbasis, and give $J^{\lambda}(\Gamma)$ the identification topology with respect to $(x,\phi) \mapsto g_x\phi$ defined on $B*\Gamma = \{(x,\phi) \in B \times \Gamma \mid x \in \mathrm{dom}(\phi)\}$. It is straightforward to show that, with this topology, $J^{\lambda}(\Gamma)$ is a topological groupoid, denoted $J^{\lambda}_{co}(\Gamma)$; the details are similar to those in Arens (1946). This structure may be more appropriate in groupoid theory itself - see 5.9. //

Example 1.16. Let Ω be a transitive algebraic groupoid on a space B. Give Ω the coarsest topology for which the anchor $[\beta,\alpha]: \Omega \to B \times B$ is continuous; that is, Ω has the sets Ω^V_U, $U,V \subseteq B$ open, as a basis. It is easy to verify that Ω is a topological groupoid on the space B, clearly the coarsest topology on Ω for which this is so. Note that each vertex group has the indiscrete topology. //

Example 1.17. Any TGB (topological group bundle - see A 1.12 for definition) is a totally intransitive topological groupoid. //

The notes by Renault (1980) contain further examples of topological groupoids; most are equivalence relation groupoids (as in I 1.5) but with topologies finer than the subspace topology from the cartesian square of the base, and all share with $J^{\lambda}_{sh}(\Gamma)$ the property that the unities form an open subset. From our point of view such examples are pathological.

Topological groupoids of the form $\frac{P \times P}{G}$ constructed in 1.12 admit an intrinsic characterization. The following definition and Proposition 1.19 are due to Dakin and Seda (1977).

Definition 1.18. Let Ω be a topological groupoid on B. Then Ω is principal if it is transitive and if for any one, and hence every, $x \in B$, the maps $\beta_x: \Omega_x \to B$ and

$\delta_x: \Omega_x \times \Omega_x \to \Omega$, $(\eta,\xi) \mapsto \eta\xi^{-1}$, are identifications. //

If δ_x is an identification then it is in fact open, for the saturation of any basic open set $V \times U \subseteq \Omega_x \times \Omega_x$ is $\bigcup_{\lambda\in\Omega_x^x} V\lambda \times U\lambda$, which is itself open. Similarly if β_x is an identification, then it is open.

If $P(B,G,\pi)$ is a Cartan principal bundle, then the groupoid $\frac{P \times P}{G}$ of 1.12 is easily seen to be principal. Conversely, if Ω is a principal topological groupoid then, for any $x \in B$, $\Omega_x(B,\Omega_x^x,\beta_x)$ is a Cartan principal bundle, and for any other $y \in B$ and $\zeta \in \Omega_x^y$, the maps $R_{\zeta^{-1}}: \Omega_x \to \Omega_y$ and $I_\zeta: \Omega_x^x \to \Omega_y^y$ form an isomorphism of Cartan principal bundles over B. These correspondences are mutually inverse, though the necessity of choosing reference points complicates the precise formulation:

<u>Proposition 1.19</u>. (i) Let $P(B,G,\pi)$ be a Cartan principal bundle. Choose $u_o \in P$ and write $x_o = \pi(u_o)$. Then the map

$$P \to \frac{P \times P}{G}\Big|_{x_o}, \qquad u \mapsto \langle u,u_o\rangle$$

is a homeomorphism, the map

$$G \to \frac{P \times P}{G}\Big|_{x_o}^{x_o}, \qquad g \mapsto \langle u_o g,u_o\rangle$$

is an isomorphism of topological groups and together they form an isomorphism of Cartan principal bundles over B.

Let $F(f,\phi): P(B,G) \to P'(B',G')$ be a morphism of Cartan principal bundles, denote by $F^*: \frac{P \times P}{G} \to \frac{P' \times P'}{G'}$ the induced morphism of groupoids, and choose $u_o \in P$, $u'_o \in P'$ such that $u'_o = F(u_o)$. Write $x_o = \pi(u_o)$, $x'_o = \pi'(u'_o)$. Then

$$\begin{array}{ccc} \frac{P \times P}{G}\Big|_{x_o} & \xrightarrow{F^*|_{x_o}} & \frac{P' \times P'}{G'}\Big|_{x'_o} \\ \uparrow & & \uparrow \\ P & \xrightarrow{F} & P' \end{array}$$

commutes, where the vertical maps are the isomorphisms corresponding to u_o and u'_o.

(ii) Let Ω be a principal topological groupoid on B, and choose $x \in B$. Then the map

$$\frac{\Omega_x \times \Omega_x}{\Omega_x^x} \to \Omega, \qquad \langle \eta, \xi \rangle \mapsto \eta\xi^{-1}$$

is an isomorphism of topological groupoids over B.

Let $\phi\colon \Omega \to \Omega'$ be a morphism of principal topological groupoids over $\phi_o\colon B \to B'$ and choose $x \in B$, $x' \in B'$ such that $x' = \phi_o(x)$. Then

$$\begin{array}{ccc} \dfrac{\Omega_x \times \Omega_x}{\Omega_x^x} & \xrightarrow{\ \phi_x^*\ } & \dfrac{\Omega'_x \times \Omega'_x}{\Omega'^x_x} \\ \Big\downarrow & & \Big\downarrow \\ \Omega & \xrightarrow{\ \phi\ } & \Omega' \end{array}$$

commutes.

<u>Proof</u>. In both cases the algebraic assertions are easily verified. To prove the continuity of the inverse of $P \to \left.\dfrac{P \times P}{G}\right|_{x_o}$ in (i), write it as

$$\begin{array}{ccc} P \times \pi^{-1}(x_o) & & \\ \Big\downarrow & \searrow & \\ \left.\dfrac{P \times P}{G}\right|_{x_o} & \longrightarrow & P \end{array}$$

where the oblique arrow is $(u, u_og) \mapsto ug^{-1}$. This map is continuous since $\delta\colon P \times_\pi P \to G$ is continuous, and the vertical map is an identification since it is the restriction of an identification to a saturated subset. In (ii) we have

$$\begin{array}{ccc} \Omega_x \times \Omega_x & & \\ \Big\downarrow & \searrow & \\ \dfrac{\Omega_x \times \Omega_x}{\Omega_x^x} & \longrightarrow & \Omega \end{array}$$

and the bottom arrow is an identification map since the other two maps are; since it is also a bijection it is a homeomorphism. //

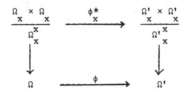

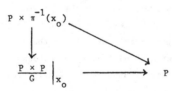

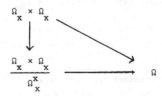

For a principal topological groupoid Ω, the bundle $\Omega_x(B,\Omega_x^x,\beta_x)$ is called the vertex bundle at x.

The choice of different reference points in 1.19(i) leads to automorphisms of P(B,G) of a specific form:

Proposition 1.20. With the notation of 1.19(i), let u_o' be a second reference point in $\pi^{-1}(x_o)$, say $u_o' = u_o h$ where $h \in G$. Then the composite automorphism

$$P \xrightarrow[(u_o)]{} \left.\frac{P \times P}{G}\right|_{x_0} \xrightarrow[(u_o')]{} P, \qquad G \xrightarrow[(u_o)]{} \left.\frac{P \times P}{G}\right|_{x_o}^{x_o} \xrightarrow[(u_o')]{} G$$

is $u \mapsto uh$, $g \mapsto h^{-1}gh$.

Proof Computation. //

In particular, G acts as a group of automorphisms of the bundle P(B,G) by $h \in G$ acting as $u \mapsto uh^{-1}$, $g \mapsto hgh^{-1}$ or, briefly, as $R_{h^{-1}}(id_B, I_h)$. Whenever a phenomenon in principal bundle theory is dependent on a reference point, one may be sure that changing the reference point within its fibre will map the phenomenon under an automorphism of this type; one may also be sure that if the phenomenon is formulated in groupoid terms then it will be an intrinsic concept, independent of reference points. The clearest example of this is the replacement of the various mutually conjugate holonomy groups and isomorphic holonomy bundles arising from a connection, by a single holonomy groupoid. See II 7.14 for this.

It is easy to verify that a transitive topological groupoid whose groupoid space is compact and Hausdorff, is principal. On the other hand, unless B is a discrete space, $J^{\lambda}_{sh}(\Gamma)$ cannot be principal, since the α-fibres $J^{\lambda}_{sh}(\Gamma)_x$ are discrete. Similarly an action groupoid $G \times B$ in which the evaluation maps $G \to B$, $g \mapsto gx_o$ are not open cannot be principal.

1.19 allows problems for principal groupoids to be reduced to problems for the vertex bundles and this technique can be extremely useful:

Proposition 1.21 Let Ω be a principal topological groupoid on B, let Ω' be any topological groupoid on B' and let $\phi\colon \Omega \to \Omega'$ be a morphism in the algebraic sense.

(i) If any one $\phi_b\colon \Omega_b \to \Omega'_{\phi_o(b)}$ is continuous, then ϕ is continuous.

(ii) If ϕ is continuous on a neighbourhood $\mathcal{U}$ of $\tilde{B}$ in Ω, then it is continuous everywhere on Ω.

Proof: (i) follows directly from 1.19. (ii) Choose $b \in B$. Now for every $\xi \in \Omega_b$, $\phi_{\beta\xi}$ is continuous in a neighbourhood of $\widetilde{\beta\xi}$ and $R_\xi: \Omega_{\beta\xi} \to \Omega_b$ maps $\widetilde{\beta\xi}$ to ξ. Since $\phi_b \circ R_\xi = R_{\phi(\xi)} \circ \phi_{\beta\xi}$ it follows that ϕ_b is continuous at ξ. Now apply (i). //

Although 1.21 (i) is easily seen to be false for arbitrary topological groupoids, it is not clear whether 1.21 (ii) is true in general.

The main examples of principal topological groupoids are locally trivial groupoids, which we treat next. In fact the main value of the concept of principal topological groupoid is that it expresses much of the force of the concept of local triviality, without using localization techniques; it also explains why all action groupoids $G \times (G/H)$ for homogeneous spaces G/H, not merely those which are locally trivial, are well-behaved and do not provide good examples of the pathology possible in the general theory. In the differentiable theory, the two concepts coincide.

§2. Local triviality.

A topological groupoid Ω is locally trivial if it is transitive and there is an open cover $\{U_i\}$ of the base such that each $\Omega_{U_i}^{U_i}$ is isomorphic to a trivial groupoid (see 2.2). For such groupoids a problem may be reduced to a local problem concerning trivial groupoids, and a globalization problem; this technique however, although it is almost universally used in principal bundle theory, is not always the most instructive, and is of course incapable of generalization to arbitrary topological groupoids. In the remainder of this book we will give intrinsic proofs, rather than use the localization-globalization technique, whenever it can be done without a great increase in length.

Locally trivial groupoids are equivalent to principal bundles under the correspondence 1.19 for principal groupoids and Cartan principal bundles. Much of the theory of principal bundles is simplified by reformulating it in groupoid terms, on account of the clearer algebraic structure of a groupoid, and because groupoid theory has a natural conceptual framework inherited from group theory. This will be especially evident in the Lie theory and connection theory of Chapter III.

In this section, after the definition and reformulations of the concept of local triviality, we examine morphisms of locally trivial groupoids in some detail, and sharpen some of the results of §1. These are chiefly technical results needed in later sections; little has been published on the algebraic analysis of morphisms of principal bundles. The section concludes with a brief account of the classification of locally trivial groupoids by cocycles; this material, well-known in the theory of principal bundles, is included here because we wish to emphasize a point about the extent to which base-preserving morphisms of locally trivial groupoids are determined by their restriction to vertex groups - this is important in the cohomology theory of locally trivial groupoids and transitive Lie algebroids, in understanding the maps $\mathcal{H}^2(\Omega,M) \to \Gamma H^2(G\Omega,M)^{B\times B}$ and $\mathcal{H}^2(A,E) \to \Gamma H^2(L,E)^{TB}$ (see Chapter IV). This classification by cocycles is one part of the theory which fits more naturally into the principal bundle formulation.

The concept of local triviality is due to Ehresmann (1959), as is the equivalence between locally trivial groupoids and principal bundles. The material from 2.17 to the end of the section is a reformulation of material standard in principal bundle theory. The remaining unsourced material in this section may be regarded as folklore.

Definition 2.1. Let Ω be a topological groupoid on B. Then Ω is locally trivial if there exists a point $b \in B$, an open cover $\{U_i\}$ of B, and continuous maps $\sigma_i: U_i \to \Omega_b$ such that $\beta_b \circ \sigma_i = \mathrm{id}_{U_i}$ for all i.

The maps σ_i will be called local sections of Ω, or local decomposing sections when it is necessary to distinguish them from the local admissible sections of §5.

The family $\{\sigma_i: U_i \to \Omega_b\}$ will be called a section-atlas for Ω.

If there is a global section $\sigma: B \to \Omega_b$ of Ω then Ω is called globally trivial or trivializable. //

A locally trivial groupoid is clearly transitive, and given any $x \in B$ there is a section-atlas $\{\sigma_i': U_i \to \Omega_x\}$ taking values in Ω_x. The significance of the concept of local triviality is shown by the following proposition, whose proof is clear.

Proposition 2.2. Let Ω be a topological groupoid on B, and let U be an open subset of B.

If $\sigma: U \to \Omega_b$ is a continuous right-inverse to β_b, for some $b \in B$, then the topological subgroupoid Ω_U^U is isomorphic to the trivial groupoid $U \times \Omega_b^b \times U$ under

$$\Sigma\colon (y,\lambda,x) \mapsto \sigma(y)\lambda\sigma(x)^{-1}.$$

Conversely, if G is a topological group and $\Sigma\colon U \times G \times U \to \Omega_U^U$ is an isomorphism of topological groupoids over U then, choosing any $b \in U$, the map $\sigma\colon U \to \Omega_b$, $x \mapsto \Sigma(x,1,b)$ is right-inverse to β_b. //

Thus locally trivial groupoids are "locally isomorphic" to trivial groupoids. The converse is not quite true, since a groupoid may be locally isomorphic to trivial groupoids without being transitive. Use of 2.2 leads to the following concept.

Definition 2.3. Let Ω be a topological groupoid on B. Then Ω is weakly locally trivial if there is an open cover $\{U_i\}$ of B, points $b_i \in B$, and continuous maps $\sigma_i\colon U_i \to \Omega_{b_i}$ such that $\beta_{b_i} \circ \sigma_i = id_{U_i}$ for all i.

The set $\{\sigma_i\colon U_i \to \Omega_{b_i}\}$ is still called a section-atlas for Ω. //

This is the concept which Ehresmann (1959) originally defined to be local triviality. It is clear that the points b_i may be assumed to lie in the corresponding sets U_i. We have chosen to include transitivity in the concept of local triviality, and to use the simpler definition 2.1 available in that case, because the transitivity components of a weakly locally trivial groupoid are easily seen to be both open and closed, and so the groupoid is topologically, as well as algebraically, the disjoint union of transitive - and locally trivial - topological subgroupoids.

Proposition 2.4. A topological groupoid which is both weakly locally trivial and transitive is locally trivial.

Proof. Trivial. //

For the last reformulation of the concept of local triviality we need the following definition from Brown et al (1976).

Definition 2.5. Let $f\colon X \to Y$ be a continuous map. Then f is a (topological) submersion if $\forall\, x_o \in X$ there is an open neighbourhood V of $f(x_o)$ in Y and a right-inverse $\sigma\colon V \to X$ to f such that $\sigma(f(x_o)) = x_o$. //

Note that this is stronger than the mere existence of a local right-inverse in some neighbourhood of any given point of Y: a map satisfying 2.5 is open, but a continuous map with a global right-inverse need not be open (project the union of the two axes in $\mathbb{R}^2$ onto one of them).

<u>Proposition 2.6</u>. Let Ω be a topological groupoid on B. The following conditions are equivalent:

(i) Ω is locally trivial;

(ii) $\beta_x : \Omega_x \to B$ is a surjective submersion for one, and hence for all $x \in B$;

(iii) $[\beta,\alpha] : \Omega \to B \times B$ is a surjective submersion;

(iv) $\beta : \Omega \to B$ is a submersion and $\delta_x : \Omega_x \times \Omega_x \to \Omega$ is a surjective submersion for one, and hence for all, $x \in B$.

<u>Proof</u>. (i) => (iii). Let $\{\sigma_i : U_i \to \Omega_b\}$ be a section-atlas for Ω and take $\xi \in \Omega$. Choose i, j such that $\alpha\xi \in U_i$, $\beta\xi \in U_j$ and define $\theta : U_j \times U_i \to \Omega$ by $\theta(y,x) = \sigma_j(y)\sigma_j(\beta\xi)^{-1}\xi\sigma_i(\alpha\xi)\sigma_i(x)^{-1}$. Clearly $[\beta,\alpha]\circ\theta = \mathrm{id}$ and $\theta(\beta\xi,\alpha\xi) = \xi$. The other parts are similar, though (iv) => (ii) is most easily proved from the diagram

$$\begin{array}{ccc} \Omega_x \times \Omega_x & \xrightarrow{\delta_x} & \Omega \\ \Big\downarrow{\scriptstyle \pi_1} & & \Big\downarrow{\scriptstyle \beta} \\ \Omega_x & \xrightarrow{\beta_x} & B \end{array} \quad . \qquad //$$

In particular, a locally trivial groupoid Ω is principal, and all the vertex bundles $\Omega_x(B,\Omega_x^x,\beta_x)$ and the bundles $\Omega_x \times \Omega_x(\Omega,\Omega_x^x,\delta_x)$ are principal bundles.

<u>Examples 2.7</u>. Obviously trivial groupoids are locally trivial. A transitive action $G \times B \to B$ gives a locally trivial groupoid iff the evaluation maps $G \to B$, $g \mapsto gx_0$, are submersions; this is always the case for a smooth (transitive) action of a Lie group (see, for example, Dieudonné (1972, 16.10.8(i))). For the standard action on a homogeneous space G/H, the action groupoid is locally trivial iff $G \to G/H$ admits local sections.

The groupoid $\frac{P \times P}{G}$ associated to a Cartan principal bundle is locally trivial iff the bundle $P(B,G,\pi)$ is locally trivial: if $\sigma : U \to P$ is a local section of the bundle, then $x \mapsto \langle\sigma(x),u_0\rangle$, $U \to \frac{P \times P}{G}\Big|_{\pi(u_0)}$ is a local section of

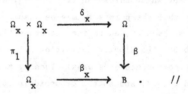

$\frac{P \times P}{G}$. If H is a subgroup of a topological group G for which $G \to G/H$ is not a submersion, then $\frac{G \times G}{H}$ is principal but not locally trivial.

That the frame groupoid $\Pi(M)$ of a fibre bundle M is locally trivial is clear from the way in which we defined the topology in $\Pi(M)$ (see 1.13). The same remark applies to the fundamental groupoid $\mathcal{T}(B)$ of a path-connected, locally path-connected and semi-locally simply-connected space B; the maps $y \mapsto \langle\theta_{i,x}(y)\rangle$ form a section-atlas of the type defined in 2.3.

A transitive topological groupoid on a discrete space is locally trivial, in fact globally trivial, and so also is a transitive groupoid with the coarsest groupoid topology (see 1.16).

If B is a Hausdorff topological manifold and $\Gamma^o(B)$ is the full pseudogroup on B, then $J^\lambda_{co}(\Gamma^o(B))$ is locally trivial - see 5.9 for a more general result.

Lastly, any inverse image of a locally trivial groupoid is locally trivial.

//

The topology of a topological group may be defined by means of a system of symmetric neighbourhoods of the identity. A neighbourhood $\mathcal{U}$ of the base Δ_B of the square topological groupoid $B \times B$ contains a neighbourhood of the form $\bigcup_i (U_i \times U_i)$, where $\{U_i\}$ is an open cover of B, and one may loosely identify open neighbourhoods of the base with open covers of it. For a general locally trivial groupoid Ω on a paracompact, second countable, Hausdorff topological manifold B, dimension theory shows that there is a finite section-atlas and it therefore follows that an open neighbourhood $\mathcal{U}$ of $\tilde{B}$ in Ω contains an open neighbourhood constructed locally from neighbourhoods of the form $U_i \times N \times U_i$, where N is a neighbourhood of the identity in a single vertex group Ω_b^b. Evidently then, the topology of Ω could be reconstructed from the system of such neighbourhoods, but this observation will not be used (compare the proof of 1.21(ii), where such methods are avoided).

We now analyze morphisms of locally trivial groupoids.

__Lemma 2.8.__ Let $\phi: \Omega \to \Omega'$ be a morphism in the algebraic sense of locally trivial groupoids, and let $\{\sigma_i : U_i \to \Omega_b\}$ and $\{\tau_j : V_j \to \Omega'_{b'}\}$ be section-atlases for Ω and Ω' with $b' = \phi_o(b)$. For any V_j and any U_i with $U_i \subseteq \phi_o^{-1}(V_j)$ define $\theta_{ij}: U_i \to \Omega'^{b'}_{b'}$ by

$\theta_{ij}(x) = \tau_j(\phi_o(x))^{-1}\,\phi(\sigma_i(x))$. Then in terms of the isomorphisms $U_i \times \Omega_b^b \times U_i \to \Omega_{U_i}^{U_i}$ and $V_j \times \Omega'^{b'}_{b'} \times V_j \to \Omega'^{V_j}_{V_j}$ induced by σ_i and τ_j, the morphism $\phi: \Omega_{U_i}^{U_i} \to \Omega'^{V_j}_{V_j}$ is represented by $(y,\lambda,x) \mapsto (\phi_o(y), \theta_{ij}(y)\phi_b^b(\lambda)\theta_{ij}(x)^{-1}, \phi_o(x))$.

Proof. Exercise. //

In particular if ϕ_o and ϕ_b^b are continuous and one can find section-atlases with respect to which the θ_{ij} are continuous, then ϕ is continuous.

The proof of the following proposition is given in full because it is a typical example of working locally with morphisms of locally trivial groupoids.

Proposition 2.9. Let Ω and Ω' be locally trivial groupoids on B and B', respectively, and let $\phi: \Omega \to \Omega'$ be a piecewise homeomorphic morphism. Then ϕ is a pullback for the category of locally trivial groupoids.

Proof. The assertion is, that if Φ is a locally trivial groupoid on B and $\psi: \Phi \to \Omega'$ is a morphism over ϕ_o, then there is a unique morphism $\bar{\psi}: \Phi \to \Omega$ over B such that $\psi = \phi\circ\bar{\psi}$.

$\bar{\psi}$ must be defined by $\bar{\psi}_x^y = (\phi_x^y)^{-1} \circ \psi_x^y$ for $x,y \in B$. Let $\{\tau_i : U_i \to \Phi_b\}$ be a section-atlas for Φ and $\{\sigma_j' : V_j \to \Omega'_{b'}\}$ a section-atlas for Ω', with $b' = \phi_o(b)$. Write $W_j = \phi_o^{-1}(V_j)$ and define $\sigma_j : W_j \to \Omega_b$ by $\sigma_j(x) = (\phi_b^x)^{-1}(\sigma_j'(\phi_o(x)))$. We prove that σ_j is continuous.

Take a continuous section $\nu_k : A_k \to \Omega_b$ of β_b with $A_k \subseteq W_j$. As in 2.8, the map $\phi_b : \Omega_b^{A_k} \to \Omega'^{V_j}_{b'}$ can be written in terms of the homeomorphisms $A_k \times \Omega_b^b \to \Omega_b^{A_k}$, $(y,\lambda) \mapsto \nu_k(y)\lambda$ and $V_j \times \Omega'^{b'}_{b'} \to \Omega'^{V_j}_{b'}$, $(y,\lambda) \mapsto \sigma_j'(y)\lambda$ as $A_k \times \Omega_b^b \to V_j \times \Omega'^{b'}_{b'}$, $(y,\lambda) \mapsto (\phi_o(y), \theta_{kj}(y)\phi_b^b(\lambda))$ where $\theta_{kj}(y) = \sigma_j'(\phi_o(y))^{-1}\phi(\nu_k(y))$. The restriction of σ_j to $A_k \to \Omega_b^{A_k}$ is mapped under $\Omega_b^{A_k} \to A_k \times \Omega_b^b$ to $y \mapsto (y, (\phi_b^b)^{-1}(\theta_{kj}(y)^{-1}))$ and is thus continuous.

Thus $\{\sigma_j : W_j \to \Omega_b\}$ is a section-atlas for Ω. Now with respect to the isomorphisms $U_i \times \Phi_b^b \times U_i \to \Phi_{U_i}^{U_i}$ and $W_j \times \Omega_b^b \times W_j \to \Omega_{W_j}^{W_j}$ induced by τ_i and σ_j (where we can assume $U_i = W_j$, by restricting τ_i and σ_j to their intersection) the algebraic morphism $\bar{\psi}$ is

$$(y,\lambda,x) \mapsto (y, (\sigma_b^b)^{-1}(\theta'_{ij}(y))\bar{\psi}_b^b(\lambda)(\phi_b^b)^{-1}(\theta'_{ij}(x)^{-1}), x)$$

where $\theta'_{ij} : U_i \to \Omega'^{b'}_{b'}$ is $y \mapsto \sigma_j'(\phi_o(y))^{-1}\ \psi(\tau_i(y))$. Since all the maps appearing in this local representation are continuous it follows that $\bar{\psi}$ is continuous. //

The following result is given by Brown et al (1976).

Proposition 2.10. Let Ω be a locally trivial groupoid on B, let Ω' be a topological groupoid on B' and let $\phi: \Omega \to \Omega'$ be a morphism. If ϕ_o is a surjective submersion then Ω' is locally trivial.

Proof. Take $b \in B$ and write $b' = \phi_o(b)$. Take $x'_o \in B'$, say $x'_o = \phi_o(x_o)$. Take a local right-inverse $\tau: V \to B$ to ϕ_o with V an open neighbourhood of x'_o and $\tau(x'_o) = x_o$. Now $\phi_o^{-1}(V)$ is an open neighbourhood of x_o; take $\sigma: U \to \Omega_b$ with $x_o \in U \subseteq \phi_o^{-1}(V)$. Define $\sigma': \tau^{-1}(U) \to \Omega'_{b'}$ by $\sigma'(x') = \phi(\sigma(\tau(x')))$. Then σ' is a continuous section of $\beta'_{b'}: \Omega'_{b'} \to B'$. //

We will mainly apply 2.10 in cases where ϕ_o is an identity map. For the remainder of this section we restrict attention to base-preserving morphisms.

Lemma 2.11. Let $\phi: \Omega \to \Omega'$ be a morphism of locally trivial groupoids over B. Let $\{\sigma_i: U_i \to \Omega_b\}$ be a section-atlas for Ω and define $\sigma'_i = \phi \circ \sigma_i$. Then, with respect to the $U_j \times \Omega_b^b \times U_i \to \Omega_{U_i}^{U_j}$ and $U_j \times \Omega'^b_b \times U_i \to \Omega'^{U_j}_{U_i}$ induced by σ_j, σ_i and σ'_j, σ'_i, the morphisms $\sigma: \Omega_{U_i}^{U_j} \times \Omega'^{U_j}_{U_i}$ is locally represented by $id_{U_j} \times \phi_b^b \times id_{U_i}$.

Proof. Exercise. //

It follows from 2.11 that base-preserving morphisms of locally trivial groupoids inherit many properties from their restrictions to vertex groups:

Definition 2.12. A continuous map $f: X \to Y$ is an embedding if it is injective and if $f: X \to f(X)$ is a homeomorphism with respect to the subspace topology on f(X). //

Proposition 2.13. Let $\phi: \Omega \to \Omega'$ be a morphism of locally trivial groupoids over B. Choose $b \in B$. Then

(i) ϕ is open iff ϕ_b^b is open;

(ii) ϕ is a surjective submersion iff ϕ_b^b is a surjective submersion;

(iii) ϕ is an embedding iff ϕ_b^b is an embedding.

Proof. (i) (=>) $\Omega_b^b = \phi^{-1}(\Omega'^b_b)$ and the restriction of an open map to a saturated subset is open. (<=) ϕ_b^b open implies that each id × ϕ_b^b × id is open.

(ii) is similar to (i) and (iii) follows from (i) since im ϕ is a topological groupoid in its own right and is locally trivial by 2.10. //

Notation 2.14. Let $\phi: \Omega \to \Omega'$ be a morphism of locally trivial groupoids over B. Then $\phi: \Omega \twoheadrightarrow \Omega'$ denotes that ϕ is a surjective submersion, and $\phi: \Omega \rightarrowtail \Omega'$ that ϕ is an embedding. //

Theorem 2.15. Let Ω be a locally trivial groupoid on a T_1 space B and let N be a totally intransitive normal subgroupoid of Ω. Then Ω/N, with the identification topology from $\natural: \Omega \to \Omega/N$, is the topological quotient groupoid, and $\natural$ is open.

Proof. Give the algebraic groupoid Ω/N the identification topology; we show that it is a topological groupoid, by showing that the bijections

$$U_j \times (\Omega/N)_b^b \times U_i \to (\Omega/N)_{U_i}^{U_j} \tag{1}$$

$$(y,[\lambda],x) \mapsto \sigma'_j(y)[\lambda]\sigma'_i(x)^{-1} = [\sigma_j(y)\lambda\sigma_i(x)^{-1}],$$

where $\{\sigma_i : U_i \to \Omega_b\}$ is a section-atlas for Ω, and $\sigma'_i = \natural\circ\sigma_i$, are homeomorphisms. Recall that if f: X → Y is an identification, and $A \subseteq Y$ is either open or closed, then the restriction f: $f^{-1}(A) \to A$ is an identification.

First take f = $\natural$ and A to be the closed subset $(\Omega/N)_b^b = [\bar{\beta},\bar{\alpha}]^{-1}(b,b)$. It follows that $\natural_b^b: \Omega_b^b \to (\Omega/N)_b^b$ is an identification and so, since it is a quotient-group projection, it is open. Hence $id_{U_j} \times \natural_b^b \times id_{U_i}$ is open, in particular an identification, and it follows that (1) is continuous.

Second, take f = $\natural$ and A to be the open subset $(\Omega/N)_{U_i}^{U_j}$. Then $\natural: \Omega_{U_i}^{U_j} \to (\Omega/N)_{U_i}^{U_j}$ is an identification and so the inverse bijection to (1) is continuous.

By working locally it is easy to show that Ω/N is a topological groupoid, obviously locally trivial. That $\natural$ is open follows from the construction, or from 2.13. //

That N is totally intransitive is crucial to the above proof; if it were not, $(\Omega/N)_{[b]}^{[b]}$ would receive from Ω/N the identification topology not from Ω_b^b, but from $\bigcup_{x\in[b]} \Omega_x^x$. Even if $\natural_o: B \to B/N$ is assumed to be a surjective submersion, it seems unlikely that the topological quotient groupoid Ω/N always receives the identification topology from Ω.

Note that in 2.15 $\natural$ need not be a surjective submersion - this already occurs for topological groups.

The following trivial result will be needed later.

Proposition 2.16. Let Ω be a locally trivial groupoid and N a totally intransitive normal subgroupoid. Let $\{\sigma_i : U_i \to \Omega_b\}$ be a section-atlas for Ω. Then the set of all $U_i \times N_b^b \to N_{U_i}^{U_i}$, $(x,\lambda) \mapsto I_{\sigma_i(x)}(\lambda) = \sigma_i(x)\lambda\sigma_i(x)^{-1}$ forms an atlas for N as a TGB. //

In particular, the kernel of a base-preserving morphism $\Omega \to \Omega'$ of locally trivial groupoids is a sub TGB of $G\Omega$.

* * * * * * *

We now wish to show that, given a locally trivial groupoid Ω on B and a group morphism $f: \Omega_b^b \to H$ for some b, there is a locally trivial groupoid Ω' on B with $\Omega'^b_b = H$ and a morphism $\phi: \Omega \to \Omega'$ over B with $\phi_b^b = f$. For this we need the concept of cocycle and the construction of locally trivial groupoids from cocycles. The treatment will be brief, since the corresponding construction for principal bundles is well-known.

Definition 2.17. Let B be a space and let G be a topological group. A cocycle on B with values in G consists of an open cover $\{U_i\}$ of B and maps $s_{ij}: U_{ij} = U_i \cap U_j \to G$ such that whenever $U_{ijk} = U_i \cap U_j \cap U_k \neq \emptyset$ we have

$$s_{ij}(x)s_{kj}(x)^{-1}s_{ki}(x) = 1 \qquad \forall\, x \in U_{ijk}\,. \tag{2}$$

Two cocycles $\{s_{ij}: U_{ij} \to G\}$ and $\{s'_{k\ell}: V_{k\ell} \to G\}$ on B are equivalent if there is a common refinement $\{W_m\}$ of $\{U_i\}$ and $\{V_k\}$ and maps $r_m: W_m \to G$ such that $s'_{mn}(x) = r_m(x)^{-1}s_{mn}(x)r_n(x)$ $\forall\, x \in W_{mn}$ and $\forall\, m,n$. (Here s_{mn} denotes the restriction of an s_{ij} to $W_{mn} \subseteq U_{ij}$.) //

Equation (2) is called the cocycle equation; clearly it implies that each s_{ii} is constant at 1 (set $i = j = k$) and that $s_{ji}(x) = s_{ij}(x)^{-1}$ $\forall\, x \in U_{ij}$ (set $j = k$). The elements s_{ij} of a cocycle may be called transition functions.

Proposition 2.18. Let Ω be a locally trivial groupoid on B and let $\{\sigma_i : U_i \to \Omega_b\}$ be a section-atlas for Ω. Then the maps $s_{ij}: U_{ij} \to \Omega_b^b$, $x \mapsto \sigma_i(x)^{-1}\sigma_j(x)$ form a cocycle. If $\{\sigma'_k: V_k \to \Omega_b\}$ is another section-atlas then the associated cocycle $\{s'_{k\ell}: V_{k\ell} \to \Omega_b^b\}$ is equivalent to $\{s_{ij}\}$.

Proof. Trivial. //

If s'_k above took values in some $\Omega_{b'}$, then, taking any $\zeta \in \Omega_b^{b'}$ and defining $\tau_k: x \mapsto \sigma'_k(x)\zeta$, the cocycles $\{t_{k\ell} = \zeta^{-1}s'_{k\ell}\zeta\}$ and $\{s_{ij}\}$ would be equivalent; that is, $\{s'_{k\ell}\}$ and $\{s_{ij}\}$, which take values in different groups, would be equivalent to within an isomorphism. The reader may work out the details of how this additional generality affects the following results.

Theorem 2.19. Let B be a space, G a topological group and $\{s_{ij}: U_{ij} \to G\}$ a cocycle.

For each i,j write $X_i^j = \{j\} \times U_j \times G \times U_i \times \{i\}$ and write X for the union of all the X_j^i. Define an equivalence relation $\sim$ in X by

$$(j,y,g,x,i) \sim (j',y',g',x',i') \iff y = y', \quad x = x' \quad \text{and} \quad g' = s_{j'j}(y)gs_{ii'}(x).$$

Denote equivalence classes by $[j,y,g,x,i]$ and $X/\sim$ by Ω. Then the following defines in Ω the structure of a groupoid on B: the source and target projections are $\alpha([j,y,g,x,i]) = x$, $\beta([j,y,g,x,i]) = y$, the object inclusion map is $\varepsilon: x \mapsto \tilde{x} = [i,x,1,x,i]$ (any i such that $x \in U_i$), and the multiplication is

$$[k,z,h,y,j_2][j_1,y,g,x,i] = [k,z,hs_{j_2 j_1}(y)g,x,i].$$

The inversion is $[j,y,g,x,i]^{-1} = [i,x,g^{-1},y,j]$.

Let Σ_i^j be the map $X_i^j \to \Omega_{U_i}^{U_j}$, $(j,y,g,x,i) \mapsto [j,y,g,x,i]$. Then each Σ_i^j is a bijection and transferring the product topologies from the X_i^j to the $\Omega_{U_i}^{U_j}$ gives a well-defined topology in Ω with respect to which it is a locally trivial topological groupoid on B.

Choose $b \in B$ and i_o such that $b \in U_{i_o}$ and define $\sigma_i: U_i \to \Omega_b$ by $x \mapsto [i,x,1,b,i_o]$. Then $\{\sigma_i\}$ is a section-atlas for Ω and the associated cocycle is $\{x \mapsto [i_o,b,s_{ij}(x),b,i_o]\}$.

Proof. The verification of the algebraic properties is an instructive exercise; the verification of the topological properties follows as in 1.13. //

Thus Ω in 2.19 is a locally trivial groupoid with a collection of distinguished section-atlases whose associated cocycles are mapped to $\{s_{ij}\}$ under the corresponding isomorphisms $[i_o,b,g,b,i_o] \mapsto g$, $\Omega_b^b \to G$. One could make this correspondence 'natural' by defining a pointed cocycle to be a cocycle $\{s_{ij}\}$ together with a point $b \in B$ and index i_o such that $b \in U_{i_o}$; or, alternatively, to be a cocycle $\{s_{ij}\}$ together with a point $b \in B$ such that $s_{ij}(b) = 1$ whenever $b \in U_{ij}$ (one may always choose a section-atlas $\{\sigma_i: U_i \to \Omega_b\}$ in such a way that $\sigma_i(b) = \tilde{b}$ whenever $b \in U_i$). The reader may work out the details.

Proposition 2.20. Let $\{s_{ij}: U_{ij} \to G\}$ and $\{s'_{ij}: U_{ij} \to G\}$ be cocycles on a space B with values in a topological group G, defined with respect to the same open cover $\{U_i\}$ and equivalent under a set of maps $\{r_i: U_i \to G\}$. Let Ω and Ω' be the groupoids constructed from $\{s_{ij}\}$ and $\{s'_{ij}\}$ in 2.19, and define a map $\phi: \Omega \to \Omega'$ by

$$[j,y,g,x,i] \mapsto [j,y,r_j(y)^{-1}gr_i(x),x,i].$$

Then ϕ is well-defined and is an isomorphism of topological groupoids over B.

Proof. Exercise. //

The condition that the two cocycles are defined on the same open cover is of course not necessary - one can always take the common refinement of the covers and all the restrictions of the elements of the cocycles.

To make the correspondence of 2.18 and 2.19 precise one must define two pointed cocycles $\{s_{ij};b,i_o\}$ and $\{s'_{ij};b',i'_o\}$ to be equivalent if $b = b'$, $i = i'_o$ and there is an equivalence $\{r_i\}$ with $r_{i_o}(b) = 1$.

2.18 and 2.20 show that there is a bijective correspondence between equivalence classes of cocycles on B with values in G and suitable isomorphism classes of locally trivial groupoids on B with vertex groups isomorphic to G, providing one defines the notions of equivalence and isomorphism with suitable care. For our purposes the results given are sufficient; we do not need a precise correspondence. Alternatively one may consider cocycles with values in TGB's and construct a locally trivial Ω from $B \times B$ and a TGB M which will be the inner subgroupoid of Ω; the cocycle must satisfy a compatibility condition with a cocycle for M, for not all TGB's are inner subgroupoids of locally trivial groupoids.

Proposition 2.21. Let Ω be a locally trivial groupoid on B, let b be a point B, and let $f: \Omega_b^b \to H$ be a morphism of topological groups. Then there is a locally trivial groupoid Φ on B together with an isomorphism $\theta: H \to \Phi_b^b$ and a morphism $\phi: \Omega \to \Phi$ over B such that $\theta \circ f = \phi_b^b$.

If Φ', θ', ϕ' are a second set of data satisfying these conditions then there is an isomorphism $\Theta: \Phi \to \Phi'$ over B such that $\Theta_b^b \circ \theta = \theta'$ and $\Theta \circ \phi = \phi'$.

<u>Proof</u>. Let $\{\sigma_i : U_i \to \Omega_b\}$ be a section-atlas for Ω and, for convenience, arrange that $\sigma_i(b) = \tilde{b}$ whenever $b \in U_i$. Denote the corresponding cocycle by $\{s_{ij}\}$, and construct Φ from the cocycle $\{f \circ s_{ij} : U_{ij} \to H\}$ as in 2.19. Define $\phi: \Omega \to \Phi$ by $\phi(\xi) = [j, \beta\xi, f(\phi_j(\beta\xi)^{-1}\xi\sigma_i(\alpha\xi)), \alpha\xi, i]$ where $\alpha\xi \in U_i$, $\beta\xi \in U_j$, and $\theta: H \to \Phi_b^b$ by $\theta(h) = [i, b, h, b, i]$ (any i with $b \in U_i$). It is straightforward to check that Φ, ϕ and θ have the required properies.

Given Φ', θ', ϕ' define $\sigma_i' = \phi' \circ \sigma_i : U_i \to \Phi_b'$ and denote the associated cocycle by $\{s_{ij}'\}$; then $s_{ij}' = \theta' \circ f \circ s_{ij}$. Define $\Theta: \Phi \to \Phi'$ by $\Theta([j,y,h,x,i]) = \sigma_j'(y)\theta'(h)\sigma_i'(x)^{-1}$. It is straightforward to check that Θ is well-defined and that $\Theta \circ \phi = \phi'$, $\Theta \circ \theta = \theta'$. //

<u>Definition 2.22</u>. In the situation of 2.21, Φ is called the <u>produced groupoid</u> of Ω along f and ϕ the <u>produced morphism</u>. //

The corresponding concept for principal bundles is usually called an 'extension' or 'prolongation'. Both terms have other meanings within bundle or groupoid theory.

<u>Proposition 2.23</u>. Let $\phi: \Omega \to \Omega'$ be a morphism of locally trivial groupoids over B. Choose $b \in B$. Then Ω' and ϕ are (isomorphic to) the produced groupoid and produced morphism of Ω along ϕ_b^b.

<u>Proof</u>. Follows from 2.21. //

We remind the reader that these results do <u>not</u> imply that an algebraic morphism of locally trivial groupoids over B is continuous if its restriction to any single vertex group is continuous (consider $B \times B \to B \times G' \times B$, $(y,x) \mapsto (y, \theta(y)\theta(x)^{-1}, x)$ for suitable $\theta: B \to G'$); and they do <u>not</u> imply that if $\phi, \phi': \Omega \to \Phi$ are morphisms of locally trivial groupoids over B and $\phi_b^b = \phi_b'^b$ for some (or all) b, then $\phi = \phi'$ (for a counterexample, in the special case of inclusions, see III 1.20 to III 1.21). These results merely reflect the fact that a locally trivial groupoid is determined by its base, a vertex group, and a cocycle:

It follows from 2.10 that if Ω is locally trivial and $\phi: \Omega \to \Omega'$ is a base-preserving morphism then Ω' is locally trivial and, what is more, any open cover $\{U_i\}$ which is the domain of a cocycle for Ω will also be the domain of a cocycle for Ω'. Loosely speaking, Ω' can be no more twisted than Ω is (and this is so even though ϕ need not be onto; indeed ϕ need not even be a morphism - one needs only that $\alpha' \circ \phi = \alpha$ and $\beta' \circ \phi = \beta$). Put differently, one cannot (with preservation of the base) map a locally trivial groupoid into a groupoid which is more twisted than itself (or that is not locally trivial at all). Given Ω and $f: \Omega_b^b \to H$, the base, a vertex group and a cocycle for the codomain groupoid Ω' are all determined and so Ω' is determined to within isomorphism.

For the benefit of a reader meeting this material for the first time, we append an example.

Example 2.24. Consider the locally trivial groupoids $\frac{SU(2) \times SU(2)}{U(1)}$ and $\frac{SO(3) \times SO(3)}{SO(2)}$ corresponding to the principal bundles $SU(2)(\mathbf{S}^2, U(1), \pi)$ and $SO(3)(\mathbf{S}^2, SO(2), \pi')$. For the first bundle, denote a typical element of $SU(2)$, $\begin{bmatrix} \alpha & \beta \\ -\bar{\beta} & \bar{\alpha} \end{bmatrix}$ with $|\alpha|^2 + |\beta|^2 = 1$, by (α, β), regard $U(1)$ as a subgroup of $SU(2)$ by $z \mapsto (z, 0)$, and let π be

$$(\alpha, \beta) \mapsto (-2\,\mathrm{Re}(\alpha\beta), -2\,\mathrm{Im}(\alpha\beta), 1 - 2|\beta|^2).$$

For the second bundle, regard $SO(2)$ as a subgroup of $SO(3)$ by $A \mapsto \begin{bmatrix} A & 0 \\ 0 & 1 \end{bmatrix}$ and let π' be $A \mapsto Ae_3$, where $\{e_1, e_2, e_3\}$ is the usual basis of $\mathbf{R}^3$.

Define a section-atlas for the first bundle by

$$U_N = \mathbf{S}^2 \setminus \{(0,0,1)\}, \quad \sigma_N(x,y,z) = \left(-\frac{x + iy}{\sqrt{2(1 - z)}}, \sqrt{\frac{1 - z}{2}}\right),$$

$$U_S = \mathbf{S}^2 \setminus \{(0,0,-1)\}, \quad \sigma_S(x,y,z) = \left(-\sqrt{\frac{1 + z}{2}}, \frac{x + iy}{\sqrt{2(1 + z)}}\right),$$

and a section-atlas for the second bundle as follows: for $i = 1,2$ let $U_i = \mathbf{S}^2 \setminus \{\pm e_i\}$ and for $x \in U_i$, let $y = \frac{x \times e_i}{\|x \times e_i\|}$ and let $\sigma_i(x)$ be the element of $SO(3)$ which maps e_1, e_2, e_3 to y, $x \times y$, x. Calculate cocycles for the two bundles.

Each $A \in SU(2)$ defines its adjoint $Ad(A): X \mapsto AXA^{-1}$, $\mathfrak{su}(2) \to \mathfrak{su}(2)$. Identify $\mathfrak{su}(2)$ with $\mathbf{R}^3$ by mapping (x_1,x_2,x_3) to

$$\begin{bmatrix} ix_3 & -x_2 + ix_1 \\ x_2 + ix_1 & -ix_3 \end{bmatrix};$$

then the corresponding map $\mathbf{R}^3 \to \mathbf{R}^3$ is an element of $SO(3)$ (see, for example, Miller (1972, p. 224)). Then $\langle A',A\rangle \mapsto \langle AdA',AdA\rangle$ is a morphism of topological groupoids $\widetilde{Ad}$: $\frac{SU(2) \times SU(2)}{U(1)} \to \frac{SO(3) \times SO(3)}{SO(2)}$ over S^2. Identify the vertex group of $\frac{SU(2) \times SU(2)}{U(1)}$ over $(0,0,1)$ with $U(1)$ using the identity matrix in $SU(2)$ as reference point, do likewise with the second groupoid, and calculate the restriction of $\widetilde{Ad}$ to $U(1) \to SO(2)$. Deduce the kernel of $\widetilde{Ad}$ and also deduce another cocycle for $\frac{SO(3) \times SO(3)}{SO(2)}$, and relate it to the one already found. //

§3. Components in topological groupoids.

We return to the study of arbitrary topological groupoids, and generalize two elementary facts about topological groups: the component of the identity is a subgroup and that subgroup is generated by any neighbourhood of the identity.

We begin however by considering the relationship between transitivity components and connectedness components. For arbitrary topological groupoids there is no relationship - any partition of any space B is the set of transitivity components for some topological groupoid on B, for example the equivalence relation corresponding to the partition itself is such a groupoid. However there is the following result:

Proposition 3.1. Let B be a locally connected space and let Ω be a topological groupoid on B for which each $\beta_x: \Omega_x \to B$, $x \in B$, is open. Then for each connectivity component C of B, Ω^C_C is transitive.

Proof. C is open so each $\beta_x: \Omega^C_x \to C$, $x \in C$, is an open map. Hence the transitivity components of Ω^C_C are open, and therefore closed, subsets of C. Since C is connected, there can be only one such transitivity component. //

As has already been noted, β_x will be open providing it is an identification.

3.1 shows that for such a topological groupoid on such a space, the transitivity components of Ω are unions of connectedness components of B. In particular each transitivity component M is open and therefore closed in B, and so Ω is algebraically and topologically the disjoint union of its transitive full subgroupoids Ω_M^M.

Now consider a transitive groupoid Ω satisfying the conditions of 3.1, and let C_i denote the connectedness components of B. If Ω is locally trivial, it can be reconstructed from the $\Omega_{C_i}^{C_i}$; we will not, however, need the details of this result. In general it seems unlikely that such a reconstruction is possible and so to restrict oneself in general to transitive groupoids on connected bases is some loss of generality. We will however often make this restriction in the locally trivial case.

<u>Proposition 3.2</u>. Let Ω be a topological groupoid on B. Let Ψ_x denote the connectedness component of $\tilde{x}$ in Ω_x, $x \in B$. Then $\Psi = \bigcup_{x \in B} \Psi_x$ is a wide subgroupoid of Ω, called the α-<u>identity-component subgroupoid</u> of Ω.

<u>Proof</u>. By definition Ψ contains each $\tilde{x}, x \in B$, so it is certainly wide. Take $\xi \in \Psi_x^y$ and $\eta \in \Psi_y^z$ and consider $\eta\xi = R_\xi(\eta) \in \Omega_x^z$. Because $R_\xi\colon \Omega_y \to \Omega_z$ is a homeomorphism, it maps components to components; since $\xi = R_\xi(\tilde{y}) \in R_\xi(\Psi_y)$ we have $\Psi_x \cap R_\xi(\Psi_y) \neq \emptyset$ and therefore $\Psi_x = R_\xi(\Psi_y)$. Hence $\eta\xi \in \Psi_x$. So Ψ is closed under multiplication. Taking $\xi \in \Psi_x^y$ again, we have $\tilde{y} \in R_{\xi^{-1}}(\Psi_x) \cap \Psi_y$ so $R_{\xi^{-1}}(\Psi_x) = \Psi_y$ and hence $\xi^{-1} = \tilde{x}\xi^{-1} \in \Psi_y$, which proves that Ψ is closed under inversion. //

Ψ need not be normal; see 3.7 below. It is implicit in the proof that the β-fibres Ψ^y are the identity components of the β-fibres Ω^y of Ω, and that, for $\xi \in \Omega_x^y$, the component of Ω_x containing ξ is $\Psi_y\xi$ and the component of Ω^y containing ξ is $\xi\Psi^x$. Clearly the various components of any one α-fibre need not be homeomorphic.

If B is connected, then $\Psi = (\bigcup_x \Psi_x) \cup \tilde{B}$ is connected, since each $\Psi_x \cap \tilde{B}$ is nonvoid. Conversely, if Ψ is connected then $B = \beta(\Psi)$ is connected.

If Ψ is transitive, then it is a connected space, since the map $\Psi_x \times \Psi_x \to \Psi$, $(\eta,\xi) \mapsto \eta\xi^{-1}$ is surjective. Thus if B is not connected then Ψ cannot be transitive. However a transitive Ω on a connected B may have $\Psi = \tilde{B}$; consider the germ groupoid $J_{sh}^\lambda(\Gamma)$ for Γ a transitive pseudogroup and B a connected space.

Proposition 3.3. Let Ω be an α-locally connected topological groupoid on a connected base B for which each $\beta_x: \Omega_x \to B$ is open. (By 3.1, Ω is transitive.) Then Ψ is transitive.

Proof. Similar to 3.1. //

Proposition 3.4. Let Ω be an α-locally connected principal topological groupoid on a connected base B. Then Ψ is a principal subgroupoid of Ω and is an open subset of Ω.

Proof. By 3.3, Ψ is transitive. Hence $\Omega_x \times \Omega_x \to \Omega$, $(\eta,\xi) \mapsto \eta\xi^{-1}$ maps $\Psi_x \times \Psi_x$ onto Ψ. This map is open, since Ω is principal, and since Ψ_x is open in Ω_x, it follows that Ψ is open in Ω, and the restriction $\Psi_x \times \Psi_x \to \Psi$ is open. //

If Ω is a differentiable groupoid (not necessarily transitive and with base not necessarily connected) then Ψ is open (see III 1.3). It would be interesting to know if, for any topological groupoid Ω on a connected base B, Ψ is the component of Ω containing $\tilde{B}$.

Proposition 3.5. Let Ω be a weakly locally trivial groupoid on a locally connected space B. Then Ψ is weakly locally trivial.

Proof. Let $\{\sigma_i: U_i \to \Omega_{b_i}\}$ be a section-atlas for Ω. Since B is locally connected we can assume the U_i are connected, and in this case each $\sigma_i(U_i)$ lies in a single component C_i of Ω_{b_i}; choose any $\xi_i \in C_i$ and define $\tau_i: U_i \to \Psi_{\beta\xi_i}$ by $x \mapsto \sigma_i(x)\xi_i^{-1}$. Then $\{\tau_i\}$ is a section-atlas for Ψ. //

Together with 3.3 this yields

Corollary 3.6. Let Ω be an α-locally connected, locally trivial groupoid on a connected base B. Then Ψ is locally trivial. //

Of course if Ψ (or any subgroupoid of Ω) is locally trivial, then Ω itself is.

Example 3.7. Let P be the space $\mathbb{R} \times \mathbb{Z}$ and let G be the discrete space $\mathbb{Z} \times \mathbb{Z}$ with the group structure

$$(m_1,n_1)(m_2,n_2) = \big(m_1 + m_2,\ (-1)^{m_2}n_1 + n_2\big).$$

Let G act on P to the right by

$$(x,p)(m,n) = (x + m,\ (-1)^m p + n)$$

and let $\pi\colon P \to B = \mathbf{S}^1$ be $\pi(x,p) = e^{2\pi i x}$. It is easy to verify that $P(B,G,\pi)$ is a principal bundle.

Let Ω denote the associated groupoid $\frac{P \times P}{G}$ on B, let $u_o = (0,0) \in P$ and let $x_o = \pi(u_o) = 1 + 0i \in \mathbf{S}^1$. Then under the identifications of P with Ω_{x_o} and G with $\Omega_{x_o}^{x_o}$ given in 2.8(i), $\Omega_{x_o}^{x_o} \subseteq \Omega_{x_o}$ corresponds to the natural inclusion $G \subseteq P$.

Therefore $\Psi_{x_o} = \mathbf{R} \times \{0\}$ and $\Psi_{x_o}^{x_o} = \mathbf{Z} \times \{0\}$. It is easy to verify that $\mathbf{Z} \times \{0\}$ is not normal in G, so Ψ is not a normal subgroupoid of Ω.

The vertex bundle $\Psi_{x_o}\big(B, \Psi_{x_o}^{x_o}\big)$ is of course the familiar example $\mathbf{R}(\mathbf{S}^1,\mathbf{Z})$ and the bundle $P(B,G)$ is the pullback of the universal cover $\mathbf{R}^2(K,G)$ of the Klein bottle K along the map $\mathbf{S}^1 = \mathbf{R}/\mathbf{Z} \to K = \mathbf{R}^2/G$ induced by $\mathbf{R} \to \mathbf{R}^2$, $x \mapsto (x,0)$. //

Let Ω be an α-locally connected principal topological groupoid on a connected base B, and assume that the vertex groups of Ω have abelian component groups. (By component group of a topological group G is meant the quotient group G/G_o where G_o is the component of the identity.) Then Ψ is normal. To see this, note first that since Ψ is transitive, it suffices to show that one Ψ_b^b is normal in Ω_b^b. Let $P(B,G)$ denote the vertex bundle of Ω at some $b \in B$, and let Q denote the component of $\tilde{b}$ in P. Then $\Psi_b^b = \{g \in G \mid R_g(Q) \subseteq Q\}$; clearly Ψ_b^b is an open subgroup of G; denote it by H. Hence H is a union of cosets of G_o in G and so H is normal in G iff H/G_o is normal in G/G_o. When G/G_o is abelian, this is always the case.

This argument shows that for locally trivial groupoids on connected bases whose vertex groups are nondiscrete Lie groups of the type encountered in many applications, the α-identity-component subgroupoid is normal.

<u>Proposition 3.8</u>. Let Ω be a topological groupoid on B. Let $\mathcal{U}$ be a symmetric set (that is, $\mathcal{U} \supseteq \tilde{B}$ and $\mathcal{U}^{-1} = \mathcal{U}$) such that each $\mathcal{U}_x$ is open in Ω_x. Then the subgroupoid Φ generated by $\mathcal{U}$ has Φ_x open in Ω_x for all $x \in B$.

<u>Proof</u>. Since $\mathcal{U}$ is symmetric, Φ is merely the set of all possible products of elements from $\mathcal{U}$. Choose $x \in B$. The set of all n-fold products $\xi_n \cdots \xi_1$ from $\mathcal{U}$

with $\alpha\xi_1 = x$ is the union of all $\mathcal{U}_{\beta\zeta}\zeta \subseteq \Omega_x$ where ζ is an $(n-1)$-fold product from $\mathcal{U}$. Since $R_\zeta : \Omega_{\beta\zeta} \to \Omega_x$ is a homeomorphism the set of all n-fold products from $\mathcal{U}$ which lie in Ω_x, is open in Ω_x. Hence Φ_x is open in Ω_x. //

As with 3.4, if Ω is principal and Φ is transitive, then Φ will itself be principal and will be an open subset of Ω. A set $\mathcal{U}$ satisfying the conditions in 3.8 will be called a <u>symmetric α-neighbourhood</u> of $\tilde{B}$ (or, of the base) in Ω.

<u>Proposition 3.9</u>. Let Ω be a topological groupoid on a connected space B for which each $\beta_x : \Omega_x \to B$ is open, and let $\mathcal{U}$ be a symmetric α-neighbourhood of $\tilde{B}$ in Ω. Then the subgroupoid Φ generated by $\mathcal{U}$ is transitive.

<u>Proof</u>. Each $\beta_x(\mathcal{U}_x)$ is open in B; denote it by U_x. Given $x, y \in B$ there is a finite chain $U_{z_0}, \ldots, U_{z_n}$ with $z_0 = x$, $z_n = y$ and $U_{z_i} \cap U_{z_{i+1}} \neq \emptyset$ $\forall\, 0 \leq i \leq n-1$; this follows from the connectivity of B. Now it is clear that there is an element $\xi \in \Phi_x^y$. //

<u>Proposition 3.10</u>. Let Ω be a topological groupoid on B, and let Φ be a wide subgroupoid of Ω. Then, if each Φ_x is open in Ω_x, $x \in B$, each Φ_x is also closed in Ω_x, $x \in B$.

<u>Proof</u>. The complement $\Omega_x \setminus \Phi_x$ is the union of all $\Phi_{\beta\xi}\xi$ as ξ ranges over $\Omega_x \setminus \Phi_x$. Since $\Phi_{\beta\xi}$ is open in $\Omega_{\beta\xi}$ it follows that $\Phi_{\beta\xi}\xi$ is open in Ω_x. //

The following result is now immediate.

<u>Proposition 3.11</u>. Let Ω be a topological groupoid on B, and let $\mathcal{U}$ be a symmetric α-neighbourhood of $\tilde{B}$ in Ω. Then $\mathcal{U}$ generates the α-identity-component subgroupoid Ψ of Ω. //

<u>Proposition 3.12</u>. Let Ω be an α-locally connected, locally trivial groupoid on a connected base B, and let $\mathcal{U}$ be a symmetric α-neighbourhood of $\tilde{B}$ in Ω. Then the subgroupoid Φ of Ω generated by $\mathcal{U}$ is locally trivial.

<u>Proof</u>. Apply 3.6 and 3.11. //

§4. Representations of topological groupoids.

This section gives the basic definitions and examples of the concept of representation (or action) of a topological groupoid, and related concepts such as isotropy subgroupoid and invariant section. The basic material is due to Ehresmann (1959, and elsewhere); see also Ngô (1967). We also give the equivalence between the concepts of action of a groupoid and covering of a groupoid; this is due to Higgins (1971) in the algebraic case and to Brown et al (1976) in the topological case.

Some deeper results for isotropy subgroupoids and the classification of locally trivial subgroupoids are given in III§1; the simplicity of their formulation there depends on the facts that all transitive smooth actions of Lie groups are homogeneous, and all closed subgroups of Lie groups admit local sections. A general topological formulation of these results would be cumbersome.

Definition 4.1. Let Ω be a topological groupoid on B and let $p: M \to B$ be a continuous map. Let $\Omega * M$ denote the subspace $\{(\xi,u) \in \Omega \times M \mid \alpha\xi = p(u)\}$ of $\Omega \times M$. An action of Ω on (M,p,B) is a continuous map $\Omega * M \to M$, $(\xi,u) \mapsto \xi u$ such that

(i) $p(\xi u) = \beta\xi, \quad \forall\ (\xi,u) \in \Omega * M$;

(ii) $\eta(\xi u) = (\eta\xi)u, \quad \forall\ (\eta,\xi) \in \Omega * \Omega, \quad (\xi,u) \in \Omega * M$;

(iii) $\widetilde{p(u)}u = u, \quad \forall\ u \in M$.

For $u \in M$, the subset $\Omega[u] = \{\xi u \mid \xi \in \Omega_{p(u)}\}$ is the orbit of u under Ω.

//

This definition goes back to Ehresmann (1959). We will be mainly concerned with two cases: (i) when (M,p,B) is a TGB and each $u \mapsto \xi u$, $M_{\alpha\xi} \to M_{\beta\xi}$ is an isomorphism of topological groups; we will then say that Ω acts on M by topological group isomorphims, (ii) when (M,p,B) is a vector bundle and each $u \mapsto \xi u$, $M_{\alpha\xi} \to M_{\beta\xi}$, is a vector space isomorphism; in this case we say that Ω acts linearly on M.

A concept of groupoid action on a groupoid is given in Brown (1972).

Definition 4.2. Let Ω be a topological groupoid on B, and let $\Omega * M^1 \to M^1$ and $\Omega * M^2 \to M^2$ be actions of Ω on continuous maps (M^i,p_i,B), $i = 1,2$. Then a continuous map $\psi: M^1 \to M^2$ such that $p^2 \circ \psi = p^1$ is Ω-equivariant if $\psi(\xi u) = \xi\psi(u)$ $\forall\ (\xi,u) \in \Omega * M^1$.

Let Ω' be a second topological groupoid with base B', let (M,p,B) and (M',p',B') be continuous maps, let $\Omega * M \to M$ and $\Omega' * M' \to M'$ be actions, let $\phi: \Omega \to \Omega'$ be a morphism of topological groupoids, and let $\psi: M \to M'$ be a continuous map such that $p' \circ \psi = \phi_o \circ p$. Then ψ is <u>ϕ-equivariant</u> if $\psi(\xi u) = \phi(\xi)\psi(u)$, $\forall\ (\xi,u) \in \Omega * M$. //

<u>Definition 4.3</u>. Let Ω be a topological groupoid on B and let (M,p,B) be a fibre bundle with locally compact, locally connected and Hausdorff fibres. Then a <u>representation</u> of Ω in (M,p,B) is a morphism $\rho: \Omega \to \Pi(M)$ of topological groupoids over B. //

For our interpretation of the term 'fibre bundle', see A§1. If (M,p,B) is a TGB we interpret $\Pi(M)$ as the groupoid of topological group isomorphisms and call ρ a <u>representation by topological group isomorphisms</u>; if (E,p,B) is a vector bundle we interpret $\Pi(E)$ as the groupoid of vector space isomorphisms and call ρ a <u>linear representation</u>.

<u>Proposition 4.4</u>. Let Ω be a locally trivial groupoid on B and let (M,p,B) be a fibre bundle whose fibres are locally compact, locally connected and Hausdorff. If $\Omega * M \to M$ is an action of Ω on M then the associated map $\Omega \to \Pi(M)$, $\xi \mapsto (u \mapsto \xi u)$ is a representation; if $\rho: \Omega \to \Pi(M)$ is a representation then $(\xi,u) \mapsto \rho(\xi)(u)$ is an action.

<u>Proof</u>. If $\Omega * M \to M$ is an action, let $\{\sigma_i: U_i \to \Omega_b\}$ be a section-atlas and define charts $\psi_i: U_i \times M_b \to M_{U_i}$ for M by $\psi_i(x,a) = \sigma_i(x)a$. Then $\Omega_{U_i}^{U_j} * M_{U_i} \to M_{U_j}$ becomes $(U_j \times \Omega_b^b \times U_i) * (U_i \times M_b) \to U_j \times M_b$, $((y,\lambda,x), (x,a)) \mapsto (y,\lambda a)$ and the result is clear. The converse is similar. //

It is not clear whether this result holds for general topological groupoids Ω and whether, using the methods of Booth and Brown (1978) a similar result can be proved without local triviality conditions on (M,p,B) or Ω.

The following examples are basic.

<u>Example 4.5</u>. Let Ω be a topological groupoid on B and let $B \times F$ be a trivial fibre bundle. Then $\xi(\alpha\xi,a) = (\beta\xi,a)$, $\xi \in \Omega$, $a \in F$, is an action of Ω on $B \times F$, called the <u>trivial action</u>. The associated representation is the <u>trivial representation</u>. //

Example 4.6. Let (M,p,B) be a fibre bundle whose fibres are locally compact, locally connected and Hausdorff. Then $\Pi(M) * M \to M$, $(\xi,u) \mapsto \xi(u)$ is an action. //

Example 4.7. Any topological groupoid Ω acts on its own β-projection through the multiplication map $\Omega * \Omega \to \Omega$. //

Example 4.8. Let $P(B,G,\pi)$ be a principal bundle, and let $M = \frac{P \times F}{G}$ be an associated fibre bundle with respect to a representation $G \to \mathrm{Homeom}(F)$. Then

$$\frac{P \times P}{G} * \frac{P \times F}{G} \dashrightarrow \frac{P \times F}{G}$$

$$(\langle v,u\rangle, \langle u,a\rangle) \mapsto \langle v,a\rangle$$

is an action. //

In fact all actions of locally trivial groupoids are of this type:

Theorem 4.9. Let Ω be a locally trivial groupoid on B, and let $\Omega * M \to M$ be an action of Ω on a continuous surjection (M,p,B) whose fibres are locally compact, locally connected, and Hausdorff. Then (M,p,B) is a fibre bundle and, for any choice of $b \in B$ and writing $P = \Omega_b$, $G = \Omega_b^b$, $F = M_b$, the map $\frac{P \times F}{G} \to M$, $\langle \xi,a\rangle \mapsto \xi a$ is a homeomorphism of continuous surjections over B and is equivariant with respect to the isomorphism $\frac{P \times P}{G} \to \Omega$ of 1.19 (ii). ($\frac{P \times F}{G}$ is constructed with respect to the representation of G on F corresponding to the restriction $\Omega_b^b \times M_b \to M_b$.)

Proof. Take a section-atlas $\{\sigma_i : U_i \to \Omega_b\}$ and use it to define charts $\psi_i: U_i \times F \to M_{U_i}$, $(x,a) \mapsto \sigma_i(x)a$, as in 4.4. This proves that (M,p,B) is a fibre bundle. Define $P \times F \to M$ by $(\xi,a) \mapsto \xi a$. In terms of the charts ψ_i for M and $(x,g) \mapsto \sigma_i(x)g$ for Ω_b, this is $U_i \times G \times F \to U_i \times F$, $(x,g,a) \mapsto (x,ga)$, which is open. Hence $\frac{P \times F}{G} \to M$, $\langle \xi,a\rangle \mapsto \xi a$ is a homeomorphism. The other statements are easily proved. //

This result should be compared with Kobayashi and Nomizu (1963, I.5.4 and subsequent discussion).

Proposition 4.10. Let $P(B,G)$ be a principal bundle and let M and M' be two associated fibre bundles corresponding to actions $G \times F \to F$ and $G \times F' \to F'$ of G on locally compact, locally connected and Hausdorff spaces F and F'.

(i) If $f: F \to F'$ is a G-equivariant map then $\tilde{f}: M \to M'$ defined by

$\langle u,a\rangle \mapsto \langle u,f(a)\rangle$ is a well-defined morphism of fibre bundles over B, and is $\frac{P \times P}{G}$-equivariant.

(ii) If ϕ: M → M' is a $\frac{P \times P}{G}$-equivariant morphism of fibre-bundles over B, then $\phi = \tilde{f}$ for some G-equivariant map f.

Proof: (i) is easy to verify. For (ii), observe that a map f: P × F → F' can be defined by the condition that

$$\phi(\langle u,a\rangle) = \langle u,f(u,a)\rangle \qquad \text{for } u \in P,\ a \in F.$$

Now it is easy to see that equivariance with respect to $\frac{P \times P}{G}$ forces f(u,a) to depend only on a. So we have f: F → F' and f must clearly be G-equivariant. //

One can formulate this result as a statement about adjoint functors (see Mackenzie (1978, 7.1)).

Example 4.11. Let Ω be a topological groupoid on B. Then the inner automorphism action is the map $\Omega * G\Omega \to G\Omega$, $(\xi,\lambda) \mapsto I_\xi(\lambda) = \xi\lambda\xi^{-1}$.

If Ω is locally trivial and its vertex groups are locally compact, locally connected and Hausdorff then I is a representation $\Omega \to \Pi(G\Omega)$. If K is a normal totally intransitive subgroupoid of Ω, whose fibres satisfy the same topological conditions, then the 'restriction' $\Omega \to \Pi(K)$ is also a representation.

Returning to GΩ itself, 4.9 shows that GΩ is equivariantly isomorphic (as a TGB) to $\frac{P \times G}{G}$ with respect to the inner automorphism representation of G on itself. In the physics literature, $\frac{P \times G}{G}$ is often called the gauge bundle associated to P(B,G). We shall call it the inner group bundle. //

The following definition introduces the last example.

Definition 4.12. Let Ω be a locally trivial groupoid on B and let (M,p,B) be a TGB with locally compact, locally connected, Hausdorff fibres. An extension of Ω by M is a sequence

$$M \overset{\iota}{\rightarrowtail} \Phi \overset{\pi}{\twoheadrightarrow} \Omega$$

in which Φ is a locally trivial groupoid on B, ι and π are groupoid morphisms over B, ι is an embedding, π is a surjective submersion, and im(ι) = ker(π). //

It is easy to see that the condition that Φ be locally trivial is superfluous.

Example 4.13. Let $M \overset{\iota}{\rightarrowtail} \Phi \overset{\pi}{\twoheadrightarrow} \Omega$ be an extension as in 4.12 with M an abelian TGB. For $\xi \in \Omega_x^y$, $x,y \in B$ choose $\xi' \in \Phi_x^y$ with $\pi(\xi') = \xi$ and define $\rho(\xi): M_x \to M_y$ as $\lambda \mapsto \xi'\lambda\xi'^{-1}$, the restriction of $I_{\xi'}$. It is clear that $\rho(\xi)$ is well-defined. Now $I: \Phi \to \Pi(M)$ is continuous (by 4.11) and π is an identification so $\rho: \Omega \to \Pi(M)$ is continuous. ρ is the representation associated to the extension $M \rightarrowtail \Phi \twoheadrightarrow \Omega$.
//

These examples are straightforward; see the discussion following III 4.14 for a representation which is not well-known in the context of principal bundles.

We now give some simple definitions and results about isotropy subgroupoids and invariant sections.

Definition 4.14. Let $\Omega * M \to M$ be an action of a transitive topological groupoid Ω on a continuous surjection (M,p,B). Then a section $\mu \in \Gamma M$ is Ω-invariant if $\xi\mu(\alpha\xi) = \mu(\beta\xi)$, $\forall\ \xi \in \Omega$. The set of Ω-invariant sections of M is denoted $(\Gamma M)^{\Omega}$.
//

If $\Omega * E \to E$ is a linear action on a vector bundle, then $(\Gamma E)^{\Omega}$ is an $\mathbb{R}$-vector space with respect to pointwise operations, but not usually a module over the ring of continuous functions on B. A general fibre bundle need not of course admit any (global) sections. In the case of a vector bundle and a linear action, $(\Gamma E)^{\Omega}$ may consist of the zero section alone (see 4.16 below).

Proposition 4.15. Let Ω be a principal topological groupoid on B and let $\Omega * E \to E$ be an action of Ω on a vector bundle (E,p,B). Choose $b \in B$, and write V for E_b and G for Ω_b^b. Then the evaluation map

$$(\Gamma E)^{\Omega} \to V^G, \quad \mu \mapsto \mu(b)$$

is an isomorphism of $\mathbb{R}$-vector spaces.

Proof. Obviously the map is injective. Given $v \in V^G$, define μ by $\mu(x) = \xi v$ where ξ is any element of Ω_b^x. Clearly $\mu(x)$ is well-defined; μ is continuous because $\beta_b: \Omega_b \to B$ is an identification. //

A different view of this result is given in Mackenzie (1978, §4).

Example 4.16. Consider the principal bundle $SO(2)(\mathbf{S}^1, \mathbf{Z}_2, p)$, where $\mathbf{Z}_2$ is embedded in $SO(2)$ as $\{1,-1\}$ and p is $z \mapsto z^2$. Let Ω be the associated groupoid and E the vector bundle $\frac{SO(2) \times \mathbf{R}}{\mathbf{Z}_2}$, where $\mathbf{Z}_2$ acts on $\mathbf{R}$ by multiplication. Then $(\Gamma E)^{\Omega} = \mathbf{R}^{\mathbf{Z}_2}$ is the zero space. (E is, of course, the Möbius band.) //

Proposition 4.17. Let $\Omega * E \to E$ be an action of a locally trivial topological groupoid Ω on a vector bundle E. For each $x \in B$, define $E^{G\Omega}|_x$ to be $E_x^{\Omega_x^x} = \{u \in E_x \mid \lambda u = u, \ \forall \lambda \in \Omega_x^x\}$. Then $E^{G\Omega}$ is a subvector bundle of E.

Proof. Let $\{\sigma_i : U_i \to \Omega_b\}$ be a section-atlas for Ω, and write $V = E_b$, $G = \Omega_b^b$. Define $\psi_i : U_i \times V^G \to E^{G\Omega}|_{U_i}$ by $(x,v) \mapsto \sigma_i(x)(v)$. Then $\psi_{i,x}$ maps V^G isomorphically onto $E^{G\Omega}|_x$. //

Proposition 4.18. With the above notation, there is a natural trivialization $B \times V^G \to E^{G\Omega}$.

Proof. For $x \in B$ and any two $\xi, \xi' \in \Omega_b^x$, the maps $V^G \to E^{G\Omega}|_x$, $v \mapsto \xi v$ and $v \mapsto \xi' v$, are identical. //

Compare Greub et al (1973, p. 384, proposition III). This result is mainly of interest because the corresponding construction for actions of transitive Lie algebroids yields sub vector bundles which are flat but not necessarily trivializable.

Definition 4.19. Let $\Omega * M \to M$ be an action of a transitive topological groupoid Ω on a continuous surjection (M,p,B). Then $\mu \in \Gamma M$ is Ω-deformable if for all $x,y \in B$ there exists $\xi \in \Omega_x^y$ such that $\xi\mu(x) = \mu(y)$.

If $\mu \in \Gamma M$ is Ω-deformable, then the isotropy subgroupoid of Ω at μ is $\Phi(\mu) = \{\xi \in \Omega \mid \xi\mu(\alpha\xi) = \mu(\beta\xi)\}$. //

The term "Ω-deformable" is adapted from Greub et al (1973, 8.2). A section μ is Ω-deformable iff its values lie in a single orbit; the condition ensures that the isotropy subgroupoid is transitive. Note that $\Phi(\mu)$ is closed in Ω providing M is Hausdorff. If Ω is locally trivial, $\Phi(\mu)$ need not be locally trivial; however for a smooth action of a locally trivial differentiable groupoid the isotropy subgroupoid at a deformable section is always locally trivial (see III 1.20).

Also in III§1 we will use the correspondence between deformable sections and their isotropy subgroupoids to give a classification of those locally trivial subgroupoids of Ω which have a preassigned vertex group at a given point $b \in B$.

* * * * * * *

The construction of an action groupoid in I 1.6 may be generalized: let Ω be a topological groupoid on B and $\Omega * M \to M$ an action of Ω on a continuous map $p: M \to B$. Give $\Omega * M$ the structure of a groupoid on M as follows: the projections are $\bar{\alpha}(\xi,u) = u$, $\bar{\beta}(\xi,u) = \xi u$, the object inclusion map is $u \mapsto (\widetilde{pu},u)$, the multiplication is $(\eta,v)(\xi,u) = (\eta\xi,u)$, defined when $v = \xi u$, the inversion is $(\xi,u)^{-1} = (\xi^{-1},\xi u)$. Then, with the subspace topology from $\Omega \times M$, $\Omega * M$ is a topological groupoid on M, and $\Omega * M \to \Omega$, $(\xi,u) \mapsto \xi$, is a continuous morphism over $p: M \to B$.

Definition 4.20. With the structure described above, $\Omega * M$ is the action groupoid associated to the action of Ω on M. //

Remarkably, action groupoids and the morphisms associated with them can be characterized intrinsically. The following discussion, including 4.21 to 4.23, is taken directly from Brown et al (1976).

Definition 4.21. Let $\phi: \Omega' \to \Omega$ be a morphism of topological groupoids over $\phi_o: B' \to B$. Then ϕ is a covering morphism if the pullback space $\Omega * B' = \{(\xi,x') \in \Omega \times B' \mid \alpha(\xi) = \phi_o(x')\}$ is homeomorphic to the space Ω' under the map $[\phi,\alpha']: \Omega' \to \Omega * B'$, $\xi' \mapsto (\phi(\xi'),\alpha'(\xi'))$. We also say that $\phi: \Omega' \to \Omega$ is a covering of Ω.

Let $\phi_1: \Omega' \to \Omega$ and $\phi_2: \Omega'' \to \Omega$ be covering morphisms with the same codomain Ω. Then a morphism of coverings $\psi: \phi_1 \to \phi_2$ over Ω is a morphism of topological groupoids $\psi: \Omega' \to \Omega''$ such that $\phi_2 \circ \psi = \phi_1$. //

Obviously each action groupoid $\Omega * M$ and its morphism $\Omega * M \to M$ form a covering Ω. It is also easy to see that if $\phi: M^1 \to M^2$ is an Ω-equivariant map of two actions $\Omega * M^1 \to M^1$ and $\Omega * M^2 \to M^2$, then $\tilde{\phi}: \Omega * M^1 \to \Omega * M^2$, $(\xi,u) \mapsto (\xi,\phi(u))$ is a morphism of coverings over Ω.

Theorem 4.22. Let $\phi: \Omega' \to \Omega$ be a covering morphism, and let $s: \Omega * B' \to \Omega'$ denote the inverse of $[\phi,\alpha']$. Then $\beta' \circ s: \Omega * B' \to B'$ is an action of Ω on $\phi_o: B' \to B$.

Let $\psi: \phi_1 \to \phi_2$ be a morphism of coverings over Ω. Then $\psi_o: B' \to B''$ is Ω-equivariant with respect to the actions induced by the coverings.

<u>Proof</u>. We will show that $\eta(\xi x') = (\eta\xi)x'$ for $(\eta,\xi) \in \Omega * \Omega$ and $x' \in B'$ with $\alpha\xi = \phi_o(x')$; the other conditions are clear. First note that each $\phi_{x'}: \Omega'_{x'} \to \Omega_{\phi_o}(x)$ is a homeomorphism - it is easy to see that $[\phi,\alpha']$ maps $\Omega'_{x'}$ onto $\Omega_{\phi_o(x')} \times \{x'\}$. Thus $s(\xi,x')$ is the unique element of $\Omega'_{x'}$ which is mapped by ϕ onto ξ. Write $y' = (\beta' \circ s)(\xi,x')$ and note that $\phi_o(y') = \beta\xi = \alpha\eta$. So $s(\eta,y')$ is defined and is the unique element of $\Omega'_{y'}$ which is mapped by ϕ onto η. Since $\alpha'(s(\eta,y')) = y' = \beta'(s(\xi,x'))$, the product $s(\eta,y')s(\xi,x')$ is defined. Obviously it belongs to $\Omega'_{x'}$ and is mapped by ϕ onto $\eta\xi$; it is therefore equal to $s(\eta\xi,x')$. That $\eta(\xi x') = (\eta\xi)x'$ now follows.

The second statement of the theorem follows from noting that $[\phi_2,\alpha''] \circ \psi = (id*\psi_o) \circ [\phi,\alpha']$, and hence $\psi(s_1(\xi,x')) = s_2(\xi,\psi_o(x'))$, $\forall\ (\xi,x') \in \Omega * B'$. //

These two constructions are indeed mutual inverses:

<u>Theorem 4.23</u>. (i) Let $\phi: \Omega' \to \Omega$ be a covering morphism, and let $\Omega * B' \to B'$ be the associated action. Then, giving $\Omega * B'$ its structure as an action groupoid, $[\phi,\alpha']: \Omega' \to \Omega * B'$ is an isomorphism of topological groupoids over B' and is an isomorphism of coverings of Ω.

(ii) Let $\Omega * M \to M$ be an action of a topological groupoid on a continuous map $p: M \to B$, and let $\pi_1: \Omega * M \to \Omega$ be the associated covering. Then the action of Ω on M induced by π_1 is the original action.

<u>Proof</u>. Straightforward exercise. //

One may express 4.23 by saying that the category of covering morphisms over a topological groupoid Ω is equivalent to the category of actions of Ω and Ω-equivariant maps. One should note too that Brown et al (1976) also prove that if $\phi_2: \Omega' \to \Omega''$ is a covering morphism and $\phi_1: \Omega \to \Omega'$ any morphism of topological groupoids, then $\phi_2 \circ \phi_1$ is a covering morphism iff ϕ_1 is.

If X, Y are path-connected, locally path-connected and semi-locally simply-connected spaces, and $p: X \to Y$ is a covering map, then $p_*: \mathcal{T}(X) \to \mathcal{T}(Y)$ is a covering morphism of topological groupoids (Brown and Danesh-Naruie (1975)). The abstract theory of covering morphisms of topological groupoids in fact models the familiar features of the theory of covering spaces: a covering $\phi: \Omega' \to \Omega$ may be called regular if $\phi^{x'}_{x'}(\Omega'^{x'}_{x'}) \triangleleft \Omega^{\phi_o(x')}_{\phi_o(x')}$, $\forall\, x' \in B'$; if Ω is a transitive topological groupoid on a Hausdorff base B, $x \in B$, and $H < \Omega^x_x$, then there is a covering $\phi: \Omega' \to \Omega$ of Ω with Ω' transitive and $\phi^{x'}_{x'}(\Omega'^{x'}_{x'}) = H$ for some $x' \in \phi_o^{-1}(x)$, which is universal in a natural sense; in particular, there is a universal covering groupoid of Ω, which is in fact the action groupoid for the action $\Omega * \Omega_x \to \Omega_x$, $(\xi,\eta) \mapsto \xi\eta$ of Ω on a chosen α-fibre Ω_x. For these results see Brown et al (1976, Theorems 6 and 13). (Presumably there is also a version of these results in which one works with subgroupoids of Ω rather than with subgroups of a particular vertex group.)

Proposition 4.24. Let $\Omega * M \to M$ be an action of a topological groupoid Ω on a continuous map $p: M \to B$.

(i) $\Omega * M$ is transitive iff $\Omega^{p(M)}_{p(M)}$ is a transitive groupoid on $p(M) \subseteq B$ and $\Omega^b_b \times M_b \to M_b$ is a transitive action for some $b \in B$;

(ii) $\Omega * M$ is locally trivial iff $\Omega_b \to M$, $\xi \mapsto \xi u$, is a surjective submersion for some $b \in B$ and $u \in M_b$.

Proof. (i) is straightforward and (ii) is merely a reformulation of the definition. //

For actions of groups, $G \times B \to B$, the action groupoid is principal iff the action is transitive and the evaluation maps $g \mapsto gx$, $x \in B$, are open. No such simple criterion seems to exist in the general case.

The concept of "homogeneous space" for a topological groupoid, and its relationship to transitive actions of groupoids, is a complicated and unsatisfactory matter and it is fortunate that we do not need to consider it here. Some results may be found in the same paper of Brown et al (1976).

§5. Admissible sections.

On a group G, the left-translations $L_g: x \mapsto gx$ form a group which is isomorphic to G itself under $g \mapsto L_g$, and the right-translations $R_g: x \mapsto xg$ likewise form a group with $g \mapsto R_g$ now an anti-isomorphism. For a topological groupoid Ω one calls a homeomorphism $\Omega \to \Omega$ a left-translation if it is the union of left-translations $L_\xi: \Omega^{\alpha\xi} \to \Omega^{\beta\xi}$; such a left-translation is not characterized by a single element of Ω, but by an admissible section. These admissible sections may be regarded as generalized elements of the groupoid, since in the generalizations of the adjoint and exponential formulas for Lie groups, they play a rôle which, in the case of groups, is taken by the group elements themselves.

The material in this section is based on Kumpera and Spencer (1972, Appendix).

Definition 5.1. Let Ω be a topological groupoid on B.

A left-translation on Ω is a pair of homeomorphisms $\phi: \Omega \to \Omega$, $\phi_o: B \to B$, such that $\beta\circ\phi = \phi_o\circ\beta$, $\alpha\circ\phi = \alpha$, and each $\phi^x: \Omega^x \to \Omega^{\phi_o(x)}$ is L_ξ for some $\xi \in \Omega_x^{\phi_o(x)}$.

An admissible section of Ω is a continuous $\sigma: B \to \Omega$ which is right-inverse to $\alpha: \Omega \to B$ and is such that $\beta\circ\sigma: B \to B$ is a homeomorphism. The set of admissible sections of Ω is denoted by $\Gamma\Omega$. //

Given an admissible section σ, define a map $L_\sigma: \Omega \to \Omega$ by $\xi \mapsto \sigma(\beta\xi)\xi$. Then L_σ and $\beta\circ\sigma: B \to B$ constitute a left-translation on Ω (the inverse of L_σ is $\eta \mapsto \sigma((\beta\circ\sigma)^{-1}(\beta\eta))^{-1}\eta)$. Conversely, let ϕ, ϕ_o constitute a left-translation on Ω. For $x \in B$ choose $\xi \in \Omega^x$ and define $\sigma(x) = \phi(\xi)\xi^{-1}$. If η is another element of Ω^x then $\eta = \xi\zeta$ for some $\zeta \in \Omega$. Now $\phi(\xi) = \theta\xi$ for some $\theta \in \Omega_x$ and $\phi(\xi\zeta) = \theta\xi\zeta$ with the same θ. Thus $\phi(\xi\zeta) = \phi(\xi)\zeta$ and so $\sigma(x)$ is well-defined. The map σ is continuous since $\beta: \Omega \to B$ is an identification. Clearly σ is an admissible section and $\phi = L_\sigma$, $\phi_o = \beta\circ\sigma$. We call L_σ (with $\beta\circ\sigma$ understood) the left-translation corresponding to σ.

Clearly the set of left-translations is a group under composition. We transfer its group structure to $\Gamma\Omega$:

Proposition 5.2. Let Ω be a topological groupoid on B. Then $\Gamma\Omega$ is a group with respect to the multiplication * defined by

$$(\sigma*\tau)(x) = \sigma((\beta\circ\tau)(x))\tau(x), \quad x \in B,$$

with identity the object inclusion map $x \mapsto \tilde{x}$, denoted in this context by id, and inversion

$$\sigma^{-1}(x) = \sigma((\beta\circ\sigma)^{-1}(x))^{-1}, \quad x \in B,$$

and $\sigma \mapsto L_\sigma$ is a group isomorphism, that is $L_{\sigma*\tau} = L_\sigma \circ L_\tau$.

Proof. Straightforward. //

Note that $\sigma \mapsto \beta\circ\sigma$ is a group morphism from $\Gamma\Omega$ to the group of homeomorphisms of B.

Example 5.3. Consider a trivial topological groupoid $B \times G \times B$. The set $\Gamma(B \times G \times B)$ can be identified with the set of pairs (ϕ,θ), where $\phi: B \to B$ is a homeomorphism and $\theta: B \to G$ is any map, by identifying (ϕ,θ) with $x \mapsto (\phi(x),\theta(x),x)$. The multiplication is then

$$(\phi_2,\theta_2) * (\phi_1,\theta_1) = (\phi_2\circ\phi_1, (\theta_2\circ\phi_1)\theta_1)$$

with inversion $(\phi,\theta)^{-1} = (\phi^{-1},\theta^{-1}\circ\phi^{-1})$; here θ^{-1} refers to the pointwise inverse of a group-valued map and ϕ^{-1} to the composition-inverse of a homeomorphism. //

Example 5.4. Consider a vector bundle (E,p,B) or, more generally, a fibre bundle whose fibres are locally compact, locally connected, and Hausdorff.
Given $\sigma \in \Gamma\Pi(E)$, define a vector bundle morphism, also denoted by σ, over $\beta\circ\sigma$, by $u \mapsto \sigma(pu)u$. Then $\sigma * \tau = \sigma\circ\tau$ and $id = id_E$, and so each $\sigma: E \to E$ is an isomorphism of vector bundles. Conversely, given an isomorphism of vector bundles $\phi: E \to E$, $\phi_o: B \to B$, the map $\sigma: x \mapsto \phi_x \in \Pi(E)_x^{\phi_o(x)}$ is an admissible section of $\Pi(E)$ (continuity is proved by using the local triviality of E) and $\sigma = \phi$.

We will use these vector bundle isomorphisms less than the maps of sections which they induce. An admissible section $\sigma: E \to E$, $\beta\circ\sigma: B \to B$ induces a map $\bar{\sigma}: \Gamma E \to \Gamma E$ by

$$\bar{\sigma}(\mu)(x) = \sigma((\beta\circ\sigma)^{-1}(x))\mu((\beta\circ\sigma)^{-1}(x)), \qquad x \in B$$

and a map $\bar{\sigma}: C(B) \to C(B)$, $f \mapsto f\circ(\beta\circ\sigma)^{-1}$ (see A§1) and these maps satisfy

$$\overline{\sigma*\tau}(\mu) = \bar{\sigma}(\bar{\tau}(\mu))$$

$$\bar{\sigma}(\mu_1 + \mu_2) = \bar{\sigma}(\mu_1) + \bar{\sigma}(\mu_2)$$

$$\bar{\sigma}(f\mu) = \bar{\sigma}(f)\bar{\sigma}(\mu) \ ,$$

as can be easily checked. For future use, we also note that

$$\overline{\sigma^{-1}}(\mu)(x) = \sigma(x)^{-1}\mu((\beta\circ\sigma)(x)). \qquad //$$

Example 5.5. Consider an action groupoid $\Omega * M$ where Ω is a topological groupoid and $p: M \to B$ is a continuous map. If $\phi: M \to M$ is a homeomorphism then an admissible section σ of $\Omega * M$ with $\bar{\beta}\circ\sigma = \phi$ may be identified with a map $f: M \to \Omega$ such that $\alpha\circ f = p$ and $f(u)u = \phi(u)$, $\forall\, u \in M$. Consider in particular a discrete group G, let B be the set G with the indiscrete topology and let $G \times B \to B$ be the group multiplication. If $\phi: B \to B$ is a permutation, but not a left-translation $g \mapsto xg$, then there is no (continuous) admissible section σ of $G \times B$ with $\bar{\beta}\circ\sigma = \phi$. For each left-translation $B \to B$ there is exactly one such admissible section. //

Example 5.6. Consider a principal topological groupoid Ω on B, and a left-translation $L_\sigma: \Omega \to \Omega$ over $\beta\circ\sigma: B \to B$. For each $x \in B$, L_σ restricts to $\Omega_x \to \Omega_x$ and $L_\sigma|_x(\beta\circ\sigma, id): \Omega_x(B, \Omega_x^x) \to \Omega_x(B, \Omega_x^x)$ is an isomorphism of Cartan principal bundles.

Conversely, let $P(B,G,\pi)$ be a Cartan principal bundle and let $\phi(\phi_o, id)$ be an isomorphism $P(B,G) \to P(B,G)$. (That is, $\pi\circ\phi = \phi_o\circ\pi$ and $\phi(ug) = \phi(u)g$, $\forall\, u \in P$, $g \in G$.) For $x \in B$ choose $u \in \pi^{-1}(x)$ and write $\sigma(x) = \langle\phi(u), u\rangle$; this is clearly well-defined and σ is continuous since π is an identification. σ is an admissible section of $\frac{P \times P}{G}$ and L_σ is $\langle v,u\rangle \mapsto \langle\phi(v),u\rangle$, which, in terms of the isomorphism of 1.19(i), corresponds to $\phi: P \to P$.

Automorphisms of principal bundles of the form $\phi(\phi_o, id)$ might thus legitimately be called left-translations. Those for which $\phi_o = id_B$ are called gauge transformations in the physics literature (for example, Atiyah et al (1978, §2)); they correspond to those admissible sections of $\frac{P \times P}{G}$ which take values in the bundle $\frac{P \times G}{G}$. //

The question of the existence of admissible sections is an extremely obscure one. Even if Ω is locally trivial and the homeomorphism group of B is transitive, it is not clear that $\sigma \mapsto \beta\circ\sigma$, $\Gamma\Omega \to \text{Homeom}(B)$, is surjective. In future however we will only be concerned with local admissible sections:

Definition 5.7. Let Ω be a topological groupoid on B. For $U \subseteq B$ open, a local admissible section of Ω on U is a map $\sigma: U \to \Omega$ which is right-inverse to α and for which $\beta\circ\sigma: U \to (\beta\circ\sigma)(U)$ is a homeomorphism from U to the open set $(\beta\circ\sigma)(U)$ in B. The set of local admissible sections of Ω on U is denoted $\Gamma_U\Omega$.

For $\sigma \in \Gamma_U\Omega$ with $V = (\beta\circ\sigma)(U)$, the local left-translation induced by σ is $L_\sigma: \Omega^U \to \Omega^V$, $\xi \mapsto \sigma(\beta\xi)\xi$. //

The set of all L_σ for $\sigma \in \Gamma_U\Omega$ and $U \subseteq B$ open is not a pseudogroup on Ω since it is not closed under restriction. The following result, which will be needed in III§4, shows that this is unimportant.

Proposition 5.8. Let Ω be a topological groupoid on B, for which $\beta\colon \Omega \to B$ is an open map. Let $\phi\colon \mathcal{U} \to \mathcal{V}$ be a homeomorphism from $\mathcal{U} \subseteq \Omega$ open to $\mathcal{V} \subseteq \Omega$ open, and let $\phi_o\colon U \to V$ be a homeomorphism from $U = \beta(\mathcal{U}) \subseteq B$ to $V = \beta(\mathcal{V}) \subseteq B$, such that $\alpha\circ\phi = \alpha$, $\beta\circ\phi = \phi_o\circ\beta$ and $\phi(\xi\eta) = \phi(\xi)\eta$ whenever $(\xi,\eta) \in \Omega*\Omega$, $\xi \in \mathcal{U}$ and $\xi\eta \in \mathcal{U}$. Then ϕ is the restriction to $\mathcal{U}$ of a unique local left-translation $L_\sigma\colon \Omega^U \to \Omega^V$ where $\sigma \in \Gamma_U\Omega$.

Proof. For $x \in U$ choose $\xi \in \mathcal{U}^x$ and define $\sigma(x) = \phi(\xi)\xi^{-1}$; clearly $\sigma(x)$ is well-defined. Since the restriction $\beta\colon \mathcal{U} \to U$ is open, σ is continuous and is therefore a local admissible section on U with $\beta\circ\sigma = \phi_o$. That $L_\sigma(\xi) = \phi(\xi)$ for $\xi \in \mathcal{U}$ is clear, as is the uniqueness. //

Thus, at least for groupoids whose projections are open maps, any local homeomorphism $\mathcal{U} \to \mathcal{V}$ which commutes with the $R_\eta\colon \Omega_{\beta\eta} \to \Omega_{\alpha\eta}$ in the sense of 5.8 is the restriction of a local left-translation $L_\sigma\colon \Omega^{\beta(\mathcal{U})} \to \Omega^{\beta(\mathcal{V})}$.

In any case, for a general topological groupoid Ω we will regard the set of local admissible sections as being in some sense a pseudogroup on B with law of composition $*$: if $\sigma \in \Gamma_U\Omega$ with $(\beta\circ\sigma)(U) = V$ and $\tau \in \Gamma_{V'}\Omega$ with $(\beta\circ\tau)(V') = W$, then $\tau * \sigma$ is the local admissible section in $(\beta\circ\sigma)^{-1}(V') \cap V$ defined by $(\tau*\sigma)(x) = \tau((\beta\circ\sigma)(x))\sigma(x)$, providing $(\beta\circ\sigma)^{-1}(V') \cap V$ is not void. We will refer to the set of all local admissible sections, together with this composition, as the pseudogroup of local admissible sections of Ω, and will denote it by $\Gamma^{\ell oc}(\Omega)$.

Let $J^\lambda(\Omega)$ denote the set of all germs of local admissible sections of Ω. Then $J^\lambda(\Omega)$ has a natural groupoid structure: the source and target projections are $\bar{\alpha}(g_x\sigma) = x$, $\bar{\beta}(g_x\sigma) = (\beta\circ\sigma)(x)$; the object inclusion map is $x \mapsto \tilde{x} = g_x\mathrm{id}$; the multiplication, denoted $*$, is $(g_y\tau) * (g_x\sigma) = g_x(\tau*\sigma)$, and the inversion is $(g_x\sigma)^{-1} = g_{(\beta\circ\sigma)(x)}(\sigma^{-1})$ where $\sigma^{\underline{-1}}$ is $y \mapsto \sigma((\beta\circ\sigma)^{-1}(y))^{-1}$, defined on $(\beta\circ\sigma)(U)$. With this structure, $J^\lambda(\Omega)$ is called the local prolongation groupoid of Ω.

One can give $J^\lambda(\Omega)$ a sheaf-type topology defined by taking as basis the sets $N_\sigma = \{g_x\sigma \mid x \in \mathrm{dom}\,\sigma\}$ for $\sigma \in \Gamma^{\ell oc}\Omega$. With this topology $J^\lambda(\Omega)$ is a topological groupoid, denoted $J^\lambda_{sh}(\Omega)$. Whether this is an interesting topology is rather uncertain; for a topological group G, $J^\lambda(G)$ is naturally isomorphic as a groupoid to the group G, but $J^\lambda_{sh}(G)$ has the discrete topology. It is thus a rather coarse invariant.

If Ω is locally compact and Hausdorff, the set of germs g_xL_σ, $x \in U$, of local left-translations $L_\sigma\colon \Omega^U \to \Omega^V$ can be given the compact-open topology of Abd-Allah and Brown (1980) (see 1.15) and this topology transferred to $J^\lambda(\Omega)$; $J^\lambda(\Omega)$ is then a topological groupoid on B, and is denoted $J^\lambda_{co}(\Omega)$. For any locally compact and Hausdorff topological group G, $J^\lambda_{co}(G)$ is naturally isomorphic to G as a topological group. Further,

Proposition 5.9. Let Ω be a locally trivial groupoid on a topological manifold B. Then $J^\lambda_{co}(\Omega)$ is locally trivial and the map $J^\lambda(\Omega) \to J^\lambda(\Gamma^o(B)) = J^\lambda(B \times B)$, $g_x\sigma \mapsto g_x(\beta\circ\sigma)$ is surjective.

Proof. Let $\{\sigma_i\colon U_i \to \Omega_b\}$ be a section-atlas for Ω; assume each U_i is the domain of a chart $\mathbb{R}^n \cong U_i$ for B. Choose some U_i which contains b and denote it by U. Choose any U_i and find a continuous $\psi_i\colon U \times U_i \to U_i$ such that each $\psi_i(-,x)\colon U \to U_i$ is a homeomorphism and $\psi_i(b,-)$ is the identity. Define $\tilde{\sigma}_i\colon U_i \to J^\lambda_{co}(\Omega)_b$ by $\tilde{\sigma}_i(x) = g_b(y \mapsto \sigma_i(\psi_i(y,x))\sigma(y)^{-1})$. It is straightforward to check that $\tilde{\sigma}_i$ is continuous.

The second assertion is proved in a similar way. //

This result is taken from Virsik (1969, 1.4). It is the only existence result for local admissible sections that we need.

Definition 5.10. Let Ω be a topological groupoid on B, and take $\sigma \in \Gamma^{\ell oc}\Omega$ with domain U and $(\beta\circ\sigma)(U) = V$.

The local right-translation defined by σ is $R_\sigma\colon \Omega_V \to \Omega_U$, $\xi \mapsto \xi\sigma((\beta\circ\sigma)^{-1}(\alpha\xi))$.

The local inner automorphism defined by σ is $I_\sigma\colon \Omega^U_U \to \Omega^V_V$, $\xi \mapsto \sigma(\beta\xi)\xi\sigma(\alpha\xi)^{-1}$. //

Clearly $R_{\sigma*\tau} = R_\tau \circ R_\sigma$ and $I_{\sigma*\tau} = I_\sigma \circ I_\tau$, wherever the products are defined. Also, $R_{\sigma^{-1}}(\xi) = \xi\sigma(\alpha\xi)^{-1}$ and $I_\sigma = L_\sigma \circ R_{\sigma^{-1}} = R_{\sigma^{-1}} \circ L_\sigma$. Note that $I_\sigma\colon \Omega^U_U \to \Omega^V_V$ is an isomorphism of topological groupoids over $\beta\circ\sigma\colon U \to V$. (Local) right-translations can be characterized intrinsically as in 5.1. Since the Lie algebroid of a differentiable groupoid will be defined using right-invariant vector fields and the flows of such fields are local left-translations, we will not use local right-translations very extensively.

Definition 5.11. Let $\phi: \Omega \to \Omega'$ be a morphism of topological groupoids over B. Then the maps $\Gamma\Omega \to \Gamma\Omega'$, $\sigma \mapsto \phi\circ\sigma$, $\Gamma^{\ell oc}\Omega \to \Gamma^{\ell oc}\Omega'$, $\sigma \mapsto \phi\circ\sigma$, and $J^{\lambda}(\Omega) \to J^{\lambda}(\Omega')$, $g_x\sigma \mapsto g_x(\phi\circ\sigma)$, are all denoted by $\tilde{\phi}$ and called induced morphisms (of groups, "pseudogroups" and groupoids, respectively). //

It is easy to see that $\tilde{\phi}: J^{\lambda}(\Omega) \to J^{\lambda}(\Omega')$ is continuous with respect to either (consistent) choice of topology.

In the case of a representation $\rho: \Omega \to \Pi(E)$ and a local admissible section $\sigma \in \Gamma_U\Omega$, the map $\tilde{\rho}(\sigma): \Gamma_U E \to \Gamma_V E$ will be denoted more simply by $\bar{\rho}(\sigma)$.

§6. The monodromy groupoid of a principal topological groupoid

Given an α-connected and principal topological groupoid Ω, whose topology is locally well-behaved, we construct a principal topological groupoid $M\Omega$ whose α-fibres are the universal covering spaces of the α-fibres of Ω and which is locally isomorphic to Ω under a canonical morphism $\psi: M\Omega \to \Omega$ which on each α-fibre is the standard covering projection. $M\Omega$ is called the monodromy groupoid of Ω and generalizes both the construction of the universal covering group of a topological group and that of the fundamental groupoid of a topological space.

The algebraic structure of $M\Omega$ is easily defined for an arbitrary α-connected toplogical groupoid; we considerably simplify the problem of defining the topology on $M\Omega$ by restricting ourselves to the case where Ω is principal and then working in terms of the vertex bundles. In fact we show that if $P(B, G)$ is a principal bundle with P connected then the universal cover $\tilde{P}$ is a principal bundle over B with respect to a group H which is locally isomorphic to G, and that the covering projection $\tilde{P} \to P$ is a principal bundle morphism over B. The corresponding result for principal groupoids then follows immediately.

For Ω a differentiable and not necessarily locally trivial groupoid, Almeida (1980), following the announcement of Pradines (1966, Théorème 2), has shown the existence of a differentiable groupoid structure on $M\Omega$ such that the covering projection $M\Omega \to \Omega$ is smooth and étale. The proof proceeds by lifting the differentiable structure of Ω back to a generating subset of $M\Omega$ and then showing that the resulting microdifferentiable structure on $M\Omega$ (see III 6.3 for definition) globalizes. This construction is vastly more complicated than the one given here, and we believe it worthwhile to have set down the simple proof available in the case of principal topological groupoids and locally trivial differentiable groupoids.

It should be noted that Almeida's construction deduces a topology on $M\Omega$ only a posteriori; it would be interesting to have a purely topological construction of the holonomy groupoid of a microtopological groupoid.

The proof given here can be reformulated to avoid use of the correspondence 1.19 between principal groupoids and Cartan principal bundles: if Ω is locally trivial and locally simple (see 6.9 for definition) 6.2 generalizes to show that $P_o^\alpha(\Omega) \to M\Omega$ is open and the continuity of the multiplication in $M\Omega$ then follows from that of the multiplication in $P_o^\alpha(\Omega)$. However it seems unlikely that $P_o^\alpha(\Omega) \to M\Omega$ is open in general.

A different approach to the construction of the universal covering principal bundle $\tilde{P}(B, H)$ of $P(B, G)$ is given by Kamber and Tondeur (1971, 6.3).

In 6.14 we show that a local morphism of topological groupoids, defined on an α-simply connected and principal domain groupoid, globalizes. In 6.11 we show that the covering projection $\psi: M\Omega \to \Omega$ has a local right-inverse morphism; for this we need to assume that Ω is locally trivial.

To avoid tedious repetition we use the following terminology.

<u>Definition 6.1</u> A topological space is <u>admissible</u> if it is Hausdorff, locally connected, locally compact and semi-locally simply connected. //

Local compactness is included here because it is used in the proof below that $C^P \times C^H \to C^P$ is continuous.

Let Ω be an α-connected and transitive topological groupoid on B, with Ω admissible. The monodromy groupoid $M\Omega$ of Ω is defined as follows.

Let $P^\alpha = P^\alpha(\Omega)$ be the set of paths $\gamma: I \to \Omega$ (where $I = [0, 1]$) for which $\alpha\circ\gamma: I \to B$ is constant; elements of P^α are called α-<u>paths</u> in Ω. Let $P_o^\alpha = P_o^\alpha(\Omega)$ be the subset of α-paths which commence at an identity of Ω; every $\gamma \in P^\alpha$ is of the form $R_\xi \circ \gamma'$ where $\gamma' \in P_o^\alpha$ and $\xi = \gamma(0)$. Define $\gamma, \delta \in P^\alpha$ to be α-<u>homotopic</u>, written $\gamma \underset{\alpha}{\simeq} \delta$, if $\gamma(0) = \delta(0)$, $\gamma(1) = \delta(1)$, and there is a continuous $H: I \times I \to \Omega$ such that $H(0, -) = \gamma$, $H(1, -) = \delta$, $H(s, 0)$ and $H(s, 1)$ are constant with respect to $s \in I$, and $H(s, -) \in P^\alpha$, $\forall s \in I$. Such a map H is called an α-<u>homotopy</u> from γ to δ. The α-homotopy class containing $\gamma \in P^\alpha$ is written $\langle\gamma\rangle$.

Define $M\Omega$ to be the set $\{\langle\gamma\rangle \mid \gamma \in P_o^\alpha\}$ with the following groupoid structure:

the projections $\bar{\alpha}$, $\bar{\beta}$: $M\Omega \to B$ are $\bar{\alpha}(\langle\gamma\rangle) = \alpha\gamma(0)$ and $\bar{\beta}(\langle\gamma\rangle) = \beta\gamma(1)$; consequently if $\bar{\alpha}(\langle\delta\rangle) = \bar{\beta}(\langle\gamma\rangle)$, then $\delta(0)\gamma(1)$ is defined and so is the standard concatenation $(R_{\gamma(1)}\circ\delta)\gamma$; we define $\langle\delta\rangle\langle\gamma\rangle$ to be $\langle(R_{\gamma(1)}\circ\delta)\gamma\rangle$. It is straightforward to verify that this product is well-defined and makes $M\Omega$ into a groupoid on B. The identities are $\tilde{x} = \langle\kappa_{\tilde{x}}\rangle$ and the inverse of $\langle\gamma\rangle$ is $\langle R_{\xi}\circ\overleftarrow{\gamma}\rangle$ where $\xi = \gamma(1)^{-1}$ and $\overleftarrow{\gamma}$ is the reverse of γ. Since Ω is α-connected and transitive, it follows that $M\Omega$ is transitive. It is clear that ψ: $M\Omega \to \Omega$, $\langle\gamma\rangle \mapsto \gamma(1)$ is a surjective morphism of groupoids over B; ψ is called the <u>covering projection</u>.

Clearly each α-fibre $M\Omega|_x$ is the set underlying the universal covering space of Ω_x, constructed from paths starting at $\tilde{x}$. This, together with the fact that $M(B \times B) = \mathcal{T}(B)$ is a topological groupoid with respect to the quotient of the compact-open topology on $C(I, B) = P_o^{\alpha}(B \times B)$, suggests that $M\Omega$ should be a topological groupoid with respect to the quotient of the compact-open topology on $P_o^{\alpha}(\Omega)$. However, this writer has not been able to give a general proof that, with this topology, the groupoid multiplication is continuous. On the one hand one may try to calculate $\pi_1(M\Omega)$ and $\pi_1(M\Omega * M\Omega)$, with the aim of lifting $M\Omega * M\Omega \to M\Omega \to \Omega$ across $M\Omega \to \Omega$; this, however, appears intractable unless $M\Omega$ has associated with it a fibre bundle structure, such as $M\Omega \cong \frac{Q \times Q}{H}$ where $Q = M\Omega|_x$, $H = M\Omega|_x^x$. On the other hand, one may try to imitate the case of $\mathcal{T}(B)$ (see 1.14), but although $C(I, B) \to \mathcal{T}(B)$ is an open map (see 6.2) it seems unlikely that $P_o^{\alpha}(\Omega) \to M\Omega$ is open in general.

These problems can be avoided when Ω is principal, by working with a single vertex bundle of Ω, and this case will suffice for our purposes. Therefore let Ω now be a principal and α-connected admissible topological groupoid on B. Choose $b \in B$ and write $P = \Omega_b$, $G = \Omega_b^b$, $\pi = \beta_b$. Then $M\Omega|_b = \tilde{P}$, the universal cover of P based at $\tilde{b}$. Write $H = \pi^{-1}(G) = M\Omega|_b^b \subseteq \tilde{P}$, and give H the subspace topology from $\tilde{P}$. Denote $\tilde{\beta}_b$: $\tilde{P} \to B$ by $\tilde{\pi}$; note that $\tilde{\pi}$ is $\langle\gamma\rangle \mapsto \pi(\gamma(1))$.

We claim that $\tilde{P}(B, H, \tilde{\pi})$ is a Cartan principal bundle. The algebraic properties follow from the groupoid stucture of $M\Omega$, and it only remains to prove that the action $\tilde{P} \times H \to \tilde{P}$ and the inversion $H \to H$ are continuous. The following result is needed.

<u>Proposition 6.2</u> Let X be a connected, locally connected and semi-locally simply connected Hausdorff space. Then the identification map

$$p: C(I,X) \to \mathcal{T}(X), \ \gamma \mapsto \langle\gamma\rangle,$$

where C(I, X) has the compact-open topology, is open.

<u>Proof</u>: The topology on C(I, X) has as sub-basis sets of the form $N(a, b, U) = \{\gamma \in C(I, X) \mid \gamma([a, b]) \subseteq U\}$, where $0 \le a \le b \le 1$ and $U \subseteq X$ is open. Let $N = \bigcap_{i=1}^{n} N([a_i, b_i], U_i)$ be a basic open set, and choose $\gamma \in N$. Write $a = \min\{a_i\}$ and $b = \max\{b_i\}$. There are now four cases.

Case $0 < a$, $b < 1$. We will show that every element of the form $\langle\delta'\rangle\langle\gamma\rangle\langle\delta\rangle$, where δ', δ are arbitary paths with $\delta'(0) = \gamma(1)$, $\delta(1) = \gamma(0)$, is in p(N). (This in fact shows that $p(N) = \pi(X)$.) For convenience, regard $\varepsilon = \delta'\gamma\delta$ as being

$$\varepsilon(t) = \begin{cases} \delta(3t) & 0 \le t \le \frac{1}{3} \\ \gamma(3t - 1) & \frac{1}{3} \le t \le \frac{2}{3} \\ \delta'(3t - 2) & \frac{2}{3} \le t \le 1. \end{cases} \tag{1}$$

Define ε' by the formula

$$\varepsilon'(t) = \begin{cases} \varepsilon(\frac{1 + a}{3a} t) & 0 \le t \le a \\ \varepsilon(\frac{t + 1}{3}) & a \le t \le b \\ \varepsilon(\frac{(2 - b)t + (1 - 2b)}{3(1 - b)}) & b \le t \le 1. \end{cases} \tag{2}$$

Then for all i we have $\varepsilon'(a_i) = \varepsilon(\frac{a_i + 1}{3}) = \gamma(a_i)$ and similarly $\varepsilon'(b_i) = \gamma(b_i)$, and $\varepsilon'([a_i, b_i]) \subseteq \gamma([a_i, b_i]) \subseteq U_i$. So $\varepsilon' \in N$. That $e' \sim \varepsilon$ should be clear; a specific homotopy is

$$H_s(t) = \begin{cases} \varepsilon(\frac{(\frac{1}{2} - a)s + a + 1}{3(\frac{1}{2} - a)s + 3a} t) \text{ for } 0 \le t \le (\frac{1}{2} - a)s + a \\ \varepsilon(\frac{t + 1}{3}) \text{ for } (\frac{1}{2} - a)s + a \le t \le (\frac{1}{2} - b)s + b \\ \varepsilon(\frac{(2 - (\frac{1}{2} - b)s - b)t + (1 - 2(\frac{1}{2} - b)s - 2b)}{3(1 - (\frac{1}{2} - b)s - b)}) \\ \quad \text{for } (\frac{1}{2} - b)s + b \le t \le 1. \end{cases}$$

Case $a = 0$, $b < 1$. Set $a^+ = \min\{a_i | a_i > 0\}$ and $U = \bigcap \{U_i | a_i = 0\}$; U is an open neighbourhood of $\gamma(0)$. We will show that every element of $\pi(X)$ of the form $\langle\delta'\rangle\langle\gamma\rangle\langle\delta\rangle$ where δ' is arbitary with $\delta'(0) = \gamma(1)$ and δ is in any path in U with $\delta(1) = \gamma(0)$, lies in $p(N)$. Clearly the set of all such products is open (see 1.14). With ε as in (1), define ε' by (2) except that a is replaced by a^+. Then $\varepsilon' \sim \varepsilon$ as before and $\varepsilon'([a_i, b_i]) \subseteq U_i$ whenever $a_i > 0$. It remains to show that $\varepsilon'([0, b_i]) \subseteq U_i$ when $a_i = 0$. Consider first the sub-case where $b_i \leqslant a^+$. Then $\varepsilon'([0, b_i]) = \varepsilon([0, \frac{1 + a^+}{3a^+} b_i])$. If $\frac{1 + a^+}{3a^+} b_i \leqslant \frac{1}{3}$, then $\varepsilon[0, \frac{1 + a^+}{3a^+} b_i] \subseteq \delta([0, 1]) \subseteq U \subseteq U_i$. And if $k = \frac{1 + a^+}{3a^+} b_i > \frac{1}{3}$, then $\varepsilon[0, k] = \varepsilon[0, \frac{1}{3}] \cup \varepsilon[\frac{1}{3}, k] = \delta([0, 1]) \cup \gamma[0, 3k - 1]$. Now $3k - 1 \leqslant b_i$, since $b_i \leqslant a^+$, so $\gamma[0, 3k - 1] \subseteq \gamma[a_i, b_i] \subseteq U_i$ as required.

Secondly, consider the sub-case where $a^+ < b_i$. Here

$$\varepsilon'[0, b_i] = \varepsilon'[0, a^+] \cup \varepsilon'[a^+, b_i]$$

$$= \varepsilon[0, \frac{1 + a^+}{3}] \cup \varepsilon[\frac{1 + a^+}{3}, \frac{1 + b_i}{3}] \quad \text{(since } b_i \leqslant b\text{)}$$

$$= \varepsilon[0, \tfrac{1}{3}] \cup \varepsilon[\tfrac{1}{3}, \frac{1 + a^+}{3}] \cup \varepsilon[\frac{1 + a^+}{3}, \frac{1 + b_i}{3}]$$

$$\subseteq \delta[0, 1] \cup \gamma[0, a^+] \cup \gamma[a^+, b_i]$$

$$\subseteq U \cup \gamma[0, b_i] \subseteq U_i, \text{ as required.}$$

The case $a > 0$, $b = 1$ is similar. In the case $a = 0$, $b = 1$, define $a^+ = \min\{a_i | a_i > 0\}$, $b^- = \max\{b_i | b_i < 1\}$ and $U = \bigcap\{U_i | a_i = 0\}$, $V = \bigcap\{V_i | b_i = 1\}$, and consider products $\langle\delta'\rangle\langle\gamma\rangle\langle\delta\rangle$ with δ' in V and δ in U and ε' defined using a^+ and b^-. //

Returning to $\tilde{P}(B, H, \tilde{\pi})$, let C^P denote the space of paths γ in $C(I, P)$ for which $\gamma(0) = b$, and C^H the space of paths γ in C^P for which $\gamma(1) \in G$. Let p be the projection $C(I, P) \to \pi(P)$. From 6.2 it follows that $p: C^P \to \tilde{P}$ is open, hence that $p: C^H \to H$ is open, and hence that $p \times p: C^P \times C^H \to \tilde{P} \times H$ is open. It therefore suffices to show that

$$C^P \times C^H \to C^P, \ (\delta, \gamma) \mapsto (R_{\gamma(1)} \circ \delta)\gamma$$

is continuous. Now $C^P \times G \to C^P$, $(\delta, g) \mapsto R_g \circ \delta$ is continuous, since the action of G on P is continuous and P is locally compact; and concatention in path spaces is continuous (see, for example, Dugundji (1966, XII.2 and XIX.1, respectively)). This completes the proof that $\tilde{P}(B, H, \tilde{\pi})$ is a Cartan principal bundle, the <u>universal covering bundle</u> of $P(B, G, \pi)$.

Since $\tilde{P}(B, H)$ is algebraically the vertex bundle at b of $M\Omega$, the topology on $\tilde{P}$ can be transferred to $M\Omega$ via the bijection $\frac{\tilde{P} \times \tilde{P}}{H} \to M\Omega$, $\langle\langle\gamma'\rangle, \langle\gamma\rangle\rangle \mapsto \langle\gamma'\rangle\langle\gamma\rangle^{-1}$ and, by 1.19, $M\Omega$ becomes a principal topological groupoid on B with $\tilde{P}(B, H)$ now the topological vertex bundle.

<u>Definition 6.3</u> Let Ω be a principal and α-connected topological groupoid on B, with Ω admissible. The topological groupoid $M\Omega$ constructed above is the <u>monodromy groupoid</u> of Ω and the morphism $\psi: M\Omega \to \Omega$, $\langle\gamma\rangle \mapsto \gamma(1)$, is the <u>covering projection</u>. //

<u>Proposition 6.4</u> Let Ω be a locally trivial α-connected, topological groupoid on B, with Ω admissible. Then $M\Omega$ is locally trivial and $\psi: M\Omega \to \Omega$ is étale.

<u>Proof</u>: Decomposing sections $U_i \to M\Omega|_b = \tilde{P}$ for $M\Omega$ can be constructed as the compositions of decomposing sections $U_j \to P$ for Ω and local sections of $\tilde{P} \to P$.

Using decomposing sections $\sigma_i: U_i \to \tilde{P}$ for $M\Omega$ and $\psi\circ\sigma_i: U_i \to P$ for Ω, define charts $\tilde{P} \times U_i \to M\Omega|_{U_i}$ and $P \times U_i \to \Omega_{U_i}$ by $(\langle\gamma\rangle, x) \mapsto \langle\gamma\rangle\sigma_i(x)^{-1}$ and $(\xi, x) \mapsto \xi\psi(\sigma_i(x)^{-1})$. Then ψ is locally $p \times id$, where $p: \tilde{P} \to P$ is the covering projection of P. //

<u>Examples 6.5</u> Clearly $M(B \times B) = \mathcal{T}(B)$ for B a connected, admissible space. For G a connected admissible topological group and H a closed subgroup, let $K = \psi^{-1}(H)$ where $\psi: \tilde{G} \to G$ is the covering projection. Then K is a closed subgroup of $\tilde{G}$, the spaces $\tilde{G}/K$ and G/H are equivariantly homeomorphic, and $\tilde{G}(\tilde{G}/K, K)$ is the universal covering bundle of $G(G/H, H)$.

In particular, the universal covering bundle of $SO(3)(\mathbf{S}^2, SO(2))$ is $SU(2)(\mathbf{S}^2, U(1))$ (see 2.24). Let the universal covering $\phi: SU(2) \times SU(2) \to SO(4)$ be realized as the map $(p, q) \mapsto (h \mapsto phq^{-1})$ mentioned in I 1.11 and let $A \in SO(3)$ act on $SO(4)$ as $\begin{bmatrix} 1 & 0 \\ 0 & A \end{bmatrix}$; then the universal covering bundle of $SO(4)(\mathbf{S}^3, SO(3))$ is $SU(2) \times SU(2)(\mathbf{S}^3, SU(2))$, where $SU(2)$ acts as the diagonal subgroup. Thus both bundles are trivializable.

More interestingly, consider $SO(4)(SO(4)/T^2, T^2)$ where T^2 is the maximal torus $\left\{\begin{bmatrix} A & 0 \\ 0 & B \end{bmatrix} \middle| A, B \in SO(2)\right\}$ in $SO(4)$. The homogeneous space $SO(4)/T^2$ is the Grassmannian $\tilde{G}_{4,2}$ of oriented 2-planes in $\mathbf{R}^4$, and a calculation shows that $\phi^{-1}(T^2)$ is the maximal torus $K = \{((e^{i\theta}, 0), (e^{i\theta'}, 0)) \mid \theta, \theta' \in \mathbf{R}\}$ where the notation for elements of $SU(2)$ is as in 2.24.

Now the action of K on $SU(2) \times SU(2)$ is the cartesian square of the action in 2.24 of $U(1)$ on $SU(2)$, so the universal covering bundle of $SO(4)(\tilde{G}_{4,2}, T^2)$ is $SU(2) \times SU(2)(\mathbf{S}^2 \times \mathbf{S}^2, U(1) \times U(1))$. Note that the restriction of $\phi: SU(2) \times SU(2) \to SO(4)$ to $K \to T^2$ is

$$((e^{i\theta}, 0), (e^{i\theta'}, 0)) \mapsto \begin{bmatrix} R_{\theta-\theta'} & 0 \\ 0 & R_{\theta+\theta'} \end{bmatrix}$$

where R_α is the rotation matrix $\begin{bmatrix} \cos\alpha & -\sin\alpha \\ \sin\alpha & \cos\alpha \end{bmatrix}$. //

Proposition 6.6 Let $P(B, G)$ be a principal bundle with P connected and admissible, and let $\tilde{P}(B, H)$ be the universal covering bundle. Then

(i) the sequence $\pi_1 P \overset{\subseteq}{\rightarrowtail} H \overset{\psi}{\twoheadrightarrow} G$ is exact;

(ii) $\pi_0 H \cong \pi_1 B$ under the boundary morphism of the long exact homotopy sequence of $\tilde{P}(B, H)$;

(iii) $\pi_1 H = \ker(\pi_1 G \to \pi_1 P)$.

Proof: (i) is immediate from the definition of H and (ii) and (iii) follow from the diagram

$$\begin{array}{ccccccccccc} \to \pi_2 B & \to & \pi_1 H & \to & 0 & \to & \pi_1 B & \to & \pi_0 H & \to & 0 \\ \| & & \downarrow & & \downarrow & & \| & & \downarrow & & \\ \to \pi_2 B & \to & \pi_1 G & \to & \pi_1 P & \to & \pi_1 B & \to & \pi_0 G & \to & 0 \end{array}$$

where the rows are the long exact homotopy sequences for $\tilde{P}(B, H)$ and $P(B, G)$ (see, for example, Hu (1959)). //

We now treat the problem of globalizing a local morphism of topological groupoids.

Definition 6.7 Let Ω and Ω' be topological groupoids on bases B and B' respectively. A local morphism of topological groupoids, denoted $\phi: \Omega \rightsquigarrow \Omega'$, consists of a continuous map $\phi: \mathcal{U} \to \Omega'$ defined on an open neighbourhood $\mathcal{U}$ of the base of Ω, together with a continuous map $\phi_o: B \to B'$, such that $\alpha' \circ \phi = \phi_o \circ \alpha$, $\beta' \circ \phi = \phi_o \circ \beta$, $\phi \circ \varepsilon = \varepsilon' \circ \phi_o$, and such that

(i) $\phi(\eta\xi) = \phi(\eta)\phi(\xi)$ whenever $\alpha\eta = \beta\xi$ and each of ξ, η, $\eta\xi$ is in $\mathcal{U}$; and

(ii) $\phi(\xi^{-1}) = \phi(\xi)^{-1}$ whenever both of ξ and ξ^{-1} are in $\mathcal{U}$.

Two local morphisms $\phi, \psi: \Omega \rightsquigarrow \Omega'$ are germ-equivalent if the maps ϕ and ψ are equal on an open neighbourhood of the base of Ω.

A local morphism $\phi: \Omega \rightsquigarrow \Omega'$, $\phi_o: B \to B'$ is a local isomorphism if there exists a local morphism $\phi': \Omega' \rightsquigarrow \Omega'$ such that $\phi \circ \phi'$ and $\phi' \circ \phi$ are germ-equivalent to $id_{\Omega'}$ and id_{Ω}. //

Example 6.8 Suppose that Ω is a locally trivial topological groupoid with a section-atlas $\{\sigma_i: U_i \to \Omega_b\}$ which has the property that each transition function $s_{ij}: U_{ij} \to G = \Omega_b^b$ is constant. Let $\mathcal{U}$ be the open neighbourhood $\bigcup_i (U_i \times U_i)$ of the base in B × B and define $\theta: \mathcal{U} \to \Omega$ by $\theta(y,x) = \sigma_i(y)\sigma_i(x)^{-1}$ whenever $(y,x) \in U_i \times U_i$. It is easy to see that because the transition functions $\{s_{ij}\}$ are constant, θ is well-defined, and so gives a local morphism $B \times B \to \Omega$ over B.

Conversely, let Ω be a transitive topological groupoid on a space B and suppose there exists a local morphism $\theta: B \times B \rightsquigarrow \Omega$ over B. Choose $b \in B$ and an open cover $\{U_i\}$ of B such that $\bigcup_i (U_i \times U_i)$ is contained in the domain of θ. In each U_i choose an x_i and for each i choose $\xi_i \in \Omega_b^{x_i}$. Define $\sigma_i: U_i \to \Omega_b$ by $\sigma_i(x) = \theta(x, x_i)\xi_i$. Then the transition functions for $\{\sigma_i\}$ are constant, and the local morphism induced by $\{\sigma_i\}$ is θ. //

6.8 applies in particular to the groupoids $\mathcal{T}(B) = M(B \times B)$ of 1.14, providing that B has a cover by canonical open sets U_i such that each nonvoid intersection $U_i \cap U_j$ is path-connected. This is the case, for example, if B is a smooth paracompact manifold, for B then possesses an open cover $\{U_i\}$ such that each U_i and each nonvoid multiple intersection $U_{i_1} \cap \dots \cap U_{i_n}$ is contractible (reference in 6.10 below). Such a cover is called simple.

In general one expects that the covering projection $\psi: M\Omega \to \Omega$ will admit a local right-inverse and in order to obtain this, one needs the following corresponding concept.

Definition 6.9 Let $P(B, G)$ be a principal bundle. A G-simple cover of P is a simple cover $\{U_i\}$ of P such that given i and given $g \in G$ there is a j such that $R_g(U_i) \subseteq U_j$. We say that $P(B, G)$ is locally simple if it admits a G-simple cover and P is admissible.

A locally trivial topological groupoid is locally simple if any one of its vertex bundles is locally simple. //

Proposition 6.10 Let $P(B, G)$ be a smooth principal bundle. Then $P(B, G)$ is locally simple.

Proof: By A 4.20, P admits a G-invariant Riemannian metric. Let $\{U_i\}$ be the set of all open subsets of P such that any two points lying in U_i can be joined by exactly one geodesic in U_i. By Helgason (1978, pp 34-36), $\{U_i\}$ covers P and since the metric is invariant under the right action of G, it follows that $\{U_i\}$ is stable under G in the sense of 6.9. By construction, $\{U_i\}$ is simple. //

Theorem 6.11 Let Ω be a locally trivial and locally simple α-connected topological groupoid. Then $\psi: M\Omega \to \Omega$ has a local right-inverse.

Proof: Choose $b \in B$ and write $P = \Omega_b$, $G = \Omega_b^b$, $Q = M\Omega|_b = \tilde{P}$, $H = M\Omega|_{\tilde{b}}^b$. Let $\{U_i\}$ be a G-simple cover of P. For each i choose $\xi_i \in U_i$, a path γ_i from $\tilde{b}$ to ξ_i within P and (as in 1.14) a function θ_i which to $\xi \in U_i$ assigns a path in U_i from ξ_i to ξ. Define $\sigma_i: U_i \to Q$ by $\sigma_i(\xi) = \langle\theta_i(\xi)\gamma_i\rangle$, the homotopy class of the concatenation of γ_i followed by $\theta_i(\xi)$. We will show that

$\phi_i: \frac{U_i \times U_i}{G} \to \frac{Q \times Q}{H}$ by $\langle\xi', \xi\rangle \mapsto \langle\sigma_i(\xi'), \sigma_i(\xi)\rangle$ is a well-defined local morphism $\frac{P \times P}{G} \rightsquigarrow \frac{Q \times Q}{H}$.

Consider $\sigma_j(\xi g)$ where $\xi \in U_i$, $g \in G$ and $R_g(U_i) \subseteq U_j$. Let

$$\lambda = R_g(\overset{\leftarrow}{\gamma_i}\,\theta_i(\xi)^{\leftarrow})\theta_j(\xi g)\gamma_j;$$

λ is a path in P from $\tilde{b}$ to g and thus defines an element $\langle\lambda\rangle$ of H. Clearly $\sigma_j(\xi g) = \sigma_i(\xi)\langle\lambda\rangle$. Let λ' denote the element obtained similarly from ξ'. To show that $\langle\lambda'\rangle = \langle\lambda\rangle$, it suffices to show that

$$R_g(\theta_i(\xi)^+)\theta_j(\xi g) \sim R_g(\theta_i(\xi')^+)\theta_j(\xi' g);$$

where both sides are paths from ξ_j to $\xi_i g$, as

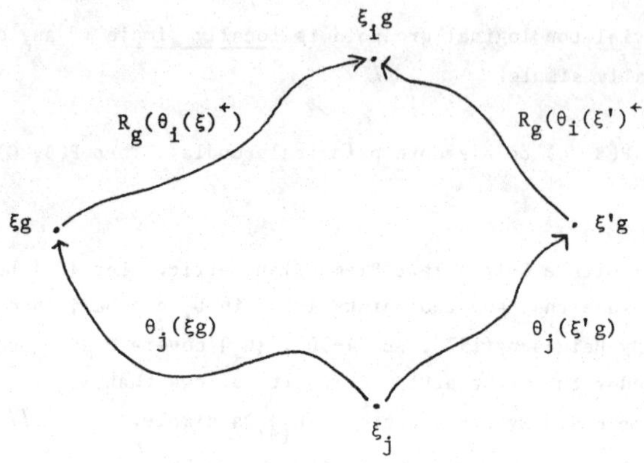

in the diagram. Since $R_g(U_i) \subseteq U_j$, it follows that all four paths in the diagram lie in U_j and so the deformation can be carried out.

Considering the σ_i as local sections of the bundle $Q(P, \pi_1 P)$, it is clear that their transition functions are constant, and the remainder of the proof follows as in 6.8. //

Corollary 6.12 Let Ω be a locally trivial and locally simple α-connected topological groupoid. Then $M\Omega$ is locally isomorphic to Ω under the covering projection ψ.

Proof: Let $\phi: \Omega \rightsquigarrow M\Omega$ be the local right-inverse to ψ constructed in 6.11. Then $\phi\circ\psi\circ\phi = \phi$ and so $\phi\circ\psi$ is equal to the identity on the image of ϕ, which is clearly open. //

We now prove that a local morphism of topological groupoids can be globalized if the domain groupoid is α-simply connected and principal. This result was proved in Mackenzie (1979) and although there it is restricted to the case of local morphisms over a fixed base, it is easy to see that the same proof applies in general. The proof given here, however, is taken directly from Almeida (1980, 4.1) and shows that the only use made of the hypothesis of principality for the domain groupoid, need be to deduce continuity of an algebraic morphism from its continuity in a neighbourhood of the base (see 1.21(ii)). It should be noted that Almeida's proof is for differentiable groupoids and that it includes a demonstration that an algebraic morphism of differentiable groupoids is smooth if it is smooth in a neighbourhood of the base; Almeida's result thus needs no hypothesis of principality or local triviality for the domain groupoid.

First recall the monodromy theorem of Chevalley (1946).

Theorem 6.13 Let P be a simply-connected and connected admissible space, and let U be an open connected neighbourhood of the diagonal Δ_P in $P \times P$. Suppose that for all $\xi \in P$ there is a (not necessarily topologized) non-empty set E_ξ and to each $(\eta, \xi) \in U$ a bijection $f_{\eta\xi}: E_\xi \to E_\eta$ such that $f_{\zeta\eta} \circ f_{\eta\xi} = f_{\zeta\xi}$ whenever all three pairs (ζ, η), (η, ξ), (ζ, ξ) are in U.

Then there exists a map $\psi: P \to \bigcup_\xi E_\xi$ such that $\psi(\xi) \in E_\xi$, $\forall \xi \in P$, and such that $\psi(\eta) = f_{\eta\xi}(\psi(\xi))$ whenever $(\eta, \xi) \in U$. Further, ψ is uniquely determined by these properties and its value at any chosen $\xi_0 \in P$. //

(This result may itself be expressed in terms of topological groupoids.)

Theorem 6.14 Let Ω be an α-connected and α-simply connected principal topological groupoid on B with Ω admissible, and let Ω' be an arbitrary topological groupoid on B'. Let $\phi: \Omega \rightsquigarrow \Omega'$, $\phi_0: B \to B'$ be a local morphism with α-connected domain $\mathcal{U}$.

Then there is a unique extension of ϕ to a global morphism $\psi: \Omega \to \Omega'$ of algebraic groupoids over ϕ_0, and it is continuous.

Proof: Let $D = \delta^{-1}(\mathcal{U}) \subseteq \Omega \underset{\alpha}{\times} \Omega$ and, for any $x \in B$, denote $D \cap (\Omega_x \times \Omega_x)$ by D_x. Clearly D_x is an open neighbourhood of the diagonal in $\Omega_x \times \Omega_x$. Now it is easy to prove that

$$D_x = \bigcup_{\xi\in\Omega_x} (R_\xi(\mathcal{U}_{\beta\xi}) \times \{\xi\})$$

and since each $R_\xi(\mathcal{U}_{\beta\xi}) \times \{\xi\}$ is connected and intersects the diagonal of $\Omega_x \times \Omega_x$, which also is connected, it follows that D_x is connected.

Write $x' = \phi_o(x)$ and for $\xi \in \Omega_x$, let E_ξ be the set $\Omega'^{\phi_o(\beta\xi)}_{x'}$. For $(\eta, \xi) \in D_x$ let $f_{\eta\xi}: E_\xi \to E_\eta$ be left-translation $L_{\phi(\eta\xi^{-1})}$; by 6.13 there is now a unique map $\psi_x: \Omega_x \to \Omega'_{x'}$, not necessarily continuous, such that $\beta' \circ \psi_x = \phi_o \circ \beta$, $\psi_x(\tilde{x}) = \widetilde{x'}$, and $\psi_x(\eta) = \phi(\eta\xi^{-1})\psi_x(\xi)$ whenever $(\eta, \xi) \in D_x$. Putting $\xi = \tilde{x}$ in this last equation shows that $\psi_x(\eta) = \phi(\eta)$ for $\eta \in \mathcal{U}_x$.

We now show that $\psi = \bigcup_x \psi_x$ is a morphism of algebraic groupoids. First take $\eta \in \mathcal{U}$ and any $\xi \in \Omega$ such that $\eta\xi$ is defined. Then $(\eta\xi, \xi) \in D_{\alpha\xi}$ and $\psi(\eta\xi) = \phi(\eta\xi\xi^{-1})\psi(\xi) = \psi(\eta)\psi(\xi)$. Next if $\eta = \eta_1\eta_2 \ldots \eta_n$ with all $\eta_i \in \mathcal{U}$, and ξ is arbitrary with $\eta\xi$ defined, then

$$\psi(\eta\xi) = \psi(\eta_1)\psi(\eta_2 \ldots \eta_n\xi) = \ldots = \psi(\eta_1) \ldots \psi(\eta_n)\psi(\xi).$$

Taking $\xi = \widetilde{\alpha\eta_1}$ in this gives $\psi(\eta) = \psi(\eta_1) \ldots \psi(\eta_n)$ and therefore $\psi(\eta\xi) = \psi(\eta)\psi(\xi)$.

Since Ω is α-connected, $\mathcal{U}$ generates Ω and this shows that ψ is an algebraic morphism. Since $\psi(\eta) = \phi(\eta)$ for $\eta \in \mathcal{U}$, ψ is continuous on a neighbourhood of the base of Ω and since Ω is principal, 1.21(ii) applies, and ψ is continuous.

The uniqueness of ψ follows from the fact that $\mathcal{U}$ generates Ω. //

Corollary 6.15 Let $\phi: \Omega \rightsquigarrow \Omega'$, $\phi_o: B \to B'$ be a local morphism of topological groupoids with Ω α-connected, locally trivial and locally simple. Then there is a unique continuous morphism $\tilde{\phi}: M\Omega \to \Omega'$ over ϕ_o such that $\tilde{\phi} = \phi \circ \psi$. //

As in the case of topological groups, it follows that two α-connected, locally trivial and locally simple topological groupoids are locally isomorphic iff their monodromy groupoids are isomorphic.

§7. Path connections in topological groupoids

Path connection is the abstraction of the differential-geometric concept of parallel translation. It differs from the concept of path lifting studied in the theory of fibrations (for example, Hu (1959, III§12), Spanier (1966, 2§7)), in that the differential geometric concept embodies a requirement of invariance under reparametrization, and most of the geometric interest follows from this condition.

This section contains some basic definitions and results which are needed in the account of C^∞-path connections in Lie groupoids given in III§7, but which are of a purely topological nature. In 7.3 we prove that a path connection induces a lifting of 1-parameter groups of homeomorphisms; this is a consequence of invariance under reparametrization and may be regarded as a continuity condition. We do not need to assume that the path connection is itself a continuous map. Continuity in this latter sense appears to be necessary only in order to deduce that the holonomy subgroupoid is locally trivial (7.7) but this is only a curiosity since in the case of C^∞-path connections, the local triviality of the holonomy subgroupoid is not necessarily with respect to the relative topology.

The main body of this section, from 7.4 on, treats the formal aspects of the concept of holonomy subgroupoid and its equivalence to the concept of holonomy bundle.

* * * * * * * *

Until 7.8 we consider a single topological groupoid Ω on a base B; we assume B to be a connected C^0-manifold and Ω to be admissible (as defined in 6.1) and to have projections $\alpha,\beta\colon \Omega \to B$ which are open. We use the notations $P^\alpha = P^\alpha(\Omega)$ and $P^\alpha_0 = P^\alpha_0(\Omega)$ of §6, although we do not assume that Ω is α-connected.

<u>Definition 7.1.</u> A <u>C^0-path connection</u> in Ω is a map $\Gamma\colon C(I,B) \to P^\alpha_0(\Omega)$, usually written $c \mapsto \bar{c}$, satisfying the following conditions:

(i) $\bar{c}(0) = \widetilde{c(0)}$ and $\beta\circ\bar{c} = c$;

(ii) If $\phi\colon [0,1] \to [a,b] \subseteq [0,1]$ is a homeomorphism then $\overline{c\circ\phi} = R_{\bar{c}(\phi(0))^{-1}} \circ (\bar{c}\circ\phi)$.

Γ is a <u>continuous C^0-path connection</u> if Γ is continuous with respect to the compact-open topologies on $C(I,B)$ and $P^\alpha_0 \subseteq C(I,\Omega)$. //

The definition of a path connection with reparametrization in a groupoid first appeared in Virsik (1971).

From (i) it follows that $\bar{c}$ is a path in $\Omega_{c(0)}$ with $\bar{c}(t) \in \Omega^{c(t)}_{c(0)}$ for all t, often called the Γ-<u>lift</u> of c.

(ii) is called the <u>reparametrization condition</u>. It is geometrically natural inasmuch as it guarantees that the images of paths in B can be meaningfully lifted to the images of α-paths in Ω. It also allows the definition of $\bar{c}$ to be extended to open paths of the form $c\colon (-\varepsilon,\varepsilon) \to B$, or $c\colon \mathbb{R} \to B$, and so to lift local 1-parameter groups of local transformations on B. This will be important in III§7. For simplicity of notation we work with the global case.

Lemma 7.2. Let Γ be a C°-path connection in Ω and let $c: \mathbf{R} \to B$ be continuous. Then there is a unique continuous $\bar{c}: \mathbf{R} \to \Omega$, with $\alpha(\bar{c}(t)) = c(0)$ for all t, $\beta \circ \bar{c} = c$, $\bar{c}(0) = \widetilde{c(0)}$ and such that for every closed interval $[a,b] \subseteq \mathbf{R}$, the restriction $\bar{c}|_{[a,b]}$ is $R_{\bar{c}(a)} \circ \Gamma(c|_{[a,b]})$, both restrictions being reparametrized by the same homeomorphism $[0,1] \to [a,b]$.

If Γ is a continuous C°-path connection, then the induced map $C(\mathbf{R},B) \to C(\mathbf{R},\Omega)$ is continuous with respect to the C° topologies.

Proof: $\bar{c}$ is most conveniently defined by lifting a suitably reparametrized $c|_{[n,n+1]}$ for each $n \in \mathbf{Z}$, and right-translating the results, so that the relevant endpoints match. The uniqueness result is then easy to see. The continuity of the associated map $C(\mathbf{R},B) \times \mathbf{R} \to \Omega$ is a local matter, and follows directly. //

We refer to $\bar{c}$, for $c: \mathbf{R} \to B$ or $c: (-\varepsilon,\varepsilon) \to B$, as the Γ-lift of c.

Proposition 7.3. Let Γ be a C°-path connection in Ω and let $\phi: \mathbf{R} \times B \to B$ be a global 1-parameter group of transformations of B. For each $x \in B$, let $\Gamma(\phi,x)$ denote the lift of $t \mapsto \phi_t(x)$ constructed in 7.2. Then $\bar{\phi}: \mathbf{R} \times \Omega \to \Omega$ defined by $\bar{\phi}_t(\xi) = \Gamma(\phi,\beta\xi)(t)\xi$ is a global 1-parameter group of transformations on Ω and $\beta \circ \bar{\phi}_t = \phi_t \circ \beta$ for all $t \in \mathbf{R}$.

Proof: Clearly $\bar{\phi}_t(\xi) \in \Omega_{\alpha\xi}^{\phi_t(\beta\xi)}$, and this establishes the last equation. Since $\phi_t: B \to B$ is continuous, and β is open, it follows that $\bar{\phi}_t$ is continuous; once we have established the group property, it will follow that $\bar{\phi}_t$ is a homeomorphism. Likewise, to prove that $\bar{\phi}: \mathbf{R} \times \Omega \to \Omega$ is continuous, observe that

$$\begin{array}{ccc} \mathbf{R} \times \Omega & \xrightarrow{\ \bar{\phi}\ } & \Omega \\ {\scriptstyle id \times \beta}\downarrow & & \downarrow{\scriptstyle \beta} \\ \mathbf{R} \times B & \xrightarrow{\ \phi\ } & B \end{array}$$

commutes.

It remains to prove the group property. Given $\xi \in \Omega$ and $s,t \in \mathbf{R}$, consider the curves $t \mapsto \bar{\phi}_{t+s}(\xi)$ and $t \mapsto \bar{\phi}_t(\bar{\phi}_s(\xi))$. Clearly both project under β to the curve $t \mapsto \phi_{t+s}(\xi) = \phi_t(\phi_s(\xi))$ in B and both have value $\bar{\phi}_s(\xi)$ at $t = 0$. It is easily checked that both are Γ-lifts (in the sense of 7.2) of the corresponding curve in B. So, by 7.2, $\bar{\phi}_{t+s}(\xi) = \bar{\phi}_t(\bar{\phi}_s(\xi))$. //

The reparametrization condition also guarantees that Γ preserves the algebraic operations on the sets of paths. The numbering of the following three results continues the numbering in 7.1.

Proposition 7.4. Let Γ be a C^o-path connection in Ω. Then

(iii) $\bar{\kappa}_x = \kappa_{\tilde{x}}$ (where κ_p, as in I 1.8, denotes the path constant at p).

(iv) $\overline{c^+} = R_{\bar{c}(1)^{-1}}(\bar{c})^+$ (where c^+ denotes the reversal, $c^+(t) = c(1 - t)$, of c).

(v) $\overline{c'c} = (R_{\bar{c}(1)} \circ \overline{c'})\bar{c}$ (where juxtaposition denotes the usual concatenation of paths).

Proof: These all follow from 7.1(ii). For (iii), take $t \in (0,1]$ and define $\rho_t: [0,1] \to [0,t]$ by $s \mapsto st$. Then $\bar{\kappa}_x(t) = (\bar{\kappa}_x \circ \rho_t)(1) =$
$(R_{\bar{\kappa}_x(0)} \circ (\overline{\kappa_x \circ \rho_t}))(1) = \overline{\kappa_x \circ \rho_t}(1)\bar{\kappa}_x(0)$ and since $\bar{\kappa}_x(0) = \tilde{x}$ and $\kappa_x \circ \rho_s = \kappa_x$, this is just $\bar{\kappa}_x(1)$. So $\overline{\kappa_x}(1) = \overline{\kappa_x}(t)$ for all $t > 0$, and since $\overline{\kappa_x}$ is continuous, it follows that $\bar{\kappa}_x(1) = \bar{\kappa}_x(0)$ also.

For (iv), use (ii) with $\phi(t) = 1 - t$. The proof of (v) is similar. //

Corollary 7.5. Let Γ be a C^o-path connection in Ω. For $c \in C(I,B)$ let $\hat{c}$ denote $\bar{c}(1)$. Then

(iii)' $\hat{\kappa}_x = \tilde{x}$

(iv)' $\widehat{c^+} = (\hat{c})^{-1}$

(v)' $\widehat{c'c} = \hat{c}'\hat{c}$. //

Definition 7.6. Let Γ be a C^o-path connection in Ω. Then

$$\Psi = \Psi(\Gamma) = \{\hat{c} \mid c \in C(I,B)\}$$

is the holonomy subgroupoid of Γ. The vertex group Ψ_x^x at $x \in B$ is the holonomy group of Γ at x. //

From 7.5 it is clear that Ψ is a wide subgroupoid of Ω; since B is path-connected, Ψ is transitive. The terminology will be justified in 7.14 below.

Proposition 7.7. Let Γ be a continuous C^o-path connection in Ω. Then Ψ is locally trivial, as a topological subgroupoid of Ω.

Proof: Let U be an open ball in B and let $\theta: U \to C(I,B)$ be a continuous map with $\theta(x)(0) = x_o$, fixed in U, and $\theta(x)(1) = x$ for all $x \in U$. Denote the composition

$$U \xrightarrow{\theta} C(I,B) \xrightarrow{\Gamma} P_o^\alpha \xrightarrow{ev} \Psi ,$$

where the last map is evaluation at 1, by σ; then $\sigma(x) \in \Psi_x^{x_o}$ for all $x \in U$. Therefore Ψ is weakly locally trivial and, since it is transitive, it follows from 2.4 that it is locally trivial. //

If Ψ is locally trivial, then Ω must be so; thus a topological groupoid which admits a continuous C^o-path connection must be locally trivial. The converse is also true: a locally trivial admissible topological groupoid on a connected C^o manifold admits a continuous C^o-path connection. This result can be proved by a modification of the usual local patching argument (Spanier (1966, 2.7.12)).

Proposition 7.8. Let Γ be a C^o-path connection in Ω. Then for each $c \in C(I,B)$, the Γ-lift $\bar{c}$ lies entirely in Ψ and, in particular, Ψ is α-connected.

Proof: For each $t \in (0,1]$ consider $c_t = c \circ \rho_t$ where ρ_t is $s \mapsto st$ as in the proof of 7.4(iii). Then $\hat{c_t} = \bar{c}(t)$ and this establishes $\bar{c}(t) \in \Psi$. //

7.8 is also, of course, true for any open path $c: (-\varepsilon,\varepsilon) \to B$ or $c: \mathbb{R} \to B$.

Example 7.9. The cartesian square groupoid $B \times B$ admits a single C^o-path connection, namely $\Gamma(c)(t) = (c(t),c(0))$.

Slightly less trivially, the fundamental groupoid $\pi(B)$ admits a single C^o-path connection, namely $\Gamma(c)(t) = \langle c\circ\rho_t\rangle$, where $\rho_t: [0,1] \to [0,t]$ is the reparametrization of 7.4 and 7.8. The easiest way to see the continuity of $\bar{c}$ is by regarding the topology on $\pi(B)$ as the quotient topology from $C(I,B)$.

Both connections are continuous and in both cases the holonomy subgroupoid is the whole groupoid. //

Definition 7.10. Let Ω and Ω' be admissible topological groupoids on B with open projections $\alpha,\beta: \Omega \to B$, $\alpha',\beta': \Omega' \to B$. Let $\phi: \Omega \to \Omega'$ be a morphism of topological groupoids over B, and let Γ be a C^0-path connection in Ω.

Then $\phi\circ\Gamma$ denotes the map $c \mapsto \phi\circ\Gamma(c)$, easily seen to be a C^0-path connection in Ω', and is called the produced (C^0-path) connection in Ω'. //

The terminology "produced" is an extension of the usage proposed after II 2.22.

Example 7.11. Let Ω be a principal, α-connected, admissible topological groupoid on B, and let Γ be a C^0-path connection in Ω. Then there is a unique C^0-path connection $\tilde{\Gamma}$ in the monodromy groupoid $M\Omega$ such that the covering projection $\psi: M\Omega \to \Omega$ maps $\tilde{\Gamma}$ to Γ.

Namely, given $c \in C(I,B)$, define $\bar{\bar{c}} = \tilde{\Gamma}(c)$ by $\bar{\bar{c}}(t) = \langle\overline{c\circ\rho_t}\rangle = \langle\overline{c\circ\rho_t}\rangle$ where, again, ρ_t is $[0,1] \to [0,t]$, $s \mapsto st$. Again, $\tilde{\Gamma}$ is continuous iff Γ is.

The uniqueness of $\tilde{\Gamma}$ follows from the uniqueness of lifts across the universal covering projections $M\Omega|_x \to \Omega_x$. //

Proposition 7.12. Let $\phi: \Omega \to \Omega'$ be a morphism of topological groupoids over B, with Ω and Ω' as in 7.10. Let Γ be a C^0-path connection in Ω and $\Gamma' = \phi\circ\Gamma$ the produced connection. Then $\phi(\Psi) = \Psi'$, where Ψ and Ψ' are the holonomy subgroupoids for Γ and Γ'.

Proof: Immediate. //

Remark: If Ω and Ω' satisfy the conditions of 6.15 and $\phi: \Omega \rightsquigarrow \Omega'$ is a local morphism of topological groupoids, then a C^0-path connection Γ in Ω induces a C^0-path connection $\tilde{\Gamma}$ in $M\Omega$ (by 7.11) and $\tilde{\Gamma}$ then induces a C^0-path connection $\Gamma' = \tilde{\phi}(\tilde{\Gamma})$ in Ω', by 6.15. In this case $\phi(\Psi) \subseteq \Psi'$, but, as 7.9 shows, equality need not hold. //

Example 7.13. Let (E,p,B) be a C^∞ vector bundle on a connected C^∞ manifold B, and let Γ be a C^0-path connection in $\Pi(E)$. Then, for $c \in C(I,B)$ and $t \in I$, $\bar{c}(t)$ is an isomorphism $E_{c(0)} \to E_{c(t)}$, generally known as parallel translation along c. //

Example 7.14. Let $P(B,G,\pi)$ be a principal bundle and let $\Omega = \frac{P \times P}{G}$ be the associated groupoid. Let Γ be a C^o-path connection in Ω. For any given $c \in C(I,B)$, $\bar{c}$ starts at the identity $\widetilde{c(0)}$ which can be written as $\langle u,u\rangle$ for any $u \in \pi^{-1}(c(0))$. Fix such a u. Since $\alpha\bar{c}(t) = c(0)$ for all t, each $\bar{c}(t)$ can be written as $\langle c^*(t),u\rangle$ with $c^*(t)$ uniquely determined by $\bar{c}(t)$ and u. From $\beta\bar{c}(t) = c(t)$, it follows that $\pi(c^*(t)) = c(t)$. Clearly $c^*(0) = u$. We call $c^* \in C(I,P)$ the Γ-lift of c starting at u, and will denote it by $\Gamma(c;u)$ when it is necessary to indicate the dependence of c^* on u. This Γ is a map $ev_o^*P \to C(I,P)$, where ev_o^*P is the pullback bundle of $P(B,G)$ over $ev_o : C(I,B) \to B$, $c \mapsto c(0)$.

If instead of $u \in \pi^{-1}(c(0))$ a second choice $u' \in \pi^{-1}(c(0))$ is made then $u' = ug$ for some $g \in G$ and from $\langle c^*(t),u\rangle = \langle c^*(t)g,ug\rangle$ it follows that $\Gamma(c;ug) = R_g \circ \Gamma(c;u)$.

Suppose that Γ is a continuous C^o-path connection in Ω, and choose a reference point $b \in B$ and $u_o \in \pi^{-1}(b)$, as in 1.19(i). Then $\langle c^*(t),u_o\rangle = \langle c^*(t),u\rangle\langle u,u_o\rangle = (R_{\langle u,u_o\rangle} \circ \bar{c})(t)$. It is easy to verify the continuity of

$$ev_o^*P \to C(I,\Omega_b), \quad (c,u) \mapsto R_{\langle u,u_o\rangle} \circ \bar{c}$$

and since $P \to \Omega_b$, $v \mapsto \langle v,u_o\rangle$ is a homeomorphism by 1.19(i) it follows that $\Gamma: ev_o^*P \to C(I,P)$ is continuous.

This proves one half of the following: There is a bijective correspondence between continuous maps $\Gamma: C(I,B) \to P_o^\alpha(\Omega)$ which satisfy (i) of 7.1 and continuous maps $\Gamma: ev_o^*P \to C(I,P)$ which satisfy $\pi \circ \Gamma(c;u) = c$, $\Gamma(c;u)(0) = u$ and $\Gamma(c;ug) = R_g \circ \Gamma(c;u)$. The other half may be proved similarly. The reparametrization condition (ii) of 7.1 becomes, in principal bundle terms,

$$\Gamma(c \circ \phi;\ \Gamma(c;u)(\phi(0))) = \Gamma(c;u) \circ \phi$$

where c and ϕ are as in 7.1(ii) and $\pi(u) = c(0)$. Compare Kobayashi and Nomizu (1963, II.3).

Returning to $P(B,G,\pi)$ and $\Omega = \frac{P \times P}{G}$, let Ψ be the holonomy groupoid of Γ. The holonomy group Ψ_x^x at $x \in B$ consists of he set of all $\hat{\ell}$ where ℓ is a loop in B at x. Choose $u \in \pi^{-1}(x)$ as reference point for all ℓ at x; then $\bar{\ell}(0) = \langle u,u\rangle$ and $\hat{\ell} = \bar{\ell}(1) = \langle \ell^*(1),u\rangle$, where $\ell^* = \Gamma(\ell;u)$. Since $\pi\ell^*(1) = \ell(1) = x$ there is a unique $g \in G$ such that $\ell^*(1) = ug$ and this g, which is the holonomy of ℓ with reference point $u \in \pi^{-1}(x)$ in the sense of Kobayashi and Nomizu (1963, II.4), corresponds to $\hat{\ell} \in \Psi_x^x$ under the isomorphism of 1.19(i), using u as reference point. Similarly it may be seen that Ψ_x corresponds, under the isomorphism of 1.19(i) with $u \in \pi^{-1}(x)$ as reference point, to the holonomy bundle of P(B,G) through u, in the sense of Kobayashi and Nomizu (1963, II.7). The variety of mutually conjugate forms of "the" holonomy group of P(B,G) and the variety of mutually isomorphic holonomy bundles, are now explained by 1.20. //

CHAPTER III LIE GROUPOIDS AND LIE ALGEBROIDS

The philosophy behind this chapter is that Lie groupoids and Lie algebroids are much like Lie groups and Lie algebras, even with respect to those phenomena - such as connection theory - which have no parallel in the case of Lie groups and Lie algebras.

We begin therefore with an introductory section, §1, which treats the differentiable versions of the theory of topological groupoids, as developed in Chapter II, §§1-6. Note that a Lie groupoid is a differentiable groupoid which is locally trivial. Most care has to be paid to the question of the submanifold structure on the transitivity components, and on subgroupoids.

§2 introduces Lie algebroids, as briefly as is possible preparatory to the construction in §3 of the Lie algebroid of a differentiable groupoid. The construction given in §3 is presented so as to emphasize that it is a natural generalization of the construction of the Lie algebra of a Lie group. One difference that might appear arbitrary is that we use right-invariant vector fields to define the Lie algebroid bracket, rather than the left-invariant fields which are standard in Lie group theory. This is for compatibility with principal bundle theory, where it is universal to take the group action to be a right action.

In §4 we construct the exponential map of a differentiable groupoid, and give the groupoid version of the standard formulas relating the adjoint maps and the exponential. The greater part of this section is concerned with the use of the exponential map to calculate the Lie algebroid of the frame groupoid $\Pi(E)$ of a vector bundle E, and of the reductions of $\Pi(E)$ defined by geometric structures on E. This calculation relies on the exponential map, in the same way as does the corresponding calculation of the Lie algebra of the general linear group.

A Lie groupoid, as well as being a generalization of a Lie group, is an alternative formulation of the concept of principal bundle, and there is therefore a version of connection theory applicable to Lie groupoids. Moreover a very great part of standard connection theory - those parts which do not refer to path-lifting or holonomy - can be presented entirely within the setting of abstract Lie algebroid theory and without reference to any groupoid. We refer to this as infinitesimal connection theory, since the Lie algebroid of a Lie groupoid is a first-order invariant. In the first part of §5 we present this infinitesimal connection theory in the setting of abstract Lie algebroids. In the second part we introduce the concept of transition form for a transitive Lie algebroid arising from a Lie groupoid. The transition forms are the right derivatives of the transition functions of the groupoid and are, in the same way, a complete invariant of the Lie algebroid. Transition forms may also be regarded as the overlap isomorphisms for a system of local isomorphisms from the given Lie algebroid to a trivial one; they thus allow problems for transitive Lie algebroids to be broken down into a local problem and a globalization problem. We will prove in IV§4 that every abstract Lie algebroid admits a classification by transition forms, and these results will then be central to the study, in Chapter V, of the integrability of Lie algebroids.

In §6 we return to the generalization of the elementary theory of Lie groups and Lie algebras and prove, firstly, that there is a bijective correspondence between α-connected reductions of a Lie groupoid and transitive Lie subalgebroids of its Lie algebroid and, secondly, that there is a bijective correspondence between germs of local morphisms of Lie groupoids over a fixed base, and morphisms of their Lie algebroids. Both results are in fact related to connection theory.

§7 treats the theory of path connections in Lie groupoids, that is, those parts of connection theory which do use the concept of path-lifting and holonomy. Here the point of view is that a path connection in a Lie groupoid is the integrated version of the corresponding infinitesimal connection in its Lie algebroid; we thus subsume connection theory under the generalization of the elementary theory of Lie groups and Lie algebras. In particular, we see that the Ambrose-Singer theorem for Lie groupoids and Lie algebroids is an immediate consequence of the correspondence between Lie subalgebroids and Lie subgroupoids (together with the correspondence between path connections and infinitesimal connections and the fact that the holonomy groupoid is Lie). In the second part of the section we give a detailed analysis of connections in vector bundles on which a Lie groupoid acts; these results are central to Chapter IV.

§1. Differentiable groupoids and Lie groupoids.

A Lie groupoid is a differentiable groupoid which is locally trivial. This usage differs from that of some authors, but has the advantage that the briefest expression is used for the most frequently occurring case.

This section treats those parts of the theory of differentiable and Lie groupoids which are refinements of the theory of topological groupoids as treated in Chapter II,§§1-6.

The concept of differentiable groupoid is due to Ehresmann; the definition used here is from Pradines (1966). In 1.4 we prove a crucial result due to Pradines, communicated to the author in 1979, that in a differentiable groupoid Ω the maps $\beta_x : \Omega_x \to B$ are subimmersions. This result sets the theory of differentiable groupoids apart from the theory of general topological groupoids. In 1.8 we use 1.4 to prove that a differentiable groupoid is locally trivial over each of its transitivity components; in particular, a transitive differentiable groupoid is a Lie groupoid. The proof depends strongly on the fact that we assume manifolds to be paracompact, Hausdorff, and of constant, and finite, dimension.

In 1.19 we apply 1.4 and 1.8 to actions of differentiable groupoids and deduce, in particular, that each orbit is a submanifold and each evaluation map is of constant rank. In 1.20 we prove that isotropy sub-groupoids for actions of Lie groupoids are closed embedded reductions. This result was given by Ngô (1967), but the proof given here appears to be the first to address the global problem. In 1.26 to 1.30 we apply 1.20 to various standard actions of frame groupoids $\Pi(E)$, where E is a vector bundle, to deduce that the frame groupoids for geometric structures defined by tensor fields are themselves Lie groupoids. This result, in terms of principal bundles, is due essentially to Greub et al (1973); the proof given here is new, and slightly more general.

Definition 1.1. A differentiable groupoid is a groupoid Ω on base B together with differentiable structures on Ω and B such that the projections $\alpha, \beta: \Omega \to B$ are surjective submersions, the object inclusion map $\varepsilon: x \mapsto \tilde{x}$, $B \to \Omega$ is smooth, and the partial multiplication $\Omega * \Omega \to \Omega$ is smooth.

A morphism of differentiable groupoids, or a smooth morphism, is a morphism of groupoids $\phi: \Omega \to \Omega'$, $\phi_o: B \to B'$ such that ϕ and ϕ_o are smooth. //

Here $\Omega * \Omega = (\alpha \times \beta)^{-1}(\Delta_B)$ is an embedded submanifold of $\Omega \times \Omega$, for since α and β are submersions, $\alpha \times \beta$ is transversal to Δ_B. The tangent bundle to $\Omega * \Omega$ is $T\Omega * T\Omega = \{Y \oplus X \in T(\Omega \times \Omega) \mid T(\alpha)(Y) = T(\beta)(X)\}$; the only formula for the tangent to the multiplication which we need is the following special case.

Lemma 1.2. Let Ω be a differentiable groupoid on B, and let $\kappa: \Omega * \Omega \to \Omega$ denote the multiplication. Then for $Y \in T(\Omega_{\alpha\eta})_\eta$, $X \in T(\Omega^{\beta\xi})_\xi$,

$$T(\kappa)(Y \oplus X) = T(R_\xi)_\eta(Y) + T(L_\eta)_\xi(X). \qquad //$$

Using this, we prove that the inversion in a differentiable groupoid Ω is a smooth map. Define $\theta: \Omega * \Omega \to \Omega \underset{\beta}{\times} \Omega$ by $(\eta,\xi) \mapsto (\eta,\eta\xi)$. Then θ is a bijection, by the algebraic properties of Ω. To see that θ is an immersion, take $Y \oplus X \in T(\Omega * \Omega)_{(\eta,\xi)}$ and suppose $T(\theta)(Y \oplus X) = 0 \oplus 0$. Since $\pi_2 \circ \theta = \pi_2$, it follows that $Y = 0$. Hence $X \in T(\Omega^{\beta\xi})_{\xi}$ and 1.2 can be applied to show that $X = 0$. Since α and β are both submersions, $\Omega * \Omega$ and $\Omega \underset{\beta}{\times} \Omega$ have the same dimension and so θ is étale everywhere on $\Omega * \Omega$. Hence it is a diffeomorphism and the composite of $\Omega \to \Omega \underset{\beta}{\times} \Omega$, $\eta \mapsto (\eta,\widetilde{\beta\eta})$, followed by θ^{-1}, followed by $\pi_2: \Omega * \Omega \to \Omega$, is smooth; this is the inversion map. Inversion is its own inverse and is therefore a diffeomorphism.

Note further that the object inclusion map ε in a differentiable groupoid is an immersion, since either projection is left-inverse to it, and is a homeomorphism onto $\tilde{B}$ by II§1. $\tilde{B}$ is therefore a closed embedded submanifold of Ω. The α-fibres and β-fibres are also (pure) closed embedded submanifolds of Ω. Lastly, in the definition of a smooth morphism, the condition that ϕ_o be smooth is superfluous, as in the topological case.

Definition 1.1 is taken from Pradines (1966). Ehresmann (1959) requires only a differentiable structure on Ω for which $\xi \mapsto \widetilde{\alpha(\xi)}$ and $\xi \mapsto \widetilde{\beta(\xi)}$ are subimmersions and multiplication $\Omega * \Omega \to \Omega$ is smooth; Kumpera and Spencer (1972) and ver Eecke (1981) require differentiable structures on Ω and B such that the projections and object inclusion map are smooth, $\Omega * \Omega$ is an embedded submanifold of $\Omega \times \Omega$ and multiplication is smooth. Ver Eecke (1981) gives a proof that even in this more general case the smoothness of the inversion map follows from the other conditions, though not so easily.

It must be admitted that the condition that α and β be submersions is a strong one; 1.9 shows that a transitive differentiable groupoid is actually locally trivial. Nonetheless, this author does not know of <u>any</u> example, much less a substantial class of interesting examples, of an α-connected, transitive groupoid which is differentiable in one of these more general senses, but for which the projections are not submersions. None of the three works cited above develops in any substantial way the theory of groupoids which are differentiable in a more general sense; nor do they give examples of the more general concept. According to Kumpera and Spencer (1972, p. 258) a simple argument shows that an α-connected groupoid which is differentiable in their sense, actually satisfies 1.1. In the absence of definite motivation for the more general concept, we accept the substantial conveniences offered by 1.1.

It may be asked why we do not carry this attitude further and restrict ourselves from the outset to differentiable groupoids which are locally trivial. Firstly, the elementary parts of the theory are not simplified by this assumption, and it is valuable to see at what points local triviality is actually necessary. Secondly, there is a substantial and important theory of microdifferentiable groupoids (see Pradines (1966) and Almeida (1980); the definition is given in 6.3) the value and interest of which lies almost entirely in the non locally trivial case. For example any foliation defines a microdifferentiable groupoid which is non locally trivial (providing the dimension and codimension are positive) - see the discussion following 6.3. Though this book does not cover that theory, it will serve as a better introduction to it if local triviality is only imposed when it is actually needed.

Consider now a differentiable groupoid Ω on B. The vertical subbundle of $T\Omega$ for $\alpha: \Omega \to B$ is denoted by $T^{\alpha}\Omega$ and called merely the vertical bundle for Ω. It is an involutive distribution on Ω whose leaves are the components of the α-fibres of Ω.

Proposition 1.3. Let Ω be a differentiable groupoid on B. Then the α-identity-component subgroupoid Ψ is open in Ω.

Proof. Let $\phi: \mathbf{R}^p \times \mathbf{R}^q \to \mathcal{U} \subseteq \Omega$ be a distinguished chart for the foliation induced by $T^{\alpha}\Omega$, where $\mathcal{U} \cap \tilde{B} \neq \emptyset$ and $\phi(\{0\} \times \mathbf{R}^q) = \mathcal{U} \cap \tilde{B}$. Then clearly $\mathcal{U} \subseteq \Psi$. Taking the union of such $\mathcal{U}$ we obtain an open neighbourhood of $\tilde{B}$ in Ω which is contained in Ψ. Now Ψ is the union of those leaves of the foliation which intersect the open neighbourhood and so is itself open. //

Theorem 1.4. Let Ω be a differentiable groupoid on B. For any $x \in B$, $\beta_x: \Omega_x \to B$ is a subimmersion.

Proof. Let Φ denote the set of values taken by the local admissible sections of Ω; clearly Φ is a wide subgroupoid of Ω. We prove that each Φ_x is open in Ω_x.

Let $\phi: \mathbf{R}^p \times \mathbf{R}^q \to \mathcal{U} \subseteq \Omega$ be a distinguished chart for the foliation induced by $T^{\alpha}\Omega$, with $\phi(0,0)$ a point of Φ. Choose ϕ so that $U = \alpha(\mathcal{U})$ is the range of a chart $\psi: \mathbf{R}^q \to U$; we identify α with the projection $\mathbf{R}^p \times \mathbf{R}^q \to \mathbf{R}^q$ and β with a submersion $\mathbf{R}^p \times \mathbf{R}^q \to \mathbf{R}^q$ ($\cong \beta(\mathcal{U})$), also denoted β. A local admissible section σ can now be identified with a map $s: \mathbf{R}^q \rightsquigarrow \mathbf{R}^p$ such that $\beta \circ (s \times \mathrm{id}): \mathbf{R}^q \rightsquigarrow \mathbf{R}^q$ is a diffeomorphism.

By assumption there exists an s with $s(0) = 0$ and $\beta\circ(s \times id)$ étale at 0 or, equivalently, such that the graph of s in transverse (in the strong sense of a direct sum) to the β-foliation of $\mathbb{R}^p \times \mathbb{R}^q$ at (0,0). Now one can see that there is a neighbourhood N of 0 in $\mathbb{R}^p$ such that $\forall\, t_0 \in N$ there is a local diffeomorphism $\chi\colon \mathbb{R}^{p+q} \rightsquigarrow \mathbb{R}^{p+q}$ near (0,0) which takes (0,0) to $(t_0,0)$, maps each α-fibre $\mathbb{R}^p \times \{u\}$ into itself, and maps the graph of s to the graph of a new map s' which is still (strongly) transverse to the β-foliation at $(t_0,0)$. This proves that Φ_x (for $x = \psi(0)$) is open in Ω_x.

By II 3.11 it follows that Φ contains Ψ, the α-identity-component subgroupoid of Ω. Fix $x \in B$ and take ξ, η in a common component of Ω_x. Then $\zeta = \xi\eta^{-1} \in \Psi$ and so there is a local admissible section $\sigma \in \Gamma_U\Omega$ with $\beta\eta \in U$ and $\sigma(\beta\eta) = \zeta$. Now $L_\sigma\colon \Omega_x^U \to \Omega_x^V$ (where $V = (\beta\circ\sigma)(U)$) maps η to ξ and $\beta_x \circ L_\sigma = (\beta\circ\sigma)\circ\beta_x$. Hence the ranks of β_x at ξ and η are equal. //

1.4 and its proof were communicated to the author by Pradines in 1979.

<u>Corollary 1.5.</u> Let Ω be a differentiable groupoid on B. Then for all $x,y \in B$, Ω_x^y is a (pure) closed embedded submanifold of Ω_x, Ω^y and Ω. In particular, each vertex group Ω_x^x is a Lie group.

<u>Proof.</u> Only the purity needs to be established and that follows from the fact that for $\xi,\eta \in \Omega_x^y$, there is a diffeomorphism $\Omega_x \to \Omega_x$ carrying ξ to η, namely R_λ, where $\lambda = \xi^{-1}\eta$. //

<u>Theorem 1.6.</u> Let Ω be a differentiable groupoid on B. Then $\forall\, x \in B$, $M_x = \beta_x(\Omega_x)$ is a submanifold of B.

<u>Proof.</u> Denote Ω_x by P and Ω_x^x by G. Then, by 1.5, the restriction of the groupoid multiplication to $P \times G \to P$ is a smooth action of a Lie group on a manifold. It is easily seen to be proper: if $K,L \subseteq P$ are compact then $\{g \in G \mid Kg \cap L \neq \emptyset\}$ is the image under $P \underset{\beta_x}{\times} P \to G$, $(\eta,\xi) \mapsto \eta^{-1}\xi$, of the closed subset $K \underset{\beta_x}{\times} L = (K \times L) \cap (P \underset{\beta_x}{\times} P)$ of the compact set $K \times L$ and is therefore compact. Since the action is also free, it follows that $\{(\xi g,\xi) \mid \xi \in P,\ g \in G\}$ is a closed embedded submanifold of $P \times P$ and so there is a quotient manifold structure on P/G (see, for example, Dieudonné (1972, 16.10.3)).

Define $j\colon P/G \to B$ by $j(\xi G) = \beta_x(\xi)$. Then j is smooth and injective. Since $P \to P/G$ is a submersion, $rk_{\xi G}(j) = rk_\xi(\beta_x)$, $\forall\, \xi \in P$, and so j is a subimmersion.

Now an injective subimmersion is an immersion. //

Corollary 1.7. $\beta_x : \Omega_x \to B$ is of constant rank.

Proof. For $\mathrm{rk}_{\xi}(\beta_x) = \dim_{\xi G}(P/G) = \dim P - \dim G$. //

Since α is a submersion, it follows that $\Omega_{M_x}^{M_x} = \Omega_{M_x} = \alpha^{-1}(M_x)$ is a submanifold of Ω. However it is not clear that the groupoid operations in $\Omega_{M_x}^{M_x}$ are smooth, since it is not clear that $\Omega_{M_x}^{M_x}$ is a quasiregular submanifold of Ω. One may define a distribution-with-singularities I on B by $I_x = \operatorname{im} T(\beta_{\alpha\xi})_{\xi}$, for any $\xi \in \Omega^x$, and the components of the M_x are clearly the leaves of this distribution, but it is not clear that the leaves of a distribution-with-singularities are necessarily quasiregular. However this awkwardness can be circumvented:

Theorem 1.8. Let Ω be a differentiable groupoid on B and let M be a transitivity component of B. Then there is a manifold structure on Ω_M^M with respect to which it is a submanifold of Ω and a differentiable groupoid on M. Further, Ω_M^M is locally trivial.

Proof. Choose $x \in M$. From 1.6, $\beta_x : \Omega_x \to M$ is a smooth surjective map. By Sard's theorem, it must be a submersion somewhere, and therefore, by 1.7, it is a submersion everywhere.

Now consider the division map $\delta_x : \Omega_x \times \Omega_x \to \Omega$. We claim it also is a subimmersion. Take η and η' in a common component of Ω_x and ξ and ξ' likewise. From the proof of 1.4 we know that there exist $\tau \in \Gamma_V\Omega$, $\sigma \in \Gamma_U\Omega$ such that $L_\tau(\eta) = \eta'$ and $L_\sigma(\xi) = \xi'$. Now the following diagram commutes where $V' = (\beta\circ\tau)(V)$, $U' = (\beta\circ\sigma)(U)$

$$\begin{array}{ccc}
\Omega_x^V \times \Omega_x^U & \xrightarrow{\ \delta_x\ } & \Omega_U^V \\
{\scriptstyle L_\tau \times L_\sigma}\downarrow & & \downarrow \\
\Omega_x^{V'} \times \Omega_x^{U'} & \xrightarrow{\ \delta_x\ } & \Omega_{U'}^{V'}
\end{array}$$

and the right-hand map is $\zeta \mapsto \tau(\beta\zeta)\zeta\sigma(\alpha\zeta)^{-1}$. Since both vertical maps are diffeomorphisms, it follows that δ_x has the same rank at (η,ξ) and (η',ξ').

Now we proceed as in 1.6, using the diagram

$$\begin{array}{ccc} \Omega_x \times \Omega_x & \xrightarrow{\ \delta_x\ } & \Omega \, . \\ \downarrow & \nearrow_{j'} & \\ \dfrac{\Omega_x \times \Omega_x}{\Omega_x^x} & & \end{array}$$

(The existence of the quotient manifold follows from the fact that $\beta_x : \Omega_x \to M$ is a submersion.) The image of δ_x is Ω_M^M and j' is, by the same argument as in 1.6, an injective immersion.

Since $\beta_x : \Omega_x \to M$ is a submersion, $\Omega_x(M, \Omega_x^x, \beta_x)$ is a principal bundle. The proof that Ω_M^M is a (locally trivial) differentiable groupoid on M is now reduced to showing that, for any principal bundle $P(M,G,\pi)$, the associated groupoid $\frac{P \times P}{G}$ is a locally trivial differentiable groupoid. This follows the same formal outline as in the topological case (II 1.12 and II 2.7). //

<u>Corollary 1.9</u>. A transitive differentiable groupoid is locally trivial. //

<u>Corollary 1.10</u>. Let Ω be a differentiable groupoid on B, and let $x \in B$. Then $\delta_x : \Omega_x \times \Omega_x \to \Omega$ is of constant rank and Ω is locally trivial iff δ_x is a submersion.

<u>Proof</u>. From 1.8, δ_x is known to be a subimmersion; the same argument as in 1.7 shows that it is actually of constant rank. The second statement follows as in II 2.6 (and does not depend on 1.8). //

1.4 and 1.9 constitute a version of Théorème 4 of Pradines (1966) appropriate to the category of pure, paracompact, Hausdorff manifolds. See also Pradines (1986).

<u>Definition 1.11</u>. A <u>Lie groupoid</u> is a locally trivial differentiable groupoid. //

It is clear from what has been said in the course of the preceding results, that the equivalence II 1.19 and II 2.7 between locally trivial topological groupoids and principal bundles remains valid for the case of Lie groupoids and (C^∞) principal bundles. Since an open subimmersion is a submersion, there is no place for a concept of "principal differentiable groupoid".

Clearly the examples of trivial groupoids (II 1.9), frame groupoids (II

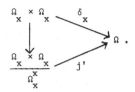

1.13), and group bundles (II 1.17) remain valid in the C^∞ case.

Example 1.12. Let B be a manifold and let $X \subseteq B \times B$ be an equivalence relation on B which is a closed embedded submanifold of $B \times B$. Then X, with its submanifold structure, is a differentiable groupoid on B iff the projection $\pi_2 : X \to B$, $(y,x) \mapsto x$, is a submersion. By Godement's criterion (Serre (1965, LG 3.27)) this is the case iff there is a quotient manifold structure on B/X. In this case X is the foliation defined by the submersion $B \to B/X$, at least if it is α-connected. //

Example 1.13. If $G \times B \to B$ is a smooth action of a Lie group on a manifold then the action groupoid $G \times B$ is a differentiable groupoid on B. 1.7 shows that each evaluation map is a subimmersion of constant rank, 1.6 that the orbits are submanifolds of B, and 1.9 that if G acts transitively, then B is equivariantly diffeomorphic to a homogeneous space; 1.9 also includes the existence of local cross-sections for closed subgroups of Lie groups.

In other words, these results are consequences of Pradines' result 1.4 and Godement's criterion. The standard proof of these results is essentially the same (see Dieudonné (1972, XVI.10)). //

Example 1.14. Let B be a connected manifold, and consider the fundamental groupoid $\pi(B)$ (see II 1.14). Since the anchor $[\beta,\alpha]: \pi(B) \to B \times B$ is a covering, there is a unique manifold structure on $\pi(B)$ for which $[\beta,\alpha]$ is smooth and it is then étale (see, for example, Dieudonné (1972, 16.8.2)). Because this manifold structure is defined locally, in terms of open sets $\pi(B)_{U_i}^{U_j}$, it is easy to see that $\pi(B)$ is a differentiable groupoid on B with respect to this structure and is locally trivial.

Similarly, the monodromy groupoid $M\Omega$ of an α-connected Lie groupoid Ω has a natural smooth structure with respect to which $M\Omega$ is a Lie groupoid and the projection $\psi: M\Omega \to \Omega$ is smooth and étale. See II 6.4. //

Example 1.15. Let Ω be a differentiable groupoid on B and denote the multiplication by $\kappa: \Omega * \Omega \to \Omega$. Then $T\Omega$ is a differentiable groupoid on TB with projections $T(\alpha)$, $T(\beta)$ and multiplication $T(\kappa): T(\Omega) * T(\Omega) \to T(\Omega)$. //

The sheaf topology on a germ groupoid $J^\lambda(\Omega)$ (see II 1.15 and II§5) is non-Hausdorff and therefore will not admit a differentiable structure in our sense. For the differentiable version of $J^\lambda_{co}(\Gamma)$, see, for example, Kumpera (1975) or ver Eecke (1981).

Much of Chapter II remains valid for differentiable and Lie groupoids and we have already used concepts such as local triviality and admissible section without comment, taking as obvious the fact that these refer to C^∞ analogues of the cncepts defined in Chapter II. Those parts of Chapter II which refer to subgroupoids and quotient groupoids cannot be assumed to remain valid; these questions are briefly treated below, as is also the theory of smooth actions and representations of Lie groupoids. With these exceptions, and excluding also the examples with sheaf topologies and example II 1.16, the material of §§1 - 6 of Chapter II is taken over without comment. The C^∞ analogue of II§7 is treated in III§7.

It is clear from the example of the groupoids $B \times B$ that a differentiable groupoid need not have a unique underlying analytic structure, and that a given topological groupoid may have several non-diffeomorphic underlying structures of differentiable groupoid. A continuous morphism of differentiable groupoids need not be smooth, even if it is a base-preserving morphism of trivial differentiable groupoids - consider $B \times B \to B \times G \times B$, $(y,x) \mapsto (y,\theta(y)\theta(x)^{-1},x)$ for suitable $\theta\colon B \to G$.

Definition 1.16. Let Ω be a differentiable groupoid on B. A differentiable subgroupoid of Ω is a differentiable groupoid Ω' on B' together with a morphism of differentiable groupoids $\phi\colon \Omega' \to \Omega$ which is an injective immersion. A differentiable subgroupoid (Ω',ϕ) of Ω is an embedded differentiable subgroupoid if ϕ is an embedding. A differentiable subgroupoid (Ω',ϕ) is wide if $B' = B$ and $\phi_o = id_B$.

Let Ω be a Lie groupoid on B. A reduction of Ω, or a Lie subgroupoid of Ω, is a wide differentiable subgroupoid (Ω',ϕ) such that Ω' is locally trivial. //

The obvious concept of equivalence for differentiable subgroupoids will be used without comment.

To generalize to differentiable or Lie groupoids the well-known results for the existence and uniqueness of Lie subgroup structures on subgroups of Lie groups seems to be difficult, and we will not treat these questions. In view of the examples $B' \times B' \subseteq B \times B$, an immediate restriction must be that the subgroupoids are wide. Conjectures can be manufactured almost indefinitely, as the choice of hypotheses varies; if one is only interested in reductions of Lie groupoids, the nature of the problems is of course much changed. Some small results will appear in the course of subsequent sections; for example, 1.21 shows that a closed reduction of a Lie groupoid is an embedded submanifold. Together with 1.9, this implies that if a closed, wide, transitive subgroupoid of a Lie groupoid has a structure of differentiable subgroupoid, then it is an embedded and locally trivial differentiable subgroupoid. This observation may help attack the existence problem, perhaps by using a closed graph theorem and approximating a continuous classifying map by a smooth one.

Proposition 1.17. Let Ω be a differentiable groupoid on B. Then $G\Omega$ is a closed embedded submanifold of Ω and is a differentiable subgroupoid of Ω.

Proof. As in 1.4, represent $\alpha: \mathcal{U} \to U = \alpha(\mathcal{U})$ locally by $\pi: \mathbb{R}^p \times \mathbb{R}^q \to \mathbb{R}^q$, for some open $\mathcal{U} \subseteq \Omega$ which intersects $G\Omega$, and represent β by a submersion $\beta: \mathbb{R}^p \times \mathbb{R}^q \to \mathbb{R}^q$. Then $G\Omega \cap \mathcal{U}$ is $\{(t,u) \in \mathbb{R}^p \times \mathbb{R}^q \mid \beta(t,u) = u\}$, that is, it is the pullback

$$\begin{array}{ccc} G\Omega \cap \mathcal{U} & \longrightarrow & \mathbb{R}^p \times \mathbb{R}^q \\ \big\downarrow & & \big\downarrow \pi \\ \mathbb{R}^p \times \mathbb{R}^q & \xrightarrow[\beta]{} & \mathbb{R}^q \end{array} .$$

Hence $G\Omega \cap \mathcal{U}$ is an embedded submanifold of $\mathcal{U}$ and since the $\mathcal{U}$ are open in Ω, $G\Omega$ is an embedded submanifold of Ω, closed since it is $[\beta,\alpha]^{-1}(\Delta_B)$.

Since $\beta: \mathbb{R}^p \times \mathbb{R}^q \to \mathbb{R}^q$ is a submersion, $G\Omega \cap \mathcal{U} \to \mathbb{R}^p \times \mathbb{R}^q$ is a submersion, and so the composite map $G\Omega \cap \mathcal{U} \to \mathbb{R}^q$ is a submersion. This is the restriction of $\alpha: G\Omega \to B$ to $G\Omega \cap \mathcal{U}$.

That $G\Omega$ satisfies the other conditions for a differentiable subgroupoid is obvious. //

For our later purposes, the most important examples of differentiable subgroupoids are closed, embedded reductions and arise as the isotropy groupoids of smooth actions:

Definition 1.18. Let Ω be a differentiable groupoid on B and let $p: M \to B$ be a smooth map. Denote the pullback of p over the submersion α by $\Omega * M$. A smooth action of Ω on (M,p,B) is a smooth map $\Omega * M \to M$ which satisfies the algebraic conditions of II 4.1. //

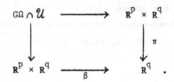

It is easy to see that the action groupoid $\Omega * M$ (see II 4.20) is a differentiable groupoid on B. From this observation comes the following result:

<u>Proposition 1.19.</u> Let $\Omega * M \to M$ be a smooth action of a differentiable groupoid Ω on a smooth map $p\colon M \to B$. Then

(i) each orbit $\Omega[u]$, $u \in M$, of Ω is a submanifold of M;

(ii) each evaluation map $\Omega_{p(u)} \to M$, $\xi \mapsto \xi u$ is a subimmersion of constant rank;

(iii) if the action is transitive, then each evaluation map is a surjective submersion;

(iv) if Ω is locally trivial, then (M,p,B) is a fibre bundle and so too is each orbit $(\Omega[u],p,B)$;

(v) the set M/Ω of orbits has the structure of a quotient manifold from M iff the graph $\Gamma = \{(u,\xi u) \in M \times M \mid u \in M,\ \xi \in \Omega_{p(u)}\}$ of the action is a closed embedded submanifold of $M \times M$.

<u>Proof.</u> (i) - (iii) follow from 1.4, 1.6, 1.7, 1.8 and 1.9.

(iv) follows as in II 4.9.

(v) follows from Godement's criterion and the observation that the following diagram commutes

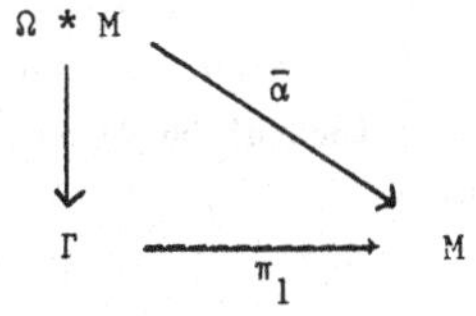

where the vertical map is $(\xi,u) \mapsto (u,\xi u)$. //

<u>Theorem 1.20.</u> Let Ω be a Lie groupoid on B and let $\Omega * M \to M$ be a smooth action of Ω on the fibre bundle (M,p,B). Let $\mu \in \Gamma M$ be an Ω-deformable section. Then the isotropy groupoid $\Phi(\mu)$ of Ω at μ is a closed embedded reduction of Ω.

<u>Proof.</u> μ takes values in a single orbit of Ω so, by 1.19, we can assume that Ω acts transitively on M.

Define $f\colon \Omega \to M \underset{p}{\times} M$ by $\xi \mapsto (\mu(\beta\xi), \xi\mu(\alpha\xi))$. Then $\Phi(\mu) = f^{-1}(\Delta_M)$, which shows that $\Phi(\mu)$ is closed in Ω. We prove that $f \pitchfork \Delta_M$ in $M \underset{p}{\times} M$. For $\xi_o \in \Phi$, choose decomposing sections $U \to \Omega_b$ and $V \to \Omega_b$ in neighbourhoods U, V of $x_o = \alpha\xi_o$, $y_o = \beta\xi_o$; then f has the local expression

$$V \times G \times U \to (V \times F) * (V \times F)$$

$$(y,g,x) \mapsto ((y,a'(y)),\ (y,ga(x)))$$

where $F = M_b$, $G = \Omega_b^b$ and $a: U \to F$, $a': V \to F$ are the local expressions of μ. Let ξ_o correspond to (y_o, g_o, x_o), and note that $a'(y_o) = g_o a(x_o)$ because $\xi_o \in \Phi(\mu)$.

Given $X,Y \in T(F)_{a'(y_o)}$, there is a $W \in T(G)_{g_o}$ such that evaluation $G \to F$, $g \mapsto ga(x_o)$ maps W to $X - Y$. This is because the action $G \times F \to F$ is smooth and transitive and therefore each evaluation is a submersion. Hence, given also $Z \in T(V)_{y_o}$, we have $f_*(0 \oplus W \oplus 0) + (Z \oplus Y) \oplus (Z \oplus Y) = (Z \oplus Y) \oplus (Z \oplus X)$ and this proves that $f \pitchfork \Delta_M$. Hence Φ is an embedded submanifold of Ω.

To show that Φ is a differentiable subgroupoid of Ω it is only necessary to show that the projections $\Phi \to B$ are submersions. In fact $\beta: \Phi \to B$ is the composite

$$\Phi \xrightarrow{f} \Delta_M \xrightarrow{\pi_1} M \xrightarrow{p} B$$

in which each map is a submersion.

That Φ is locally trivial follows from 1.9. //

This theorem is taken from Ngô (1967, I.3.a); the proof given there, however, seems to address only the local problem. There is a converse, which goes back to Ehresmann (1959) and whose principal bundle formulation is well-known from Kobayashi and Nomizu (1963, I 5.6):

<u>Proposition 1.21</u>. Let Ω be a Lie groupoid on B and Φ a reduction of Ω whose vertex groups are closed (and hence embedded) subgroups of the vertex groups of Ω. Choose $b \in B$, write $G = \Omega_b^b$, $H = \Phi_b^b$, $P = \Omega_b$ and let M be the fibre bundle $\dfrac{P \times (G/H)}{G}$ corresponding to the standard action of G on G/H, and $\Omega * M \to M$ the associated action $\xi\langle\eta,gH\rangle = \langle\xi\eta,gH\rangle$ (see II 4.8, II 4.9).

For $x \in B$ choose $\zeta \in \Phi_b^x$ and define $\mu(x) = \langle\zeta,H\rangle$. Then μ is a well-defined, smooth, Ω-deformable section of M, and $\Phi(\mu) = \Phi$. In particular, Φ is a closed, embedded reduction of Ω.

Proof. That μ is smooth follows from the facts that $\beta_b: \Phi_b \to B$ is a surjective submersion, and Φ_b is a submanifold of Ω_b. That $\Phi(\mu) = \Phi$ is a trivial algebraic manipulation. //

1.21 shows that a closed reduction of a Lie groupoid is an embedded submanifold. It would be very interesting to know if every closed transitive wide subgroupoid of a Lie groupoid is an embedded submanifold, and a differentiable subgroupoid.

1.20 and 1.21 give a classification of those closed embedded reductions of a Lie groupoid which have a specified vertex group at a chosen point of the base. A closed embedded reduction may of course fail to be trivializable over open subsets of the base over which the larger groupoid is trivializable, as the following example shows.

Example 1.22. Let G be a Lie group and H a closed subgroup of G. Then

$$\frac{G \times G}{H} \to (G/H) \times G \times (G/H)$$

$$\langle g_2, g_1 \rangle \mapsto (g_2 H, g_2 g_1^{-1}, g_1 H)$$

is a smooth morphism over G/H and an embedding. //

Proposition 1.23. Let Ω be a differentiable groupoid. Then $\Omega * G\Omega \to G\Omega$, $(\xi, \lambda) \mapsto I_\xi(\lambda) = \xi\lambda\xi^{-1}$ is a smooth action of Ω on the inner subgroupoid $G\Omega$.

Proof. Since $G\Omega$ is an embedded submanifold of Ω (1.17), this is trivial. //

The following construction is needed in 1.25.

Proposition 1.24. Let Ω and Ω' be differentiable groupoids on B, and let Ω be Lie. Then the product groupoid $\Omega \underset{B\times B}{\times} \Omega'$ defined in I§3 has a unique differentiable structure with respect to which it is a differentiable groupoid on B, the projections $\pi: \Omega \underset{B\times B}{\times} \Omega' \to \Omega$, $\pi': \Omega \underset{B\times B}{\times} \Omega' \to \Omega'$ are smooth, and π' is a submersion. $\Omega \underset{B\times B}{\times} \Omega'$ is Lie iff Ω' (in addition to Ω) is Lie, and this is the case iff π (in addition to π') is a submersion.

Proof. The differentiable structure on $\Omega \underset{B\times B}{\times} \Omega'$ is of course the pullback structure of $[\beta, \alpha]$ over $[\beta', \alpha']$. The stated properties are easily verified. //

The following proposition and examples are basic to connection theory.

<u>Proposition 1.25</u>. Let (E,p,B) be a vector bundle.

(i) The action $\Pi(E) * \mathrm{Hom}^n(E; B \times \mathbf{R}) \to \mathrm{Hom}^n(E; B \times \mathbf{R})$ defined by $\xi\phi = \phi\circ(\xi^{-1})^n$ is smooth.

(ii) The action $\Pi(E) * \mathrm{Hom}^n(E;E) \to \mathrm{Hom}^n(E;E)$ defined by $\xi\phi = \xi\circ\phi\circ(\xi^{-1})^n$ is smooth.

(iii) Let (E',p',B) be another vector bundle on the same base. Then the action $(\Pi(E) \underset{B\times B}{\times} \Pi(E')) * \mathrm{Hom}(E;E') \to \mathrm{Hom}(E;E')$ defined by $(\xi,\xi')\phi = \xi'\circ\phi\circ\xi^{-1}$ is smooth.

<u>Proof</u>. The bundles $\mathrm{Hom}^n(E; B \times \mathbf{R})$, etc., are commonly defined in terms of charts induced from charts for E (for example, Dieudonné (1972, XVI.16)). Since charts for E are decomposing sections for $\Pi(E)$, the results reduce locally to the smoothness of the corresponding actions of Lie groups on vector spaces, $GL(V) \times \mathrm{Hom}^n(V;\mathbf{R}) \to \mathrm{Hom}^n(V;\mathbf{R})$, etc. //

$\mathrm{Hom}^n(E;E')$ of course denotes the vector bundle on B whose fibre over $x \in B$ is the space of n-multilinear maps $E_x \times \cdots \times E_x \to E'_x$ and whose bundle structure is induced from the bundle structure of E and E' as in the reference cited above. The actions (i) and (ii) clearly restrict to the subbundles $\mathrm{Alt}^n(E;E')$ and $\mathrm{Sym}^n(E;E')$ of the alternating and symmetric multilinear maps; further, Hom(E;E') in (iii) could be replaced by $\mathrm{Hom}^n(E;E')$ and the obvious action. Lastly, there are analogous actions of $\Pi(E)$ on the tensor bundles $\otimes^r_s E$. We take all these variations of 1.25 to be included in its statement.

<u>Example 1.26</u>. Let (E,p,B) be a vector bundle, and let $\langle\ ,\ \rangle$ be a Riemannian structure in E, regarded as a section of $\mathrm{Hom}^2(E; B \times \mathbf{R})$ (see, for example, Greub et al (1972, 2.17)). Then $\langle\ ,\ \rangle$ is $\Pi(E)$-deformable with respect to the action of 1.25(i), since any two vector spaces of the same dimension with any positive-definite inner products, are isometric. Let $\Pi\langle E\rangle$ denote the isotropy groupoid of $\langle\ ,\ \rangle$. By 1.20 it is a Lie groupoid on B, the <u>Riemannian</u> or <u>orthonormal frame groupoid</u> of (E, $\langle\ ,\ \rangle$). A decomposing section $\sigma: U \to \Pi\langle E\rangle_b$ of $\Pi\langle E\rangle$ is a moving frame for E, and the local triviality of $\Pi(E)$ is in fact equivalent to the existence of moving frames in E. //

Example 1.27. Similarly, if Δ is a determinant function in a vector bundle (E,p,B), regarded as a never-zero section of $Alt^r(E; B \times \mathbf{R})$ (r = rank E), then Δ is $\Pi(E)$-deformable and the groupoid of orientation-preserving isomorphisms between the fibres of E is a closed embedded reduction of $\Pi(E)$, denoted $\Pi^+(E)$. //

Example 1.28. Let (L,p,B) be a vector bundle and let $[\ ,\ ]$ be a section of $Alt^2(L;L)$ such that each $[\ ,\ ]_x: L_x \times L_x \to L_x$ is a Lie algebra bracket. We call such a section a field of Lie algebra brackets in L.

A field of Lie algebra brackets need not be $\Pi(L)$-deformable. For example, let $\mathfrak{g}$ be a non-abelian Lie algebra with bracket $[\ ,\ ]$ and in $L = \mathbf{R} \times \mathfrak{g}$ define $[\ ,\ ]_t = t[\ ,\ ]$. However 1.20 implies that if $[\ ,\ ]$ is a field of Lie algebra brackets in a vector bundle L and if the fibres of L are pairwise-isomorphic as Lie algebras, then L admits an atlas of charts which fibrewise are Lie algebra isomorphisms. In this case, $(L,[\ ,\ ])$ is a Lie algebra bundle, as defined in 2.3 below, and we denote the isotropy groupoid by $\Pi[L]$. //

Example 1.29. Let μ be a section of a vector bundle E on a connected base B. Then, by a similar argument, E has an atlas of charts $U \times V \to E_U$ such that the local representatives $U \to V$ of μ are constant, iff μ is either never-zero or always zero. This (trivial) result is well-known in the case of tangent vector fields. //

Greub et al (1973, Chapter VIII) introduce a concept of Σ-bundle, defined as a vector bundle E together with a finite set Σ of sections σ_i of various tensor bundles $\otimes^{r_i}_{s_i}(E)$, such that E admits an atlas with respect to which all the σ_i are constant. Their Theorem 1 (loc.cit.) follows from 1.20 by considering the natural action of $\Pi(E)$ on the direct sum of the relevant tensor bundles. It is interesting to compare the proof of their result with that of 1.20.

It is worth noting that in 1.28 and 1.29 the condition of pairwise isomorphism, or of being never-zero or always zero, need hold only on each component of the base separately. A similar comment applies to the next, and last, example.

Example 1.30. Let (E^ν,p_ν,B), $\nu = 1,2$, be vector bundles on base B, and let $\phi: E^1 \to E^2$ be a morphism, considered as a section of $Hom(E^1;E^2)$. Then ϕ is $\Pi(E) \underset{B\times B}{\times} \Pi(E')$-deformable iff it is of constant rank. Now 1.20 shows that if this is the case, there are atlases $\{\psi^\nu_i: U_i \times V^\nu \to E^\nu_{U_i}\}$, $\nu = 1,2$, and a linear map $f: V^1 \to V^2$ such that $\phi: E^1_{U_i} \to E^2_{U_i}$ is represented by $(x,v) \mapsto (x,f(v))$.

This is of course a vector-bundle version of the standard characterization of a subimmersion; note, however, that it does not apply to general subimmersions $M \to N$ since we are working with a fixed base. The result, nonetheless, is important in the abstract theory of transitive Lie algebroids (see IV§1). //

The following result is often used.

<u>Proposition 1.31</u>. Let $\phi: \Omega \to \Omega'$ be a morphism of Lie groupoids over a connected base B. Then

(i) ϕ is of constant rank;

(ii) if ϕ^b_b is injective for some $b \in B$, then ϕ is an injective immersion;

(iii) if ϕ^b_b is surjective for some $b \in B$, then ϕ is a surjective submersion.

<u>Proof</u>. Follows from II 2.13 and the corresponding results for Lie group morphisms. //

<u>Proposition 1.32</u>. Let $\phi: \Omega \to \Omega'$ be a morphism of Lie groupoids over B. Then $K = \ker \phi$ is a closed embedded submanifold of $G\Omega$, and a sub Lie group bundle of $G\Omega$. Further, $\mathrm{im}(\phi)$ is a submanifold of Ω' and a reduction of Ω'.

<u>Proof</u>. Let $\{\sigma_i : U_i \to \Omega_b\}$ be a section-atlas for Ω, and denote by $\psi_i = I_{\sigma_i}$ the induced charts for $G\Omega$. Since ϕ^b_b is of constant rank, K_b is an embedded submanifold of Ω^b_b. Clearly $\psi_i(U_i \times K_b) = K_{U_i}$, so K_{U_i} is an embedded submanifold of $G\Omega|_{U_i}$. Since the $G\Omega|_{U_i}$ are open in $G\Omega$, it follows that K is an embedded submanifold of $G\Omega$. Since $K = \phi^{-1}(\tilde{B})$, it is obviously closed.

Let X denote the equivalence relation $\{(\xi\lambda, \xi) \mid \xi \in \Omega, \lambda \in K, \alpha\xi = \beta\lambda\}$ on Ω induced by K. As in 1.10, it is easy to see that $\delta': \Omega \underset{\beta}{\times} \Omega \to \Omega$, $(\eta, \xi) \mapsto \eta^{-1}\xi$ is a submersion; since $\delta'^{-1}(K) = X$ it follows that X is a closed embedded submanifold of $\Omega \underset{\beta}{\times} \Omega$, and hence of $\Omega \times \Omega$. The projection $X \xrightarrow{\pi_2} \Omega$ is a submersion, since

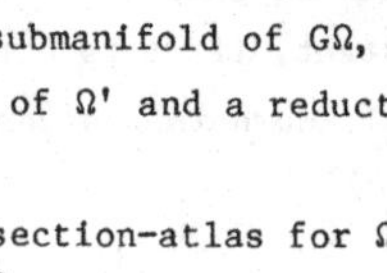

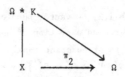

commutes; here $\Omega * K \to X$ is $(\xi,\lambda) \mapsto (\xi\lambda,\xi)$ and $\Omega * K \to \Omega$ is $(\xi,\lambda) \mapsto \xi$; this latter map is a submersion because in the pullback square defining $\Omega * K$, the projection $K \to B$ is a submersion.

So, by Godement's criterion, the quotient manifold Ω/X - that is, Ω/K - exists. It is easy to adapt the proof of II 1.6 to show that Ω/K is a differentiable groupoid; since $\natural: \Omega \to \Omega/K$ is now a smooth morphism over B, it follows that Ω/K is Lie. Now the induced morphism $\bar{\phi}: \Omega/K \to \Omega'$ is smooth and injective; since ϕ is of constant rank (by 1.31(i)) it follows (as in 1.6) that $\bar{\phi}$ is also. It is therefore an immersion. This completes the proof. //

1.32 is the only result concerning quotient differentiable groupoids that we will need. A more general result is stated in Pradines (1966).

2. Lie algebroids.

This section introduces the concept of Lie algebroid, preparatory to the construction in §3 of the Lie algebroid of a differentiable groupoid.

The concept of Lie algebroid, as a generalization of the concept of Lie algebra, and obtained from a differentiable groupoid by a process clearly generalizing that by which a Lie algebra is obtained from a Lie group, first appeared in Pradines (1967). It was not used by Ehresmann (1951, 1956) in his definitions of a connection. Atiyah (1957) had earlier constructed from a principal bundle an exact sequence of vector bundles which is the Lie algebroid of the corresponding groupoid (see 3.20); see also Nickerson (1961).

Related to the concept of Lie algebroid is the purely algebraic concept of Lie pseudo-algebra over a (commutative and unitary) ring, which stands in the same relationship to the concept of Lie algebroid as does that of module over a ring to the concept of vector bundle. The concept of Lie pseudo-algebra has been introduced by Herz (1953), Palais (1961b), Rinehart (1963), Hermann (1967), Nelson (1967) and Pradines (1967), and by other writers since Pradines, usually with a view to studying de Rham cohomology and Lie algebra cohomology simultaneously.

Rinehart (1963) proves a Poincaré-Birkhoff-Witt theorem for projective Lie pseudo-algebras and this is the only result of substance in the theory before the announcement in Pradines (1968b) of the integrability of Lie algebroids. A treatment of the material of this section, as well as parts of §5 and §7, and much of IV§1, in terms of Lie pseudo-algebras, was given in Mackenzie (1979). We manage to avoid here the explicit use of the concept of Lie pseudo-algebra.

The basic examples of Lie algebroids first appeared in Ngô (1968).

Definition 2.1. Let B be a manifold. A Lie algebroid on B is a vector bundle (A,p,B) together with a vector bundle map $q: A \to TB$ over B, called the anchor of A, and a bracket $[\ ,\ ]: \Gamma A \times \Gamma A \to \Gamma A$ which is $\mathbb{R}$-bilinear and alternating and satisfies the Jacobi identity, and is such that

(1) $$q([X,Y]) = [q(X),q(Y)] \qquad X,Y \in \Gamma A$$

(2) $$[X,uY] = u[X,Y] + q(X)(u)Y \qquad X,Y \in \Gamma A,\ u \in C(B).$$

The Lie algebroid A is transitive if q is a submersion, regular if q is of locally constant rank, and totally intransitive if q = 0. B is the base of A.

Let A' be a second Lie algebroid, on the same base B. Then a morphism of Lie algebroids $\phi: A \to A'$ over B is a vector bundle morphism such that $q' \circ \phi = q$ and $\phi[X,Y] = [\phi(X),\phi(Y)]$, $\forall X,Y \in \Gamma A$. //

The simplest examples of Lie algebroids are Lie algebras, Lie algebra bundles, and the tangent bundle to a manifold. The reader familiar with principal bundles should at this point read A§2-§3 where it is shown that the Atiyah sequence of a principal bundle is a transitive Lie algebroid.

Pradines (1967) named the map $q: A \to TB$ the "flèche" of the Lie algebroid. The usual translation, "arrow", is a much-used word and we propose to call this map the "anchor" of the Lie algebroid. It ties, or fails to tie, the bracket structure on ΓA to the Poisson bracket on ΓTB.

The anchor of a Lie algebroid A encodes its geometric properties. If A is transitive then right inverses to the anchor are connections in A (see §5). If A is regular then the image of the anchor defines a foliation of the base manifold and over each leaf of this foliation, the Lie algebroid is transitive. Compare the situation with groupoids, where local right-inverses to the anchor correspond to decomposing sections.

Proposition 2.2. Let A be a Lie algebroid on B, and $U \subseteq B$ an open subset. Then the bracket $\Gamma A \times \Gamma A \to \Gamma A$ "restricts" to $\Gamma_U A \times \Gamma_U A \to \Gamma_U A$ and makes A_U a Lie algebroid on U, called the restriction of A to U.

Proof: It suffices to show that if $X,Y \in \Gamma A$ and Y vanishes on an open set $U \subseteq B$, then $[X,Y]$ vanishes on U. For $x_o \in U$ take $u: B \to \mathbb{R}$ with $u(x_o) = 0$, $u(B \setminus U) = \{1\}$; then $[X,Y](x_o) = [X,uY](x_o) = u(x_o)[X,Y](x_o) + q(X)(u)(x_o)Y(x_o) = 0$. //

For the restriction of Lie algebroids to more general submanifolds of the base, see Almeida and Kumpera (1981).

The following examples are basic.

Definition 2.3. A Lie algebra bundle, or LAB, is a vector bundle (L,p,B) together with a field of Lie algebra brackets $[\ ,\ ]: \Gamma L \times \Gamma L \to \Gamma L$ (see 1.28) such that L admits an atlas $\{\psi_i: U_i \times \mathfrak{g} \to L_{U_i}\}$ in which each $\psi_{i,x}$ is a Lie algebra isomorphism.

A morphism of LAB's $\phi: L \to L'$, $\phi_o: B \to B'$, is a morphism of vector bundles such that each $\phi_x: L_x \to L'_{\phi_o(x)}$ is a Lie algebra morphism. //

An LAB is clearly a totally intransitive Lie algebroid. On the other hand a totally intransitive Lie algebroid is merely a vector bundle with a field of Lie algebra brackets, as in 1.28, and need not be an LAB.

Example 2.4. Let B be a manifold and let $\mathfrak{g}$ be a Lie algebra. On $TB \oplus (B \times \mathfrak{g})$ define an anchor $q = \pi_1 : TB \oplus (B \times \mathfrak{g})$ and a bracket

$$[X \oplus Y, Y \oplus W] = [X,Y] \oplus \{X(W) - Y(V) + [V,W]\}.$$

Then $TB \oplus (B \times \mathfrak{g})$ is a transitive Lie algebroid on B, called the trivial Lie algebroid on B with structure algebra $\mathfrak{g}$.

Let $\phi: TB \oplus (B \times \mathfrak{g}) \to TB \oplus (B \times \mathfrak{g}')$ be a morphism of trivial Lie algebroids on B. Then the condition $q' \circ \phi = q$ implies that ϕ has the form

$$\phi(X \oplus V) = X \oplus (\omega(X) + \phi^+(V)) \tag{3}$$

where $\omega: TB \to B \times \mathfrak{g}'$ is a $\mathfrak{g}'$-valued 1-form on B and $\phi^+: B \times \mathfrak{g} \to B \times \mathfrak{g}'$ is a vector bundle morphism. Writing out the equation

$$[\phi(X \oplus V), \phi(Y \oplus W)] = \phi[X \oplus V, Y \oplus W]$$

and setting firstly $X = Y = 0$, then $V = W = 0$, and lastly $V = 0$, we obtain, successively

$$[\phi^+(V), \phi^+(W)] = \phi^+([V,W]) \tag{4}$$

$$\delta\omega(X,Y) + [\omega(X),\omega(Y)] = 0 \tag{5}$$

$$X(\phi^+(W)) - \phi^+(X(W)) + [\omega(X),\phi^+(W)] = 0. \tag{6}$$

(4) is the condition that ϕ^+ be an LAB morphism, (5) is the condition that ω be a Maurer-Cartan form. We call (6) the compatibility condition.

It is easy to see that, conversely, a Maurer-Cartan form $\omega \in A^1(B,\mathfrak{g}')$ and an LAB morphism $\phi^+: B \times \mathfrak{g} \to B \times \mathfrak{g}'$ which satisfy (6), define by (3) a morphism of Lie algebroids $TB \oplus (B \times \mathfrak{g}) \to TB \oplus (B \times \mathfrak{g}')$.

This decomposition should be contrasted with the decomposition I 2.13 of morphisms of trivial groupoids. Given a morphism $\phi: B \times G \times B \to B \times G' \times B$ over B, if we define $F: B \times G \to B \times G'$ by $(x,g) \mapsto (x, \pi_2 \circ \phi(x,g,x))$ and $\theta: B \to G'$ by $\theta(x) = \pi_2 \circ \phi(x,1,b)$, where b is fixed, then θ and F must satisfy a compatibility condition comparable to (6). See 3.21 and 7.30.

Similarly, it is possible to describe $\phi\colon TB \oplus (B \times \mathfrak{g}) \to TB \oplus (B \times \mathfrak{g}')$ in terms of ω and a single morphism of Lie algebras $f = \phi^{+}\big|_{b}\colon \mathfrak{g} \to \mathfrak{g}'$. Here, however, f and ω must still obey a compatibility condition; see 7.30. //

<u>Example 2.5.</u> Let E be a vector bundle on B and consider the jet bundle exact sequence (see, for example, Palais (1965))

$$\mathrm{End}(E) \overset{\subseteq}{\rightarrowtail} \mathrm{Diff}^1(E) \overset{\sigma}{\twoheadrightarrow} \mathrm{Hom}(T^*(B),\mathrm{End}(E)).$$

Here $\mathrm{Diff}^1(E)$ is the vector bundle $\mathrm{Hom}(J^1(E),E)$, whose sections can be naturally regarded as first or zeroth order differential operators from E to itself, and the symbol map, σ, maps $D \in \Gamma\mathrm{Diff}^1(E)$ to

$$\delta f \mapsto (\mu \mapsto D(f\mu) - fD(\mu)) \qquad f \in C(B),\ \mu \in \Gamma E.$$

Map TB injectively into $\mathrm{Hom}(T^*(B),\mathrm{End}(E))$ by

$$X \mapsto (\omega \mapsto (\mu \mapsto \omega(X)\mu)) \qquad \omega \in \Gamma T^*(B),\ \mu \in \Gamma E$$

and construct the inverse image vector bundle over B (see C.5),

$$\begin{array}{ccc} \mathrm{CDO}(E) & \longrightarrow & TB \\ \big\downarrow & & \big\downarrow \\ \mathrm{Diff}^1(E) & \xrightarrow{\ \sigma\ } & \mathrm{Hom}(T^*(B),\mathrm{End}(E))\ . \end{array} \tag{7}$$

The inverse image exists because σ is a surjective submersion and since the right-hand vertical arrow is an injective immersion, it follows as usual that the left-hand arrow is also and we can therefore regard CDO(E) as a subvector bundle of $\mathrm{Diff}^1(E)$. Similarly, because σ is a surjective submersion, it follows that the top arrow is also; we denote it by q. Clearly the kernel of q is still End(E), and we have an exact sequence of vector bundles over B,

$$\mathrm{End}(E) \rightarrowtail \mathrm{CDO}(E) \overset{q}{\twoheadrightarrow} TB$$

where the sections of CDO(E) are those first- or zeroth-order differential operators $D\colon \Gamma E \to \Gamma E$ for which

(8) $$\sigma(D)(\delta f)(\mu) = X(f)\mu$$

for some $X = q(D) \in \Gamma TB$ and all $f \in C(B)$, $\mu \in \Gamma E$. Equivalently,

(8a) $$D(f\mu) = fD(\mu) + q(D)(f)\mu \qquad f \in C(B),\ \mu \in \Gamma E.$$

A (first- or zeroth-order) differential operator D with the property (8) for some vector field X is usually called a derivation in E. We will not use this term because when E = L is an LAB it would unavoidably lead to confusion with the concept of derivation which refers to the Lie bracket. Instead we call such a D a covariant differential operator in E, since for any connection ∇ in E and vector field X on B, the covariant derivative ∇_X obeys (8).

For $D, D' \in \Gamma\mathrm{Diff}^1(E)$, the bracket

$$[D,D'] = D\circ D' - D'\circ D$$

is also in $\Gamma\mathrm{Diff}^1(E)$ and it is easy to check from (8a) that if $D, D' \in \Gamma CDO(E)$, then $[D,D'] \in \Gamma CDO(E)$, and $q([D,D']) = [q(D),q(D')]$. Lastly, one can also easily check, again from (8a), that

$$[D,fD'] = f[D,D'] + q(D)(f)D' \qquad D,D' \in \Gamma CDO(E),\ f \in C(B)$$

and so CDO(E) is a transitive Lie algebroid on B, the Lie algebroid of covariant differential operators on E.

Let $E = B \times V$ be a trivial vector bundle, and define a morphism from the trivial Lie algebroid $TB \oplus (B \times \mathfrak{gl}(V))$ into $CDO(B \times V)$ by

$$(X \oplus \phi)(\mu) = X(\mu) + \phi(\mu) \qquad \mu: B \to V$$

where $X(\mu)$ is the Lie derivative. It is easy to check that this is an isomorphism of Lie algebroids over B.

In general, CDO(E) plays the role for E that is played for a vector space V by the general linear Lie algebra $\mathfrak{gl}(V)$. This will become apparent in the course of the following sections.

See also the original construction in Atiyah (1957, §4), where the bundle D(E) is a form of the dual of CDO(E). //

Example 2.6. An involutive distribution (without singularities) Δ on a manifold B is a regular Lie algebroid on B with respect to the inclusion $\Delta \subseteq TB$ as anchor and the standard bracket of vector fields. //

Example 2.7. Let $B = \mathbb{R}$ and define a bracket $[\ ,\ ]'$ on TB by

$$[\xi \frac{d}{dt}, \eta \frac{d}{dt}]' = t(\dot{\xi}\eta - \xi\dot{\eta}) \frac{d}{dt}, \qquad \xi, \eta : B \to \mathbb{R}$$

and an anchor $q: TB \to TB$ by $q(\xi \frac{d}{dt}) = t\xi \frac{d}{dt}$. It is straightforward to check that TB is a Lie algebroid on B with this structure, and that it is not regular. //

Let A be a transitive Lie algebroid on B. The kernel L of $q: A \twoheadrightarrow TB$ inherits the bracket structure of A (by (1) in 2.1) and is itself a totally intransitive Lie algebroid on B. We usually write a transitive Lie algebroid in the form

$$L \overset{j}{\rightarrowtail} A \overset{q}{\twoheadrightarrow} TB.$$

The notation j allows L to be any totally intransitive Lie algebroid isomorphic to the kernel of q; for example, in the case of the Atiyah sequence of a principal bundle $P(B,G,\pi)$, the kernel $\frac{T^{\pi}P}{G}$ of π_* is usually replaced by $\frac{P \times \mathfrak{g}}{G}$ (see A§3).

We call L the adjoint bundle of A. The reason for this terminology will gradually become clear - see, for example, A§3, 3.18 and 3.20, and 5.8. In IV§1 we will prove that L is actually an LAB.

A morphism $\phi: A \to A'$ of transitive Lie algebroids over B obeys $q' \circ \phi = q$ and therefore induces a morphism of the adjoint bundles $\phi^+: L \to L'$. This is a morphism of totally intransitive Lie algebroids. In IV§1 we show that ϕ^+ (and hence ϕ) is of locally constant rank.

The following version of the 5-lemma is extremely useful.

Proposition 2.8. Let $\phi: A \to A'$ be a morphism of transitive Lie algebroids over B. Then ϕ is a surjection, injection or bijection iff $\phi^+: L \to L'$ is, respectively, a surjection, injection or bijection.

Proof: This is a diagram-chase in

$$\begin{array}{ccccc} L & \rightarrowtail & A & \overset{q}{\twoheadrightarrow} & TB \\ {\scriptstyle \phi^+}\downarrow & & \downarrow{\scriptstyle \phi} & & \| \\ L' & \rightarrowtail & A' & \overset{q'}{\twoheadrightarrow} & TB . \end{array}$$ //

If A is a regular Lie algebroid on B then I = im q is an involutive distribution on B and L = ker q is still defined. Such a Lie algebroid could be written in the form

$$L \xrightarrow{j} A \xrightarrow{q} I$$

and a version of 2.8 would continue to hold. However in this case L need not be an LAB and morphisms need not be of locally constant rank (as the totally intransitive case shows).

The regular case may to some extent be reduced to the transitive case, since the restriction of a regular Lie algebroid to a leaf of the foliation defined by I = im q is transitive. However there remains the question of how the restrictions are bound together, and we will not address this problem.

We conclude this section with some basic algebraic definitions. These are kept to the minimum needed in the present chapter, since little of substance can be said except in the transitive case and then only by using results which will not be proved until IV§1.

Definition 2.9. Let A be a Lie algebroid on B and let E be a vector bundle, also on B. A representation of A on E is a morphism of Lie algebroids over B,

$$\rho\colon A \to CDO(E).$$

Let $\rho'\colon A \to CDO(E')$ be a second representation of A. Then a vector bundle morphism $\phi\colon E \to E'$ is A-equivariant if $\phi(\rho(X)(\mu)) = \rho(X)(\phi(\mu))$ for $X \in A$, $\mu \in E$.

Let $\rho^i\colon A^i \to CDO(E^i)$, i = 1,2 be representations of Lie algebroids over B on vector bundles over B, let $\phi\colon A^1 \to A^2$ be a morphism of Lie algebroids over B, and let $\psi\colon E^1 \to E^2$ be a morphism of vector bundles. Then ψ is ϕ-equivariant if $\psi(\rho^1(X)(\mu)) = \rho^2(\phi(X))(\psi(\mu))$, $\forall X \in A^1$, $\mu \in E^1$. //

In IV§1 we will see that if A is transitive, then equivariant morphisms are of locally constant rank.

Example 2.10. Let A be a Lie algebroid on B and let V be a vector space. The trivial representation of A on B × V is

$$\rho^0(X)(f) = q(X)(f) \qquad f\colon B \to V,\ X \in \Gamma A. \qquad //$$

Example 2.11. Let A be a transitive Lie algebroid on B. The adjoint representation of A is the representation

$$\mathrm{ad}: A \to CDO(L)$$

of A on its adjoint bundle L defined by

$$\mathrm{ad}(X)(V) = [X,V] \qquad X \in \Gamma A,\ V \in \Gamma L. \qquad //$$

We return to the adjoint representation in §4 and §5. It plays a crucial role in the developments of Chapter IV.

Example 2.12. Let P(B,G) be a principal bundle and let $E = \frac{P \times V}{G}$ be an associated vector bundle. Given $X \in \Gamma(\frac{TP}{G})$ denote by $\bar{X}$ the corresponding G-invariant vector field on P; and given $\mu \in \Gamma E$ denote by $\tilde{\mu}: P \to V$ the corresponding G-equivariant map. Then the Lie derivative $\bar{X}(\tilde{\mu})$ is also G-equivariant (see A 4.4); denote the corresponding element of ΓE by $X(\mu)$. Then A 4.6 shows that $X \mapsto (\mu \mapsto X(\mu))$ is a representation of $\frac{TP}{G}$ on E. //

Example 2.13. Let E be a vector bundle on B, and let ∇ be a flat connection in E. Then $X \mapsto \nabla_X$ is a representation of TB on E. //

Definition 2.14. An exact sequence of Lie algebroids over B is a sequence of morphisms of Lie algebroids over B

$$A' \overset{\iota}{\rightarrowtail} A \overset{\pi}{\twoheadrightarrow} A''$$

which is exact as a sequence of vector bundles over B. //

The Lie algebroid A' must be totally intransitive, for $q' = q \circ \iota = (q'' \circ \pi) \circ \iota = 0$. The basic example is of course $L \overset{j}{\rightarrowtail} A \overset{q}{\twoheadrightarrow} TB$ for A a transitive Lie algebroid.

Definition 2.15. Let A be a transitive Lie algebroid. A is abelian if its adjoint bundle is, that is, if

$$[V,W] = 0 \qquad \forall V,W \in L. \qquad //$$

Proposition 2.16. Let $E \overset{\iota}{\rightarrowtail} A' \overset{\pi}{\twoheadrightarrow} A$ be an exact sequence of Lie algebroids over B with E abelian. Then

$$\iota(\rho(X)(\mu)) = [X',\iota(\mu)] \qquad X \in \Gamma A,\ \mu \in \Gamma E,$$

where $X' \in \Gamma A'$ has $\pi(X') = X$, defines a representation of A on E.

Proof: Exercise. //

In particular, if A is transitive and abelian, there is a natural representation of TB on the adjoint bundle L (compare the final remarks in I§3).

Definition 2.17. Let A be a Lie algebroid on B. A Lie subalgebroid of A is a Lie algebroid A' on B together with an injective morphism $A' \rightarrowtail A$ of Lie algebroids over B.

If A is transitive, a reduction of A is a Lie subalgebroid of A which is itself transitive. //

Proposition 2.18. Let A and A' be Lie algebroids on B and let A be transitive. Let $A \oplus_{TB} A'$ denote the inverse image vector bundle over B

$$\begin{array}{ccc} A \oplus_{TB} A' & \longrightarrow & A' \\ \big\downarrow & & \big\downarrow q' \\ A & \xrightarrow[q]{} & TB \end{array}$$

(see C.5). Let $\bar{q}\colon A \oplus_{TB} A' \to TB$ be the diagonal composition and define a bracket on $\Gamma(A \oplus_{TB} A')$ by

$$[X \oplus X',\ Y \oplus Y'] = [X,Y] \oplus [X',Y'].$$

Then $A \oplus_{TB} A'$ is a Lie algebroid on B, and the diagram above is now a pullback in the category of Lie algebroids over B. Lastly, $A \oplus_{TB} A'$ is transitive iff A' (as well as A) is transitive and this is so iff $A \oplus_{TB} A' \to A$ (as well as $A \oplus_{TB} A' \to A'$) is surjective.

Proof: Straightforward. //

$A \oplus_{TB} A'$ is called the direct sum Lie algebroid of A and A' over TB. Note that the trivial Lie algebroid $TB \oplus (B \times \mathfrak{g})$ is not a direct sum of TB and $B \times \mathfrak{g}$.

Concepts of kernel, ideal and quotient Lie algebroid are defined in IV§1.

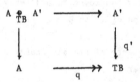

3. The Lie algebroid of a differentiable groupoid.

This section gives the construction of the Lie algebroid of a differentiable groupoid and works the basic examples. The construction follows very closely the construction of the Lie algebra of a Lie group. However since the right translations on a groupoid, the $R_\xi \colon \Omega_{\beta\xi} \to \Omega_{\alpha\xi}$, are diffeomorphisms of the α-fibres only and not of the whole groupoid, a vector field X on a differentiable groupoid Ω can be defined to be right-invariant only if it is tangent to the α-fibres. Having noted this, a right-invariant vector field on Ω is determined by its values on the unities $\tilde{x}$, $x \in B$, and we define the Lie algebroid of Ω to be $A\Omega = \bigcup_{x \in B} T(\Omega_x)_{\tilde{x}}$ with the natural vector bundle structure over B which it inherits from $T\Omega$. The Lie bracket is placed on the module of sections of $A\Omega$ (not on $A\Omega$ itself) via the correspondence between sections of $A\Omega$ and right-invariant vector fields on Ω. This bracket is not bilinear with respect to the module structure on $\Gamma A\Omega$ but obeys

$$[X, fY] = f[X,Y] + q(X)(f)Y \qquad f \in C(B),\ X,Y \in \Gamma A\Omega$$

where $q \colon A\Omega \to TB$ is a vector bundle morphism over B which maps each $X \in \Gamma A\Omega$ to the β-projection of the corresponding right-invariant vector field. The map q, which we are calling the anchor of $A\Omega$, ties the bracket structure on $\Gamma A\Omega$ to its module structure, and is the only feature of the Lie algebroid concept which does not appear in the case of Lie algebras. The anchor $q \colon A\Omega \to TB$ is, further, essential to the connection theory of Ω; for this see §5.

The construction of the Lie algebroid of a differentiable groupoid is due to Pradines (1967).

Let Ω be a differentiable groupoid on a manifold B.

Definition 3.1. The vector bundle $A\Omega \to B$ is the inverse image of $T^\alpha\Omega \to \Omega$ across the embedding $\varepsilon \colon B \to \Omega$. Thus

(1)
$$\begin{array}{ccc} A\Omega & \longrightarrow & T^\alpha\Omega \\ \big\downarrow & & \big\downarrow \\ B & \xrightarrow{\ \varepsilon\ } & \Omega \end{array}$$

is a pullback. //

Since ε is an embedding we will usually regard $A\Omega$ as the restriction of $T^\alpha\Omega$ to B and we identify the fibres $A\Omega|_x$ with the tangent spaces $T(\Omega_x)^{\sim}_x$, $x \in B$, and regard $A\Omega \to T^\alpha\Omega$ as an inclusion.

<u>Definition 3.2</u>. A <u>vertical vector field</u> on Ω is a vector field X which is vertical with respect to α; that is, $X_\xi \in T(\Omega_{\alpha\xi})_\xi$ for all ξ in the domain of X.

A global vector field X on Ω is <u>right-invariant</u> if it is vertical and $X(\eta\xi) = T(R_\xi)_\eta(X(\eta))$ for all $(\eta,\xi) \in \Omega * \Omega$. //

A trivial manipulation shows that a vertical vector field X is right-invariant iff $X(\xi) = T(R_\xi)(X(\widetilde{\beta\xi}))$ for all $\xi \in \Omega$. Thus a right-invariant vector field is determined by its values on the submanifold of identity elements. We denote the section of $A\Omega$ corresponding to the composite $X\circ\varepsilon\colon B \to T^\alpha\Omega$ by $X\!\downarrow_B$. The next result provides an inverse for $X \mapsto X\!\downarrow_B$.

<u>Proposition 3.3</u>. The vector bundle morphism

$$\text{(2)} \qquad \begin{array}{ccc} T^\alpha\Omega & \xrightarrow{\ \mathcal{R}\ } & A\Omega \\ \big\downarrow & & \big\downarrow \\ \Omega & \xrightarrow{\ \beta\ } & B \end{array}$$

where $\mathcal{R}_\xi = T(R_{\xi^{-1}})_\xi\colon T(\Omega_{\alpha\xi})_\xi \to T(\Omega_{\beta\xi})_{\beta\xi}$, is a pullback.

<u>Proof</u>: The composite of $T(\delta)\colon T(\Omega \underset{\alpha}{\times} \Omega) \to T\Omega$ with the tangent to $\Omega \to \Omega \underset{\alpha}{\times} \Omega$, $\xi \mapsto (\widetilde{\alpha\xi},\xi)$, restricts to $T^\alpha\Omega \to T^\alpha\Omega$; factoring this map over the pullback $A\Omega \to T^\alpha\Omega$ gives $\mathcal{R}$, which is therefore smooth.

Each $\mathcal{R}_\xi$ is clearly an isomorphism of vector spaces, so $\mathcal{R}$ is a pullback by C.2. //

<u>Corollary 3.4</u>. Given $X \in \Gamma A\Omega$, the formula

$$\vec{X}(\xi) = T(R_\xi)(X(\beta\xi)), \qquad \xi \in \Omega,$$

defines a right-invariant vector field on Ω.

<u>Proof</u>: For, in the notation of C.3, $\vec{X} = \mathcal{R}^{\#}(X)$. //

Corollary 3.5 The map

$$C(\Omega) \underset{C(B)}{\otimes} \Gamma A\Omega \to \Gamma T^{\alpha}\Omega, \quad f \otimes X \mapsto f\vec{X}$$

is an isomorphism of $C(\Omega)$-modules.

Proof: See C.4. //

For a Lie group G it is well-known that $\Gamma TG \cong C(G) \underset{\mathbb{R}}{\otimes} \mathfrak{g}$ and since a Lie algebra is free as a vector space it follows that any vector field X on G can be written as

(3) $$X = f_1\vec{X_1} + \cdots + f_n\vec{X_n}$$

where $\{X_1,\ldots,X_n\}$ is any basis for $\mathfrak{g}$. For a vertical vector field $X \in \Gamma T^{\alpha}\Omega$, 3.5 states that there are $X_i \in \Gamma A\Omega$ for which (3) holds, but since $\Gamma A\Omega$ is in general only a projective $C(B)$-module, and not free, the X_i and n may vary with X.

The two pullback diagrams (1) and (2) show that $T^{\alpha}\Omega \to \Omega$ is trivializable iff $A\Omega \to B$ is so; this generalizes the fact that the tangent bundle of a Lie group is trivializable.

Denote the set of right-invariant vector fields on Ω by $\Gamma^{RI}T^{\alpha}\Omega$. It is a $C(B)$-module under the multiplication $fX = (f\circ\beta)X$ and the maps

(4) $$\Gamma^{RI}T^{\alpha}\Omega \to \Gamma A\Omega, \quad X \mapsto X\big|_{B}; \qquad \Gamma A\Omega \to \Gamma^{RI}T^{\alpha}\Omega, \quad X \mapsto \vec{X}$$

are mutually inverse $C(B)$-module isomorphisms.

Lemma 3.6. $\Gamma^{RI}T^{\alpha}\Omega$ is closed under the Poisson bracket.

Proof: $X \in \Gamma T\Omega$ is vertical iff $X \overset{\alpha}{\sim} 0 \in \Gamma TB$ and $X \in \Gamma T^{\alpha}\Omega$ is right-invariant iff $X\big|_{\Omega_y} \overset{R_\xi}{\sim} X\big|_{\Omega_x}$ for each $\xi \in \Omega_x^y$, $x, y \in B$. The result now follows from the fact that for any smooth map $\phi: M \to N$, and $X, Y \in \Gamma TM$, $X', Y' \in \Gamma TN$, $X \overset{\phi}{\sim} X'$ and $Y \overset{\phi}{\sim} Y'$ imply $[X,Y] \overset{\phi}{\sim} [X',Y']$. //

3.6 permits the Poisson bracket on $\Gamma^{RI}T^{\alpha}\Omega$ to be transferred to $\Gamma A\Omega$, that is, we define

(5) $$[X,Y] = [\vec{X},\vec{Y}]\big\downarrow_B \qquad X,Y \in \Gamma A\Omega.$$

This bracket is alternating, $\mathbb{R}$-bilinear and satisfies the Jacobi identity; these follow immediately from the corresponding properties of the Poisson bracket. For a function $f \in C(B)$ and $X,Y \in \Gamma A\Omega$ we have

$$\begin{aligned}\overrightarrow{[X,fY]} &= [\vec{X},(f\circ\beta)\vec{Y}]\\ &= (f\circ\beta)[\vec{X},\vec{Y}] + \vec{X}(f\circ\beta)\vec{Y}\\ &= \overrightarrow{f[X,Y]} + \vec{X}(f\circ\beta)\vec{Y}.\end{aligned}$$

Now because $\beta\colon \Omega \to B$ is a surjective submersion and $\beta\circ R_\xi = \beta$, $\forall\xi \in \Omega$, every right-invariant vector field $\vec{X}$ is β-projectable; that is, there is a vector field X' on B such that

$$X'(f)\circ\beta = \vec{X}(f\circ\beta) \qquad \forall f \in C(B),$$

and in terms of X', we obtain

$$[X,fY] = f[X,Y] + X'(f)Y.$$

X' is the β-projection of the right-invariant vector field associated to X, and is described more simply as follows.

Definition 3.7. The anchor $q = q^\Omega\colon A\Omega \to TB$ of $A\Omega$ is the composite of the vector bundle morphisms

$$\begin{array}{ccc} A\Omega & \longrightarrow & T^\alpha\Omega \\ \downarrow & & \downarrow \\ B & \xrightarrow{\ \varepsilon\ } & \Omega \end{array}$$

and

$$\begin{array}{ccccc} T^\alpha\Omega & \longrightarrow & T\Omega & \xrightarrow{T(\beta)} & TB \\ \downarrow & & \downarrow & & \downarrow \\ \Omega & = & \Omega & \xrightarrow{\ \beta\ } & B \end{array} . \qquad //$$

Since $\beta\circ\varepsilon = id_B$, q is a morphism over B.

Lemma 3.8. For $X \in \Gamma A\Omega$, $\vec{X}$ is β-related to q(X).

Proof: $T(\beta)(\vec{X}(\xi)) = T(\beta\circ R_\xi)(X(\beta\xi)) = T(\beta)(X(\beta\xi))$, and clearly $q_x = T(\beta_x)_x^\sim : T(\Omega_x)_x^\sim \to T(B)_x$. //

It now follows that for any $X,Y \in \Gamma A\Omega$ and any $f \in C(B)$,

(6) $$[X,fY] = f[X,Y] + q(X)(f)Y.$$

Further, since $\vec{X}$ is β-related to q(X) and $\vec{Y}$ is β-related to q(Y), it follows that $[\vec{X},\vec{Y}]$ is β-related to $[q(X),q(Y)]$. But $[\vec{X},\vec{Y}] = \overrightarrow{[X,Y]}$ is also β-related to $q([X,Y])$ and since β is surjective it follows that

(7) $$q([X,Y]) = [q(X),q(Y)].$$

These results permit the

Definition 3.9. The Lie algebroid of Ω is the vector bundle $A\Omega \to B$ defined in 3.1 together with the bracket [,] defined in (5) and the anchor q defined in 3.7. //

It is worth noting that the vector bundle $A\Omega \to B$ is always locally trivial and that this is not related to the local triviality of Ω. The local triviality of $A\Omega$ goes back ultimately to the assumption that β is a submersion.

We now need to make a few comments on the local version of the correspondence between sections of $A\Omega$ and right-invariant vector fields on Ω. For $\mathcal{U} \subseteq \Omega$ open, a local right-invariant vector field on $\mathcal{U}$ is a vertical vector field on $\mathcal{U}$ such that $X(\eta\xi) = T(R_\xi)(X(\eta))$ whenever $\alpha\eta = \beta\xi$ and both $\eta\xi$ and η are in $\mathcal{U}$. If $X \in \Gamma_U A\Omega$, where $U \subseteq B$ is open, then 3.3 shows that $\vec{X}(\xi) = T(R_\xi)(X(\beta\xi))$ defines a smooth local right-invariant vector field on Ω^U. On the other hand, we have

Lemma 3.10. Let $X \in \Gamma^{RI}T^\alpha\Omega$ be a local right-invariant vector field on an open set $\mathcal{U} \subseteq \Omega$. For $x \in \beta(\mathcal{U})$ choose any $\xi \in \mathcal{U}^x$ and define

$$\underset{\sim}{X}(x) = T(R_{\xi^{-1}})(X(\xi)).$$

Then $\underset{\sim}{X}$ is a well-defined and smooth local section of $A\Omega$ on $\beta(\mathcal{U})$.

Proof: That $\underset{\sim}{X}$ is well-defined follows from the right-invariance of X. Let $U = \beta(\mathcal{U})$ and note that $\mathcal{R}$ restricts to $T^{\alpha}\Omega|_{\mathcal{U}} \to A\Omega|_{U}$. Now $\underset{\sim}{X}\circ\beta = \mathcal{R}\circ X$ is smooth, and since β is a submersion it follows that $\underset{\sim}{X}$ is smooth. //

If we now take $X \in \Gamma^{RI}_{\mathcal{U}} T^{\alpha}\Omega$, where $\mathcal{U} \subseteq \Omega$ is open, and apply first 3.10 to get $\underset{\sim}{X} \in \Gamma_{\beta(\mathcal{U})} A\Omega$ and then 3.3 to get $(\vec{\underset{\sim}{X}}) \in \Gamma^{RI}_{\beta^{-1}(\beta(\mathcal{U}))} T^{\alpha}\Omega$, then we obtain a right-invariant vector field defined on the β-saturation of $\mathcal{U}$ which is equal to X on $\mathcal{U}$. We call $(\vec{\underset{\sim}{X}})$ the β-saturation of X and any right-invariant vector field defined on a β-saturated open subset of Ω will be called a β-saturated right-invariant vector field. The above shows that we need not consider right-invariant vector fields defined on more general open sets. Clearly $(\vec{\underset{\sim}{X}})$ is the only right-invariant vector field on $\beta^{-1}(\beta(\mathcal{U}))$ which coincides with X on $\mathcal{U}$.

If $\mathcal{U}$ is itself β-saturated, say $\mathcal{U} = \beta^{-1}(U)$ where U is open in B, and $X \in \Gamma^{RI}_{\mathcal{U}} T^{\alpha}\Omega$, then $\underset{\sim}{X}$ is actually the restriction of X to $\mathcal{U} \cap B = U$. In this case we write $X\!\downarrow_{U}$, rather than $\underset{\sim}{X}$, and there are mutually inverse C(U)-module isomorphisms

(8a) $$X \mapsto \vec{X}, \qquad \Gamma_U A\Omega \to \Gamma^{RI}_{\beta^{-1}(U)} T^{\alpha}\Omega$$

and

(8b) $$X \mapsto X\!\downarrow_{U}, \qquad \Gamma^{RI}_{\beta^{-1}(U)} T^{\alpha}\Omega \to \Gamma_U A\Omega.$$

It is straightforward to show that the bracket on $\Gamma_U A\Omega$ transported from $\Gamma^{RI}_{\beta^{-1}(U)} T^{\alpha}\Omega$ via (8) coincides with the bracket induced from the bracket on $\Gamma A\Omega$ by the method of 2.2.

Now consider a morphism $\phi: \Omega \to \Omega'$ of differentiable groupoids over B. We construct the induced morphism $\phi_*: A\Omega \to A\Omega'$ of Lie algebroids over B.

Since $\alpha'\circ\phi = \alpha$, the vector bundle morphism $T(\phi)$ restricts to $T^{\alpha}(\phi): T^{\alpha}\Omega \to T^{\alpha}\Omega'$. Since $\phi\circ\varepsilon = \varepsilon'$ the composition

(9) $$\begin{array}{ccccc} A\Omega & \longrightarrow & T^{\alpha}\Omega & \xrightarrow{T^{\alpha}(\phi)} & T^{\alpha}\Omega' \\ \downarrow & & \downarrow & & \downarrow \\ B & \xrightarrow{\varepsilon} & \Omega & \xrightarrow{\phi} & \Omega' \end{array}$$

is a vector bundle morphism over $\varepsilon': B \to \Omega'$ and so there is a unique vector bundle morphism $\phi_*: A\Omega \to A\Omega'$ over B such that the composition

$$\begin{array}{ccccc} A\Omega & \xrightarrow{\phi_*} & A\Omega' & \longrightarrow & T^{\alpha}\Omega' \\ \downarrow & & \downarrow & & \downarrow \\ B & = & B & \xrightarrow{\varepsilon'} & \Omega' \end{array}$$

is equal to (9).

From $\beta' \circ \phi = \beta$ and the definition of the anchors on $A\Omega$ and $A\Omega'$ it follows immediately that $q' \circ \phi_* = q$. It remains to prove that ϕ_* preserves the brackets.

<u>Lemma 3.11.</u> Let $X \in \Gamma A\Omega$, $X' \in \Gamma A\Omega'$. Then $X' = \phi_*(X)$ iff $\vec{X} \overset{\phi}{\sim} \vec{X'}$.

<u>Proof:</u> This is a straightforward manipulation. //

Now take $X, Y \in \Gamma A\Omega$ and write $X' = \phi_*(X)$, $Y' = \phi_*(Y)$. From 3.11 it follows that $\vec{X} \overset{\phi}{\sim} \vec{X'}$ and $\vec{Y} \overset{\phi}{\sim} \vec{Y'}$, so $\overrightarrow{[X,Y]} = [\vec{X},\vec{Y}] \overset{\phi}{\sim} [\vec{X'},\vec{Y'}] = \overrightarrow{[X',Y']}$ and by 3.11 again, we have $[X',Y'] = \phi_*([X,Y])$.

The constructions $\Omega \mapsto A\Omega$ and $\phi \mapsto \phi_*$ constitute a functor from the category of differentiable groupoids on a given base B and smooth morphisms of differentiable groupoids over B to the category of Lie algebroids on B and Lie algebroid morphisms over B. It is called the <u>Lie functor.</u>

The construction of the induced morphism of Lie algebroids $\phi_*: A\Omega \to A\Omega'$ can be extended to the case where $\phi: \Omega \to \Omega'$ is a morphism of differentiable groupoids over an arbitary smooth map $\phi_o: B \to B'$. For this see Almeida and Kumpera (1981); the difficulty lies not in the construction of ϕ_* but in giving an abstract formulation of the bracket-preservation property of ϕ_*. The case where ϕ_o is a diffeomorphism does not present this difficulty and is treated in §4.

<u>Example 3.12.</u> The Lie algebroid of a cartesian square groupoid $B \times B$ is naturally isomorphic to TB. For any differentiable groupoid Ω, the anchor map $[\beta,\alpha]: \Omega \to B \times B$, considered as a morphism over B, induces the anchor $q: A\Omega \to TB$ of $A\Omega$. //

We now need a series of results concerning the relationship between properties of ϕ and properties of ϕ_*.

Proposition 3.13. Let $\phi\colon \Omega \to \Omega'$ be a morphism of differentiable groupoids over B. Then

(i) ϕ is a submersion iff each $\phi_x\colon \Omega_x \to \Omega'_x$ is a submersion, and iff $\phi_*\colon A\Omega \to A\Omega'$ is a fibrewise surjection;

(ii) ϕ is an immersion iff each ϕ_x is an immersion, and iff ϕ_* is a fibrewise injection;

(iii) ϕ is of locally constant rank iff ϕ_* is.

Proof: We prove (iii); the other results can be proved by the same method.

($\Rightarrow$) Let the components of Ω be C_i and let the rank of ϕ on C_i be k_i. Take $\xi \in \Omega$ and write $x = \alpha\xi$, $y = \beta\xi$. From the diagram

$$\begin{array}{ccccc} T(\Omega_x)_\xi & \rightarrowtail & T(\Omega)_\xi & \twoheadrightarrow & T(B)_y \\ \Big\downarrow T(\phi_x)_\xi & & \Big\downarrow T(\phi)_\xi & & \Big\| \\ T(\Omega_y)_{\phi(\xi)} & \rightarrowtail & T(\Omega)_{\phi(\xi)} & \twoheadrightarrow & T(B)_y \end{array}$$

it follows that $rk_\xi(\phi_{\alpha\xi}) = k_i - \dim B$, for $\xi \in C_i$. Now for any $\xi \in \Omega$ we have $\phi_{\beta\xi}\circ R_\xi = R_{\phi(\xi)}\circ\phi_{\alpha\xi}$ so $rk_{\widetilde{\beta\xi}}(\phi_{\beta\xi}) = rk_\xi(\phi_{\alpha\xi})$. Therefore, $rk_{\tilde{x}}(\phi_x) = k_i - \dim B$ is constant for $x \in \beta(C_i)$ and since β is open, this shows that $rk_{\tilde{x}}(\phi_x)$ is a locally constant function of x.

($\Leftarrow$) Let the components of B be B_i and let the rank of ϕ_* on B_i be k_i. By the same argument it follows that $rk_\xi(\phi_{\alpha\xi}) = rk_{\widetilde{\beta\xi}}(\phi_{\beta\xi}) = k_i$ for $\xi \in \Omega^{B_i}$ and so $rk_\xi(\phi) = k_i + \dim B$ for $\xi \in \Omega^{B_i}$. Since Ω^{B_i} is open in Ω, this completes the proof. //

Recall from 1.31 that every base-preserving morphism of Lie groupoids is of locally constant rank, and that such a morphism is a submersion if it is surjective and is an immersion if it is injective. It is not true that every base-preserving morphism of differentiable groupoids is of locally constant rank, even if the base is connected: let Ω be an action groupoid $G \times B$ where $G \times B \to B$ is a smooth action with a fixed point x_0 and all other orbits of dimension ≥ 1. Then $[\beta,\alpha]$ is $G \times B \to B \times B$, $(g,x) \mapsto (gx,x)$ and the rank is not constant in any neighbourhood of $(1,x_o)$. Since $[\beta,\alpha]_x = \beta_x$ is known, by 1.4, to be of locally constant rank for any differentiable groupoid this example also shows that 3.13 (iii) cannot be strengthened to make it similar in form to (i) and (ii).

The following result is used in §6.

Proposition 3.14. Let $\phi\colon \Omega \to \Omega'$ be a morphism of differentiable groupoids over B. If Ω' is α-connected and $\phi_*\colon A\Omega \to A\Omega'$ is fibrewise surjective, then ϕ is surjective.

Proof: By 3.13, ϕ is a submersion and so $\phi(\Omega)$ is a symmetric α-neighbourhood of the base in Ω'. By II 3.11, $\phi(\Omega)$ generates Ω'; since $\phi(\Omega)$ is a subgroupoid of Ω' it follows that $\phi(\Omega) = \Omega'$. //

Proposition 3.15. Let $M \xrightarrow{\iota} \Omega' \xrightarrow{\pi} \Omega$ be an exact sequence of differentiable groupoids over B; that is, ι and π are morphisms of differentiable groupoids over B, ι is an embedding, π is a surjective submersion, and $\mathrm{im}(\iota) = \ker(\pi)$. Then $AM \xrightarrow{\iota_*} A\Omega' \xrightarrow{\pi_*} A\Omega$ is an exact sequence of Lie algebroids over B.

Proof: All that need be proved is that the sequence of vector spaces

$$T(M_x)_{\tilde{x}} \xrightarrow{T(\iota_x)_{\tilde{x}}} T(\Omega'_x)_{\tilde{x}} \xrightarrow{T(\pi_x)_{\tilde{x}}} T(\Omega_x)_{\tilde{x}}$$

is exact, and this follows immediately from the fact that $\pi_x\colon \Omega'_x \to \Omega_x$ is a submersion with $(\pi_x)^{-1}(\tilde{x}) = M_x$. //

3.15 asserts that the Lie functor is exact.

Corollary 3.16. Let Ω be a differentiable groupoid on B. If Ω is locally trivial then $A\Omega$ is transitive. If $A\Omega$ is transitive and B is connected, then Ω is locally trivial.

Proof: Since $[\beta,\alpha]_* = q\colon A\Omega \to TB$, the first result follows from 3.13 (i) and the second from 3.14. //

If Ω is a Lie groupoid on B then

$$G\Omega \rightarrowtail \Omega \xrightarrow{[\beta,\alpha]} B \times B$$

is an exact sequence of differentiable groupoids over B, and $G\Omega$ is a Lie group bundle on B; if $\{\sigma_i : U_i \to \Omega_b\}$ is a section-atlas for Ω then $\{I_{\sigma_i} : U_i \times \Omega_b^b \to G\Omega|_{U_i}\}$ is an LGB-atlas for $G\Omega$. Since $[\beta,\alpha]_* = q\colon A\Omega \to TB$, the adjoint bundle of $A\Omega$ is the Lie algebroid of $G\Omega$ as a differentiable groupoid; the following calculation confirms that this coincides with the Lie algebra bundle associated to $G\Omega$.

Example 3.17. Let (M,p,B) be a Lie group bundle. (See A§1.) The Lie algebra bundle asociated to M is denoted M_* and defined as follows (Douady and Lazard (1966)):

Let M_* be the inverse image of $T^pM \to M$ across the identity section $B \to M$, $x \mapsto 1_x$, of M. The fibre of M_* over x is then the Lie algebra of M_x. Let $\{\psi_i : U_i \times G \to M_{U_i}\}$ be an LGB atlas for M; then the induced chart for T^pM is the composite of $\psi_i^{-1} \times \mathrm{id}\colon M_{U_i} \times \mathfrak{g} \to U_i \times G \times \mathfrak{g}$ followed by $U_i \times G \times \mathfrak{g} \to T(U_i \times G)$, $(x,g,X) \mapsto 0_x \oplus T(R_g)(X)$, followed by $T(\psi_i)\colon T(U_i \times G) \to TM$. The restriction of this chart to the identity section $U_i \subseteq M_{U_i}$ is therefore

$$(x, X) \mapsto T(\psi_i)(0_x \oplus X) = T(\psi_{i,x})(X)$$

and is denoted $\psi_i{}^*\colon U_i \times \mathfrak{g} \to M_*|_{U_i}$; fibrewise $\psi_i{}^*$ is the Lie algebra isomorphism $\mathfrak{g} \to M_*|_x$ induced by $\psi_{i,x}\colon G \to M_x$. Thus M_* with the Lie algebra bracket on each fibre induced from the Lie group structures on the fibres of M, is a Lie algebra bundle.

It is clear that the construction of M_* as an LAB coincides with the construction of the Lie algebroid AM of M considered as a totally intransitive differentiable groupoid. //

Definition 3.18. Let Ω be a Lie groupoid. The LAB associated to $G\Omega$ is denoted $L\Omega$ and is called the adjoint Lie algebra bundle of Ω. //

Proposition 3.19. Let Ω be a Lie groupoid. For $\xi \in \Omega$ define $\mathrm{Ad}(\xi)\colon L\Omega|_{\alpha\xi} \to L\Omega|_{\beta\xi}$ to be $T(I_\xi)_{\alpha\xi}$. Then $\mathrm{Ad}\colon \Omega \to \Pi[L\Omega]$ is a smooth representation of Ω on $L\Omega$, called the adjoint representation.

Proof: The proof is straightforward, using either local triviality to reduce to the case of Lie groups, or 1.23. //

If Ω is abelian then, as in I§3, $Ad(\xi)$ depends only on $\alpha\xi$ and $\beta\xi$ and so there is a global chart $B \times \mathfrak{g} \to L\Omega$, where $\mathfrak{g}$ is the Lie algebra of a vertex group of Ω.

For an arbitrary differentiable groupoid Ω, the adjoint representation can be defined and shown to be smooth as a map $\Omega * A(G\Omega) \to A(G\Omega)$, by using 1.23.

The construction of the Lie algebroid of a differentiable groupoid, as we have presented it here, emphasizes that it is a direct and natural generalization of the concept of Lie algebra of a Lie group. The next result shows that, for a Lie groupoid Ω, the Lie algebroid $L\Omega \rightarrowtail A\Omega \twoheadrightarrow TB$ is the Atiyah sequence of any of the principal bundles associated to Ω. See A§3 for the construction of the Atiyah sequence of a principal bundle.

Proposition 3.20. Let Ω be a Lie groupoid on B, let $b \in B$ and write $P = \Omega_b$, $G = \Omega_b^b$ and $p = \beta_b : P \to B$. Then the restriction of $\mathcal{R}: T^{\alpha}\Omega \to A\Omega$ to $TP \to A\Omega$ induces an isomorphism

$$R: \frac{TP}{G} \to A\Omega$$

of Lie algebroids over B. The induced morphism

$$R^{+}: \frac{P \times \mathfrak{g}}{G} \to L\Omega$$

is $\langle \xi, X\rangle \mapsto Ad(\xi)(X)$.

Proof: Clearly $\mathcal{R}: TP \to A\Omega$, $p: P \to B$ is constant on each orbit of the action of G so by A 2.2 it quotients to a vector bundle map $R: \frac{TP}{G} \to A\Omega$ over B, given by $R(\langle X_\xi\rangle) = T(R_{\xi^{-1}})(X)$. Since $\mathcal{R}$ is fibrewise bijective, it follows that R is also, and it is therefore a vector bundle isomorphism. That $q \circ R = p_*$ is a straightforward consequence of $\beta \circ R_{\xi^{-1}} = \beta$, $\forall \xi \in \Omega$; it remains to verify that R is bracket-preserving.

Take $X, Y \in \Gamma(\frac{TP}{G})$; it suffices to verify that $[\overrightarrow{R(X)}, \overrightarrow{R(Y)}]$ and $\overrightarrow{R[X,Y]}$ are equal on P; that they are equal on Ω then follows from the transitivity of Ω. Now $[\overrightarrow{R(X)}, \overrightarrow{R(Y)}]\big|_P = [\overrightarrow{R(X)}\big|_P, \overrightarrow{R(Y)}\big|_P]$, because $\overrightarrow{R(X)}\big|_P$ and $\overrightarrow{R(Y)}\big|_P$ are tangent to P, and $[\overrightarrow{R(X)}\big|_P, \overrightarrow{R(Y)}\big|_P] = [\overline{X}, \overline{Y}] = \overline{[X,Y]} = \overrightarrow{R[X,Y]}\big|_P$.

Lastly, the map $\frac{P \times \mathfrak{g}}{G} \rightarrowtail \frac{TP}{G}$, $\langle \xi, X\rangle \mapsto \langle T(m_\xi)(X)\rangle$ of A 3.2 can be rewritten as $\langle \xi, X\rangle \mapsto \langle T(L_\xi)(X)\rangle$ since $m_\xi: G \to P$ is precisely the restriction of

$L_\xi: \Omega^b \to \Omega^{\beta\xi}$ to $G = \Omega_b^b \to \Omega_b^{\beta\xi} \subseteq P$. Consequently, $R^+(\langle\xi,X\rangle) = T(R_{\xi^{-1}} L_\xi)(X) = \mathrm{Ad}(\xi)(X)$. //

We now give some simple examples of the construction of the Lie algebroid of a differentiable groupoid. Further examples are developed in the following sections.

Example 3.21. Let B be a manifold, G a Lie group, and Ω the trivial Lie groupoid $B \times G \times B$.

α is the projection $\pi_3: B \times G \times B \to B$, so $T^\alpha\Omega$ is the Whitney sum $\pi_1{}^*TB \oplus \pi_2{}^*TG$ of the inverse image bundles $\pi_1{}^*TB$ and $\pi_2{}^*TG$ over $B \times G \times B$. Since the compositions $\pi_1\circ\varepsilon$ and $\pi_2\circ\varepsilon$ are $x \mapsto x$ and $x \mapsto 1$ respectively, the inverse image of $T^\alpha\Omega$ over ε is $(\pi_1\circ\varepsilon)^*TB \oplus (\pi_2\circ\varepsilon)^*TG = TB \oplus (B \times \mathfrak{g})$, where the Whitney sum is now over B. This is the vector bundle $A\Omega$.

We will write a general vector field on Ω in the form $X \oplus V \oplus Y$ where X, V, Y are sections of $\pi_1{}^*TB$, $\pi_2{}^*TG$, $\pi_3{}^*TB$, respectively. Such a field is vertical iff $Y = 0$. The right-translation $R_{(y,h,z)}: B \times G \times \{y\} \to B \times G \times \{z\}$ has tangent

$$T(R_{(y,h,z)})_{(x,g,y)}: X_x \oplus V_g \oplus 0_y \mapsto X_x \oplus T(R_h)(V_g) \oplus 0_z$$

and so $X \oplus V \oplus 0$ is right-invariant iff $V(x,gh,z) = T(R_h)_g(V(x,g,y))$ and $X(x,gh,z) = X(x,g,y)$ identically in x, y, z, g, h. This is so iff $V(x,g,y)$ is independent of y and right-invariant in g and $X(x,g,y)$ depends only on x. When this is so, V can be identified with the function $x \mapsto V(x,1,x)$, $B \to \mathfrak{g}$, also denoted by V, and X can be identified with the vector field $x \mapsto X(x,1,x)$ on B, also denoted by X; with this notation $X \oplus V$ is the section $X \oplus V \oplus 0|_B$ of $TB \oplus (B \times \mathfrak{g})$. Conversely, given $X \in \Gamma TB$ and $V: B \to \mathfrak{g}$, the right-invariant vector field $\overrightarrow{X \oplus V}$ is $(x,g,y) \mapsto X_x \oplus T(R_g)(V(x)) \oplus 0_y$.

To simplify the calculation of the bracket on $A\Omega$, temporarily denote by $\vec{V}$ and $\vec{X}$ the right-invariant vector fields $\overrightarrow{0 \oplus V}$ and $\overrightarrow{X \oplus 0}$. With this notation, for any $V,W: B \to \mathfrak{g}$ and $X,Y \in \Gamma TB$,

$$[\overrightarrow{X \oplus V}, \overrightarrow{Y \oplus W}] = [\vec{X} + \vec{V}, \vec{Y} + \vec{W}]$$

$$= [\vec{X},\vec{Y}] + [\vec{X},\vec{W}] - [\vec{Y},\vec{V}] + [\vec{V},\vec{W}].$$

It is easy to verify that $\vec{V}$ has the global flow $\psi_t(y,g,x) = (y, \exp tV(y)g, x)$ and that if $\{\phi_t\}$ is a local flow for X on $U \subseteq B$ then $\{\vec{\phi}_t\}$ defined by $\vec{\phi}_t(y,g,x) = (\phi_t(y),g,x)$

is a local flow for $\vec{X}$ on $U \times G \times B$. Using these flows it is straightforward to show that $[\vec{V},\vec{W}] = \overrightarrow{[V,W]}$, where $[V,W]$ is the map $x \mapsto [V(x),W(x)]$, $B \to \mathfrak{g}$; that $[\vec{X},\vec{W}] = \overrightarrow{X(W)}$, where $X(W)$ is the Lie derivative of the vector-valued function W: and that $[\vec{X},\vec{Y}] = \overrightarrow{[X,Y]}$. Consequently the bracket on $TB \oplus (B \times \mathfrak{g})$ is given by

$$[X \oplus V,\ Y \oplus W] = [X,Y] \oplus \{X(W) - Y(V) + [V,W]\}.$$

The anchor $q\colon TB \oplus (B \times \mathfrak{g}) \to TB$ is clearly the projection $X \oplus V \mapsto X$ and the adjoint bundle is, by the formula for the bracket in $TB \oplus (B \times \mathfrak{g})$, the trivial Lie algebra bundle $B \times \mathfrak{g}$. Thus $A(B \times G \times B)$ is the trivial Lie algebroid on B with structure algebra $\mathfrak{g}$ constructed in 2.4.

Now consider a morphism $\phi\colon B \times G \times B \to B \times H \times B$ of trivial Lie groupoids over B. By I 2.13, ϕ can be written in the form

$$\phi(y,g,x) = (y,\ \theta(y)f(g)\theta(x)^{-1},\ x)$$

for some morphism of Lie groups $f\colon G \to H$ and smooth map $\theta\colon B \to H$. In terms of this description $T(\phi)(X \oplus V \oplus Y)$, where $X \in T(B)_x$, $V \in T(G)_g$, $Y \in T(B)_y$, is given by

$$X \oplus \{T(R_{f(g)\theta(y)^{-1}})T(\theta)(X) + T(L_{\theta(x)})T(R_{\theta(y)^{-1}})T(f)(V)$$

$$- T(L_{\theta(x)f(g)\theta(y)^{-1}})T(R_{\theta(y)^{-1}})T(\theta)(Y)\} \oplus Y.$$

Setting $Y = 0$ and $y = x$, $g = 1$, this reduces to $X \oplus \{\Delta(\theta)(X) + Ad(\theta(x))f_*(V)\} \oplus 0$, where Δ is the right-derivative (see B§2). Hence $\phi_*\colon TB \oplus (B \times \mathfrak{g}) \to TB \oplus (B \times \mathfrak{h})$ is

$$\phi_*(X \oplus V) = X \oplus \{\Delta(\theta)(X) + Ad(\theta)f_*(V)\}.$$

In terms of the description (3) in 2.4, ϕ_* is formed from the Maurer-Cartan form $\Delta(\theta)$ and the LAB morphism $Ad(\theta)\circ f_*$. The compatibility condition for $\Delta(\theta)$ and $Ad(\theta)\circ f_*$ is proved directly in B 2.1. //

Example 3.22. Let $m\colon G \times B \to B$ be a smooth action of a Lie group G on a manifold B, and Ω the corresponding action groupoid (see I 1.6). Then, as in the preceding example, $T^{\alpha}\Omega = \pi_1{}^*TG \oplus 0$ and $A\Omega = B \times \mathfrak{g}$ as a vector bundle on B. A vertical vector field $V\colon G \times B \to TG$ is right-invariant iff $V(gh,y) = T(R_h)(V(g,hy))$ identically in g, h, y. For $V\colon B \to \mathfrak{g}$ the corresponding right-invariant vector field $\vec{V}$ is $\vec{V}(g,x) = T(R_g)(V(gx))$. The anchor $q\colon B \times \mathfrak{g} \to TB$ is $(x,X) \mapsto T(m(-,x))_1(X)$.

It would be interesting to have a simple formula to describe the bracket on $\Gamma A\Omega$ in terms of the isomorphism $A\Omega = B \times \mathfrak{g}$: it is certainly not the case that the Lie algebroid bracket on $\Gamma(B \times \mathfrak{g})$ is the pointwise bracket of $\mathfrak{g}$-valued maps, since if the action is transitive and abelian, the pointwise bracket on $\Gamma(B \times \mathfrak{g})$ is identically zero and the anchor surjective, but the bracket on ΓTB will only be zero in degenerate cases. //

<u>Example 3.23.</u> Let G be a Lie group and H a closed subgroup of G. The Atiyah sequence of the principal bundle G(G/H,H) is calculated in A 3.9 and, by 3.20, is the Lie algebroid of the Lie groupoid $\frac{G \times G}{H}$ (see II 1.12).

By II 1.12, the Lie groupoid $\frac{G \times G}{H}$ is isomorphic to the Lie groupoid $G \times (G/H)$ of 3.22, where G acts on G/H in the standard fashion; the isomorphism is $\langle g_2, g_1\rangle \mapsto (g_2 g_1^{-1}, g_1 H)$. The induced isomorphism of Lie algebroids $\frac{TG}{H} \to (G/H) \times \mathfrak{g}$ is described in A 3.9. //

The following result is a straightforward exercise.

<u>Proposition 3.24.</u> Let Ω and Ω' be differentiable groupoids on B and let Ω be Lie. Denote by π and π' the projections $\Omega \underset{B\times B}{\times} \Omega' \to \Omega$ and $\Omega \underset{B\times B}{\times} \Omega' \to \Omega'$. Then $A(\Omega \underset{B\times B}{\times} \Omega')$ is naturally isomorphic to $A\Omega \underset{TB}{\oplus} A\Omega'$ under $X \mapsto \pi_* X \oplus \pi'_* X$. //

* * * * * * * *

We insert here some remarks about differential forms on differentiable groupoids which will be needed in later sections. Until 3.28, let Ω be a differentiable groupoid on B.

<u>Definition 3.25.</u> (i) Let $\mathcal{E}$ be a vector bundle on Ω. Then $\Gamma\mathrm{Alt}^n(T^\alpha\Omega;\mathcal{E})$ is denoted $A^n(\Omega,\mathcal{E})$ and elements of $A^n(\Omega,\mathcal{E})$ are called <u>fibred</u> <u>n-forms</u> on Ω with values in $\mathcal{E}$.

(ii) Let E be a vector bundle on B; recall that β^*E is the inverse image of E over $\beta\colon \Omega \to B$. A fibred n-form $\omega \in A^n(\Omega,\beta^*E)$ is called <u>right-invariant</u> if

$$\omega(T(R_\xi)_\eta(V_1),\dots,T(R_\xi)_\eta(V_n)) = \omega(V_1,\dots,V_n)$$

for all $V_i \in T^\alpha(\Omega)_\eta$ and $\xi \in \Omega$ such that $\eta\xi$ is defined.

The subspace of right-invariant fibred n-forms is denoted $A^n_{RI}(\Omega,\beta^*E)$. //

Whereas $A^n(\Omega,\beta^*E)$ is a $C(\Omega)$-module, $A^n_{RI}(\Omega,\beta^*E)$ is only a $C(B)$-module, with

$u \in C(B)$ acting by $u \circ \beta \in C(\Omega)$.

Definition 3.26. Let E be a vector bundle on B. Then the vector bundle $\mathrm{Alt}^n(A\Omega;E)$ is denoted $C^n(A\Omega,E)$ and elements of $\Gamma C^n(A\Omega,E)$ are called n-cochains on $A\Omega$ with values in E. //

Proposition 3.27. There is an isomorphism of C(B)-modules

$$\Gamma C^n(A\Omega,E) \xrightarrow{\cong} A^n_{RI}(\Omega,\beta^*E)$$

which to $\omega \in \Gamma C^n(A\Omega,E)$ assigns $\vec{\omega}$ defined by

$$\vec{\omega}(V_1,\ldots,V_n) = \omega_{\beta\xi}\big(T(R_{\xi^{-1}})(V_1),\ldots,T(R_{\xi^{-1}})(V_n)\big)$$

for $V_1,\ldots,V_n \in T^\alpha\Omega|_\xi$, $\xi \in \Omega$, and which to $\omega \in A^n_{RI}(\Omega,\beta^*E)$ assigns $\omega\!\downarrow_B$ defined by

$$\omega\!\downarrow_B(X_1,\ldots,X_n) = \omega_{\tilde{x}}(X_1,\ldots,X_n)$$

for $X_1,\ldots,X_n \in A\Omega|_x = T^\alpha\Omega|_{\tilde{x}}$.

Proof: Straightforward. //

Taking $E = B \times \mathbb{R}$ we obtain the vector bundle dual $C^1(A\Omega, B \times \mathbb{R}) = \mathrm{Hom}(A\Omega, B \times \mathbb{R}) = A\Omega^*$. By 3.27, the module of global sections of this dual can be identified with the C(B)-module of real-valued right-invariant forms $A^1_{RI}(\Omega, \Omega \times \mathbb{R})$. Given $\omega \in \Gamma C^1(A\Omega, B \times \mathbb{R})$ and $X \in \Gamma A\Omega$ one can form $\langle X,\omega\rangle = \omega(X)\colon B \to \mathbb{R}$ and $\langle\vec{X},\vec{\omega}\rangle = \vec{\omega}(\vec{X})\colon \Omega \to \mathbb{R}$ and $\vec{\omega}(\vec{X})(\xi) = \vec{\omega}_\xi(T(R_\xi)(X(\beta\xi))) = \omega_{\beta\xi}(X(\beta\xi))$, for all $\xi \in \Omega$, so $\vec{\omega}(\vec{X}) = \omega(X)\circ\beta$. One could call the elements of $A^1_{RI}(\Omega, \Omega \times \mathbb{R})$ the Maurer-Cartan forms of Ω.

Taking E to be the vector bundle $A\Omega$ itself, let $\theta \in \Gamma C^1(A\Omega,A\Omega)$ be the identity map $A\Omega \to A\Omega$. Then $\vec{\theta} \in A^1_{RI}(\Omega,\beta^*A\Omega)$ maps $V \in T^\alpha\Omega|_\xi$ to $T(R_{\xi^{-1}})(V)$ and is essentially the map $\mathcal{R}$ of 3.3. One could call $\vec{\theta}$ the Maurer-Cartan form of Ω.

If now (M,p,B) is a fibred manifold and $f\colon M \to \Omega$ is a smooth map with $\alpha\circ f = p$, one can define a right derivative

$$\begin{array}{ccc} T^P M & \xrightarrow{\Delta(f)} & \beta^*A\Omega \\ \downarrow & & \downarrow \\ M & \xrightarrow{f} & \Omega \end{array}$$

by $\Delta(f)_u = T(R_{f(u)^{-1}})\,T(f_x)_u$, where $u \in M_x$, $x \in B$. Then $\Delta(f)$ is, in an obvious extension of 3.25, a fibred form on M, and $\Delta(id_\Omega) = \vec{\theta}$. We will not pursue this concept here.

Pradines (1967) obtains $\vec{\theta}$ as $A(\delta)$, where δ is the division map $\Omega \underset{\alpha}{\times} \Omega \to \Omega$, $(\eta,\xi) \mapsto \eta\xi^{-1}$, regarded as a morphism of differentiable groupoids over $\beta\colon \Omega \to B$ (see I 2.14).

__Proposition 3.28.__ Let Ω be a Lie groupoid on B. Choose $b \in B$ and write $P = \Omega_b$, $G = \Omega_b^b$. Let V be a vector space and $G \times V \to V$ a linear action of G on V; let $E = \frac{P \times V}{G}$ be the associated vector bundle.

Then there are natural isomorphisms of C(B)-modules

$$\Gamma C^n(A\Omega,E) \xrightarrow{\cong} A^n(P,V)^G$$

where $A^n(P,V)^G = \{\omega \in A^n(P,V) \mid (R_g)_*(\omega) = g^{-1}\omega,\ \forall g \in G\}$ is the C(B)-module of G-equivariant V-valued forms on P.

__Proof:__ See A 4.12(i). //

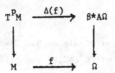

§4. The exponential map and adjoint formulas.

The exponential map with which this section deals associates to a local section X of the Lie algebroid of a differentiable groupoid Ω, a 1-parameter family ExptX of local admissible sections of Ω. Namely, the right-invariant vector field $\vec{X}$ has flows consisting of local left-translations, and the local admissible sections corresponding to these left-translations are the ExptX. There is a second concept of exponential map, which maps $A\Omega$ into Ω itself and is étale on a neighbourhood of the zero section of $A\Omega$; it however depends on a choice of connection in the vector bundle $A\Omega$. For this see Pradines (1968a) or Almeida (1980, §9).

The exponential map for a Lie groupoid is used in the same way, and for the same purposes, as the exponential map of a Lie group. The first major use (4.5) is to identify the Lie algebroid $A\Pi(E)$, for a vector bundle E, with CDO(E). This result is from Kumpera (1971), though the proof has been simplified. Using 4.5, we calculate in 4.7 the Lie algebroid of an isotropy subgroupoid; this result is central in the connection theory of §7. In 4.8 we use the exponential to differentiate the standard representations of the frame groupoid $\Pi(E)$ of a vector bundle E on the associated bundles $Hom^n(E; B \times \mathbb{R})$, $Hom^n(E;E)$, etc. From 4.7 and 4.8 it follows that, for a vector bundle E and a geometric structure on E defined by tensor fields, the Lie algebroid of the frame groupoid consists of those covariant differential operators with respect to which the tensor fields are constant (or parallel). This result encapsulates a number of calculations usually regarded as part of connection theory. Here we need only the cases of Riemannian bundles and Lie algebra bundles.

In the remaining part of the section we give formulas for the adjoint representations of a differentiable groupoid and its Lie algebroid which generalize the well-known formulas for Lie groups and Lie algebras.

The definition and basic properties of the exponential, the fundamental theorem 4.5, and the adjoint formulas in 4.11 are due to Kumpera (1971). (Except for 4.5, this material also appears in Kumpera and Spencer (1972, Appendix).) The remainder of the section is due to the author.

Let Ω be a differentiable groupoid on B and take $X \in \Gamma A\Omega$. Let $\{\phi_t : \mathcal{U} \to \mathcal{U}_t\}$ be a local flow for $\vec{X} \in \Gamma T\Omega$. Since $\vec{X}$ is α-vertical, we have $\alpha \circ \phi_t = \alpha$ so for $x \in B$ with $\mathcal{U} \cap \Omega_x \neq \emptyset$, each ϕ_t restricts to $\mathcal{U} \cap \Omega_x \to \mathcal{U}_t \cap \Omega_x$. For $\xi \in \Omega_x^y$ where $\mathcal{U} \cap \Omega_x \neq \emptyset$, $\mathcal{U} \cap \Omega_y \neq \emptyset$, we know $(R_\xi)_*(\vec{X}|_{\Omega_y}) = \vec{X}|_{\Omega_x}$ so $\phi_t \circ R_\xi = R_\xi \circ \phi_t$ for all t.

Now define $U = \beta(\mathcal{U})$, $U_t = \beta(\mathcal{U}_t)$ and $\psi_t\colon U \to U_t$ so that

$$\begin{array}{ccc} \mathcal{U} & \xrightarrow{\ \phi_t\ } & \mathcal{U}_t \\ \beta\downarrow & & \downarrow\beta \\ U & \xrightarrow[\ \psi_t\ ]{} & U_t \end{array}$$

commutes; that is, $\psi_t(x) = \beta(\phi_t(\eta))$ for any $\eta \in \mathcal{U}$ with $\beta\eta = x$. Since $\phi_t(\eta\xi) = \phi_t(\eta)\xi$, this is well-defined, and since β is a submersion ψ_t is smooth. Since $(\phi_t)^{-1} = \phi_{-t}$ has the same properties as ϕ_t, it follows that $(\psi_{-t} \circ \psi_t)\circ\beta = \beta$ and ψ_t is therefore a local diffeomorphism. Now, for $x \in U$,

$$\frac{d}{dt}\psi_t(x)\Big|_0 = \frac{d}{dt}\beta(\phi_t(\eta))\Big|_0 = \beta_*(\vec{X}(\eta))$$

for any $\eta \in \mathcal{U} \cap \Omega^x$, and so $\{\psi_t\colon U \to U_t\}$ is a local flow for $q(X) = \beta_*(\vec{X})$.

Lastly, $\phi_t\colon \mathcal{U} \to \mathcal{U}_t$ and $\psi_t\colon U \to U_t$ satisfy the conditions of II 5.8 and so each ϕ_t is the restriction of a unique local left-translation $L_{\sigma_t}\colon \Omega^U \to \Omega^{U_t}$ where $\sigma_t \in \Gamma_U\Omega$ is defined by $\sigma_t(x) = \phi_t(\eta)\eta^{-1}$, where η is any element of $\mathcal{U} \cap \Omega^x$. This proves the following result.

<u>Proposition 4.1.</u> Let Ω be a differentiable groupoid on B, let $W \subseteq B$ be an open subset, and take $X \in \Gamma_W A\Omega$. Then for each $x_0 \in W$ there is an open neighbourhood U of x_0 in W, called a <u>flow neighbourhood</u> for X, an $\varepsilon > 0$, and a unique smooth family of local admissible sections $\mathrm{Exp}tX \in \Gamma_U\Omega$, $|t| < \varepsilon$, such that;

(i) $\dfrac{d}{dt}\mathrm{Exp}tX\Big|_0 = X$,

(ii) $\mathrm{Exp}0X = id \in \Gamma_U\Omega$,

(iii) $\mathrm{Exp}(t + s)X = (\mathrm{Exp}tX)*(\mathrm{Exp}sX)$, whenever $|t|, |s|, |t + s| < \varepsilon$,

(iv) $\mathrm{Exp}{-t}X = (\mathrm{Exp}tX)^{-1}$,

(v) $\{\beta\circ\mathrm{Exp}tX\colon U \to U_t\}$ is a local 1-parameter group of transformations for $q(X) \in \Gamma_W TB$ in U. //

(iii) may be expressed more fully as follows: if $x \in U$, if $|s|, |t|, |s + t| < \varepsilon$ and if $\beta(\mathrm{Exp}sX(x)) \in U$ (which is true for all $|s| < \delta$ for some $\delta > 0$), then $(\mathrm{Exp}(t + s)X)(x) = ((\mathrm{Exp}tX)(\beta(\mathrm{Exp}sX(x)))\mathrm{Exp}sX(x)$.

The family $t \mapsto \mathrm{Exp}tX$ is smooth in the sense that $\mathbb{R} \times U \leadsto \Omega$, $(t,x) \mapsto \mathrm{Exp}tX(x)$ is smooth; this follows from the smoothness of the local flow $\mathbb{R} \times \Omega \leadsto \Omega$, $(t,\xi) \mapsto \phi_t(\xi) = \mathrm{Exp}tX(\beta\xi)\xi$ for $\vec{X}$.

A well-defined exponential map $X \mapsto \mathrm{Exp}X$ may be defined on the sheaf of germs of local sections of $A\Omega$ with values in the sheaf of germs of local admissible sections of Ω. We will use the term "exponential map" in the obvious loose sense.

The following result is proved in Kumpera and Spencer (1972, Appendix).

Theorem 4.2. Let Ω be a differentiable groupoid on B, let $X \in \Gamma A\Omega$, and let $\xi_0 \in \Omega$ with $\beta\xi_0 = y_0$. Then the integral curve for $\vec{X}$ through ξ_0 is infinitely extendable in both directions iff the integral curve for $q(X)$ through y_0 is infinitely extendable in both directions. In particular, $\vec{X}$ is complete iff $q(X)$ is complete. //

Proposition 4.3. Let $\phi: \Omega \to \Omega'$ be a morphism of differentiable groupoids over B. Then if U is a flow neighbourhood for $X \in \Gamma A\Omega$, it is also a flow neighbourhood for $\phi_*(X)$ and

$$\tilde{\phi}(\mathrm{Exp}tX) = \mathrm{Exp}t\phi_*(X)$$

for all t for which $\mathrm{Exp}tX$ is defined.

Proof: It is easy to verify that $t \mapsto \tilde{\phi}(\mathrm{Exp}tX)$ has the properties which characterize $t \mapsto \mathrm{Exp}t\phi_*(X)$. //

Examples 4.4. Let $\Omega = B \times B$ and let $X \in \Gamma A\Omega = \Gamma TB$ have a local flow $\{\phi_t : U \to U_t\}$. Then $\{\phi_t \times \mathrm{id}_B : U \times B \to U_t \times B\}$ is a local flow for $\vec{X} = X \oplus 0$ and $\mathrm{Exp}tX$ can be identified with $\phi_t \in \Gamma_U(B \times B)$ (see II 5.3).

Let M be an LGB on B, and let $X \in \Gamma(M_*)$. Then $\mathrm{Exp}tX$ is the global section $x \mapsto \mathrm{exp}tX(x)$ of M whose value at $x \in B$ is the Lie algebra exponential of $X(x) \in M_*|_x$.

Let $P(B,G,\pi)$ be a principal bundle and let $X \in \Gamma(\frac{TP}{G})$. It is shown in A 3.7 that the local flows of $\bar{X} \in \Gamma^G TP$ are of the form $\phi_t(\psi_t, id_G)$, where $\{\psi_t: U \to U_t\}$ is a local flow for $\pi_*(X)$ and $\{\phi_t: \pi^{-1}(U) \to \pi^{-1}(U_t)\}$ is defined on π-saturated open sets. It is easy to see that if X is regarded as in $\Gamma A(\frac{P \times P}{G})$ (see 3.20) then $\phi_t(\psi_t, id_G)$ corresponds to ExptX under II 5.6. //

Theorem 4.5. Let E be a vector bundle on B. For $X \in \Gamma A\Pi(E)$ define $\mathcal{D}(X); \Gamma E \to \Gamma E$ by

$$\mathcal{D}(X)(\mu)(x) = -\frac{d}{dx}\overline{\mathrm{Expt}X(\mu)(x)}\Big|_0$$

where $\mu \in \Gamma E$, $x \in B$, and the exponential is taken in a flow neighbourhood of x. (The bar notation is defined in II 5.4.)

Then $\mathcal{D}(X) \in \Gamma CDO(E)$, and $\mathcal{D}: \Gamma A\Pi(E) \to \Gamma CDO(E)$ defines an isomorphism $A\Pi(E) \to CDO(E)$ of Lie algebroids over B.

Proof: Choose $b \in B$, write $P = \Pi(E)_b$ and $V = E_b$ and for $\mu \in \Gamma E$ define $\tilde{\mu}: P \to V$ by $\tilde{\mu}(\xi) = \xi^{-1}\mu(\beta\xi)$. For $X \in \Gamma A\Pi(E)$, let $\bar{X}$ denote the restriction of $\vec{X} \in \Gamma^{RI}T^{\alpha}\Pi(E)$ to P. Then it is straightforward to verify that

$$\widetilde{\mathcal{D}(X)(\mu)} = \bar{X}(\tilde{\mu}).$$

Now for $f: B \to \mathbb{R}$ it follows that

$$\widetilde{\mathcal{D}(X)(f\mu)} = \bar{X}(f\circ\beta_b)(\tilde{\mu})$$

$$= (f\circ\beta_b)\bar{X}(\tilde{\mu}) + \bar{X}(f\circ\beta_b)\tilde{\mu}$$

and hence that $\mathcal{D}(X)(f\mu) = f\mathcal{D}(X)(\mu) + q(X)(f)\mu$. This shows that $\mathcal{D}(X)$ is a first or zeroth order differential operator and, further, an element of $\Gamma CDO(E)$.

Similarly it is easy to verify that

$$\mathcal{D}(X + Y) = \mathcal{D}(X) + \mathcal{D}(Y), \qquad X,Y \in \Gamma A\Pi(E),$$

and

$$\mathcal{D}(fX) = f\mathcal{D}(X), \qquad f: B \to \mathbb{R},$$

and so $\mathcal{D}$ induces a morphism $\mathcal{D}: A\Pi(E) \to CDO(E)$ of vector bundles over B. It follows from what we have already done that $\mathcal{D}$ respects the anchors on $A\Pi(E)$ and $CDO(E)$. As for the bracket condition, for $X,Y \in \Gamma A\Pi(E)$ and $\mu \in \Gamma E$,

$$\widetilde{\mathcal{D}([X,Y])(\mu)} = \overline{[X,Y]}(\tilde{\mu})$$

$$= [\bar{X},\bar{Y}](\tilde{\mu})$$

$$= \bar{X}(\bar{Y}(\tilde{\mu})) - \bar{Y}(\bar{X}(\tilde{\mu}))$$

$$= \widetilde{\mathcal{D}(X)(\mathcal{D}(Y)(\mu)) - \mathcal{D}(Y)(\mathcal{D}(X)(\mu))}$$

and so $\mathcal{D}$ is a morphism of Lie algebroids over B.

Lastly, $\mathcal{D}^+$: $L\Pi(E) \to End(E)$ is fibrewise the realization of $T(GL(E_x))_{id}$ as $\mathfrak{gl}(E_x)$, $x \in B$, (see B§1), and is therefore a vector bundle isomorphism. By 2.8 it follows that $\mathcal{D}$ is an isomorphism. //

A proof of 4.5 first appeared in Kumpera (1971). The proof given here is a simplification of Kumpera's. A similar construction occurs in Kobayashi and Nomizu (1963, p. 115).

In what follows we will identify $A\Pi(E)$ with CDO(E) via this isomorphism without comment.

Definition 4.6. Let $\rho: \Omega \to \Pi(E)$ be a representation of a differentiable groupoid Ω on a vector bundle E. Then the induced representation ρ_* of $A\Omega$ on E is

$$\rho_*(X)(\mu) = -\frac{d}{dt}\,\bar{\rho}(\text{Exp}tX)(\mu)\Big|_0,$$

$X \in \Gamma A\Omega$, $\mu \in \Gamma E$. //

The notation $\bar{\rho}$ is introduced at the end of II§5. The next result generalizes a simple formula for Lie groups and Lie algebras. In the present generality it is essentially a part of connection theory.

Theorem 4.7. Let $\rho: \Omega \to \Pi(E)$ be a representation of a Lie groupoid Ω on a vector bundle E. Let $\mu \in \Gamma E$ be Ω-deformable and let Φ be the isotropy groupoid of μ. Then

$$\Gamma A\Phi = \{X \in \Gamma A\Omega \mid \rho_*(X)(\mu) = 0\}.$$

Proof: If $X \in \Gamma A\Phi$ then each ExptX takes values in Φ. Hence $\bar{\rho}(\text{Exp}tX)(\mu) = \mu$ for all t and $\rho_*(X)(\mu) = 0$ follows.

Conversely, take $X \in \Gamma A\Omega$ and suppose $\rho_*(X)(\mu) = 0$. Since $X = \frac{d}{dt}\mathrm{Exp}tX\big|_0$, and Φ is an embedded submanifold of Ω, it suffices to show that each ExptX takes values in Φ. Take $x \in B$ and ExptX defined in a neighbourhood of x. Consider the curve

$$c(t) = \rho((\mathrm{Exp}tX(x))^{-1})\mu((\beta\circ\mathrm{Exp}tX)(x))$$

in E_x; that is,

$$c(t) = \bar{\rho}(\mathrm{Exp}{-tX})(\mu)(x).$$

Now, fixing t_o, we have

$$\frac{d}{dt}c(t_o + t)\Big|_{t=0} = \frac{d}{dt}\bar{\rho}(\mathrm{Exp}{-(t_o + t)X})(\mu)(x)\Big|_0$$

$$= \bar{\rho}(\mathrm{Exp}{-t_oX})\left(\frac{d}{dt}\bar{\rho}(\mathrm{Exp}{-tX})(\mu)\Big|_0\right)(x)$$

$$= \bar{\rho}(\mathrm{Exp}{-t_oX})(\rho_*(X)(\mu))(x)$$

$$= 0$$

so $c(t_o) = 0$ for all t_o and hence c is constant at $c(0)$. Therefore

$$\rho(\mathrm{Exp}tX(x)^{-1})\mu((\beta\circ\mathrm{Exp}tX)(x)) = \mu(x),$$

which shows that $\mathrm{Exp}tX(x) \in \Phi$, as required. //

The proof of 4.7 of course relies on the fact that Φ is already known to be a differentiable (in fact, a Lie) subgroupoid of Ω. In the applications of 4.7 we need the following formulas for induced representations.

<u>Theorem 4.8.</u> Let E be a vector bundle on B.

(i) Let $\Pi(E) * \mathrm{Hom}^n(E; B \times \mathbb{R}) \to \mathrm{Hom}^n(E; B \times \mathbb{R})$ be the action of 1.25(i). Then the induced representation of $CDO(E)$ on $\mathrm{Hom}^n(E; B \times \mathbb{R})$ is given by

$$X(\phi)(\mu_1,\ldots,\mu_n) = q(X)(\phi(\mu_1,\ldots,\mu_n)) - \sum_{i=1}^{n}\phi(\mu_1,\ldots,X(\mu_i),\ldots,\mu_n).$$

(ii) Let $\Pi(E) * \mathrm{Hom}^n(E;E) \to \mathrm{Hom}^n(E;E)$ be the action of 1.25(ii). Then the induced representation of $CDO(E)$ on $\mathrm{Hom}^n(E;E)$ is given by

$$X(\phi)(\mu_1,\ldots,\mu_n) = X(\phi(\mu_1,\ldots,\mu_n)) - \sum_{i=1}^{n} \phi(\mu_1,\ldots,X(\mu_i),\ldots,\mu_n).$$

(iii) Let E' be a second vector bundle on the same base, and let $(\Pi(E) \underset{B\times B}{\times} \Pi(E')) * \text{Hom}(E;E') \to \text{Hom}(E;E')$ be the action of 1.25(iii). Then the induced representation of $CDO(E) \underset{TB}{\oplus} CDO(E')$ on Hom(E;E') is given by $(X \oplus X')(\phi)(\mu) = X'(\phi(\mu)) - \phi(X(\mu))$.

<u>Remark</u>: These formulas generalize results which are well-known in the case of general linear groups of vector spaces. As with 1.25, we take 4.8 to include the restrictions of (i) and (ii) to Alt^n and Sym^n and the corresponding formulas for general tensor bundles, exterior algebra bundles, and symmetric bundles.

<u>Proof</u>: To illustrate the use of the groupoid exponential, we prove (ii) with n = 1. The adaptation of the proof to the other cases follows as in the case of general linear groups of vector spaces.

Take $X \in \Gamma CDO(E)$, $\phi \in \Gamma\text{Hom}(E;E)$, $\mu \in \Gamma E$ and $x \in B$. Let ExptX be defined in a neighbourhood of x and write $x_t = (\beta\circ\text{Expt}X)^{-1}(x)$. Then (all limits are taken as $t \to 0$)

$$X(\phi)(x) = -\frac{d}{dt}\overline{\text{Expt}X}(\phi)(x)\Big|_0$$

$$= -\lim \frac{1}{t}\{\text{Expt}X(x_t)(\phi(x_t)) - \phi(x)\}$$

$$= -\lim \frac{1}{t}\{\text{Expt}X(x_t)\circ\phi(x_t)\circ(\text{Expt}X(x_t))^{-1} - \phi(x)\}$$

so

$$X(\phi)(\mu)(x) = -\lim \frac{1}{t}\{\text{Expt}X(x_t)\circ\phi(x_t)\circ(\text{Expt}X(x_t))^{-1})(\mu(x)) - \phi(x)(\mu(x))\}.$$

On the other side we have

$$X(\phi(\mu))(x) = -\lim \frac{1}{t}\{(\text{Expt}X(x_t)\circ\phi(x_t))(\mu(x_t)) - \phi(x)(\mu(x))\}$$

and

$$\phi(X(\mu))(x) = -\lim \frac{1}{t}\{(\phi(x)\circ\text{Expt}X(x_t))(\mu(x_t)) - \phi(x)(\mu(x))\}.$$

Define a curve $c(t)$ for t near $0 \in \mathbb{R}$ with values in $Hom(E_x;E_x)$ by

$$c(t) = ExptX(x_t) \circ \phi(x_t) \circ (ExptX(x_t))^{-1}$$

and a curve $f(t)$ for t near $0 \in \mathbb{R}$ with values in E_x by

$$f(t) = ExptX(x_t)(\mu(x_t)).$$

Note $c(0) = \phi(x)$ and $f(0) = \mu(x)$. Then the left-hand side of our equation is

$$-\lim \frac{1}{t} \{c(t)f(0) - c(0)f(0)\}$$

and the right-hand side is

$$-\lim \frac{1}{t} \{c(t)f(t) - c(0)f(t)\};$$

both limits being taken in the one vector space E_x. It is elementary that these limits are equal. //

Corollary 4.9. (i) Let E be a vector bundle on B, and $\langle\ ,\ \rangle$ a Riemannian structure in E. Then the Lie algebroid of the Riemannian frame groupoid $\Pi\langle E\rangle$ (see 1.26) is given by

$$\Gamma A\Pi\langle E\rangle = \{X \in \Gamma A\Pi(E) \mid q(X)(\langle u,\nu\rangle) = \langle X(\mu),\nu\rangle + \langle\mu,X(\nu)\rangle,\ \ \forall\mu,\nu \in \Gamma E\}.$$

In particular, the fibres of the adjoint bundle $L\Pi\langle E\rangle$ of $A\Pi\langle E\rangle$ are the Lie algebras $\mathfrak{so}(E_x)$, $x \in B$.

(ii) Let L be a Lie algebra bundle on B. Then the Lie algebroid of the LAB frame groupoid $\Pi[L]$ (see 1.28) is given by

$$\Gamma A\Pi[L] = \{X \in \Gamma A\Pi(L) \mid X([\mu,\nu]) = [X(\mu),\nu] + [\mu,X(\nu)],\ \ \forall\mu,\nu \in \Gamma L\}.$$

In particular the fibres of the adjoint bundle $L\Pi[L]$ of $A\Pi[L]$ are the Lie algebras $Der(L_x)$, $x \in B$. //

4.9 and 1.20 (on which 4.9 strongly depends) show that (i) given a Riemannian vector bundle $(E,\langle\ ,\ \rangle)$ there is a well-defined transitive Lie algebroid, which we will denote

$$End\langle E\rangle \rightarrowtail CDO\langle E\rangle \twoheadrightarrow TB,$$

which is the reduction of $CDO(E)$ characterized by the equation in 4.9(i), and (ii) given an LAB L there is a well-defined transitive Lie algebroid, which we will denote

$$Der(L) \rightarrowtail CDO[L] \twoheadrightarrow TB$$

which is the reduction of CDO(L) characterized by the equation in 4.9(ii). We will in what follows identify AΠ<E> with CDO<E> and AΠ[L] with CDO[L] without comment. It is possible to construct CDO<E> and CDO[L] without use of the underlying groupoids if one grants the existence of Riemannian and Lie connections in E and L respectively - this method is used in 7.22 in another case. It is also possible to construct CDO<E> and CDO[L] from transition forms - this method will be clear after §5 and IV§4. The method used above seems the most natural.

If (E,Σ) is a Σ-bundle in the sense of Greub et al (1973, Chapter VIII) then clearly 4.7 and 4.8 may be used to calculate the Lie algebroid of the Lie groupoid of Σ-preserving isomorphisms. Its adjoint bundle is the Lie algebra bundle which is constructed in Greub et al (op. cit., 8.4).

It would be interesting to know of useful conditions under which, given a representation $\rho: A \to CDO(E)$ of an abstract transitive Lie algebroid A and a section $\mu \in \Gamma E$, there is a reduction A' of A characterized by the condition $X \in \Gamma A' \iff \rho(X)(\mu) = 0$.

We turn now to the adjoint representations of Lie groupoids and Lie algebroids. The first result is an immediate consequence of the Jacobi identity.

Proposition 4.10. Let Ω be a Lie groupoid on B. Then the adjoint representation $ad: A\Omega \to CDO(L\Omega)$ of its Lie algebroid (defined in 2.11) takes values in $CDO[L\Omega]$. //

This is actually true for all transitive Lie algebroids (see IV 1.5) but we have not yet established that the adjoint bundle of a transitive Lie algebroid is an LAB.

We are now concerned to establish that the representation $Ad: \Omega \to \Pi[L\Omega]$ induces $ad: A\Omega \to CDO[L\Omega]$. We in fact prove a stronger version of this result, and for this we need to generalize the construction of induced morphisms.

Let $\phi: \Omega \to \Omega'$ be a morphism of differentiable groupoids over $\phi_o: B \to B'$. As in §3, $\alpha'\circ\phi = \phi_o\circ\alpha$ implies that $T(\phi)$ restricts to $T^{\alpha}\Omega \to T^{\alpha'}\Omega'$ and $\phi\circ\varepsilon = \varepsilon'\circ\phi_o$ (where ε and ε' are the object inclusion maps $B \to \Omega$ and $B' \to \Omega'$) implies that

$$
\begin{array}{ccccc}
A\Omega & \xrightarrow{\subseteq} & T^{\alpha}\Omega & \longrightarrow & T^{\alpha'}\Omega' \\
\downarrow & & \downarrow & & \downarrow \\
B & \xrightarrow{\varepsilon} & \Omega & \xrightarrow{\phi} & \Omega'
\end{array}
\tag{1}
$$

is a vector bundle morphism over $\varepsilon' \circ \phi_o$. Now the composite

$$
\begin{array}{ccccc}
\phi_o^{*}A\Omega' & \longrightarrow & A\Omega' & \longrightarrow & T^{\alpha}\Omega' \\
\downarrow & & \downarrow & & \downarrow \\
B & \xrightarrow{\phi_o} & B' & \xrightarrow{\varepsilon'} & \Omega'
\end{array}
$$

is a pullback because each factor is, and so there is a unique vector bundle morphism $A\Omega \to \phi_o^{*}A\Omega'$ over B such that the composite

$$
\begin{array}{ccccccc}
A\Omega & \longrightarrow & \phi_o^{*}A\Omega' & \longrightarrow & A\Omega' & \longrightarrow & T^{\alpha}\Omega' \\
\downarrow & & \downarrow & & \downarrow & & \downarrow \\
B & = & B & \xrightarrow{\phi_o} & B' & \xrightarrow{\varepsilon'} & \Omega'
\end{array}
\tag{2}
$$

is equal to (1). We denote the composition of the first two squares in (2) by $A(\phi)\colon A\Omega \to A\Omega'$, or by ϕ_*.

It is easy to see that $\beta' \circ \phi = \phi_o \circ \beta$ implies that $q' \circ \phi_* = T(\phi_o) \circ q$. Further, just as in 3.11, it is easy to see that if $X \in \Gamma A\Omega$ and $X' \in \Gamma A\Omega$, then

$$\phi_* \circ X = X' \circ \phi_o \iff \vec{X} \overset{\phi}{\sim} \vec{X}' \tag{3}$$

and from this it follows, also as in §3, that for $X,Y \in \Gamma A\Omega$ and $X',Y' \in \Gamma A\Omega'$,

$$\phi_* \circ X = X' \circ \phi_o \quad \text{and} \quad \phi_* \circ Y = Y' \circ \phi_o \quad \text{imply} \quad \phi_* \circ [X,Y] = [X',Y'] \circ \phi_o. \tag{4}$$

If $\phi_o\colon B \to B'$ is a diffeomorphism, then (4) may be expressed by saying that the map

$$\phi_*\colon \Gamma A\Omega \to \Gamma A\Omega \qquad X \mapsto \phi_* \circ X \circ \phi_o^{-1}$$

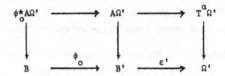

which is semi-linear with respect to

$$C(B) \to C(B') \qquad f \mapsto f\circ\phi_o^{-1},$$

satisfies the equation

$$(5) \qquad \phi_*([X,Y]) = [\phi_*(X),\phi_*(Y)] \qquad X,Y \in \Gamma A\Omega$$

and in this sense we will say that $\phi_*: A\Omega \to A\Omega'$ is the <u>morphism of Lie algebroids</u> over $\phi_o: B \to B'$ <u>induced by</u> $\phi: \Omega \to \Omega'$. It is easy to modify the definition of a morphism of abstract Lie algebroids in 2.1 to include this case.

When $\phi_o: B \to B'$ is allowed to be an arbitrary smooth map a different approach is needed to the definition of a morphism of Lie algebroids $A \to A'$ over $\phi_o: B \to B'$. For this see Almeida and Kumpera (1981). The construction of $\phi_*: A\Omega \to A\Omega'$ from $\phi: \Omega \to \Omega'$ is however as we have given it here, and (4) still essentially expresses the fact that ϕ_* is a morphism of Lie algebroids. We will use only the case in which ϕ_o is a diffeomorphism.

Let Ω be a differentiable groupoid on B, until we reach 4.14. Let $\sigma \in \Gamma_U\Omega$ be a local admissible section with $(\beta\circ\sigma)(U) = V$. Then $I_\sigma: \Omega_U^U \to \Omega_V^V$ is a morphism of differentiable groupoids over $\beta\circ\sigma: U \to V$ and we define

$$Ad(\sigma) = (I_\sigma)_* : A\Omega|_U \to A\Omega|_V.$$

<u>Proposition 4.11.</u> With the notation just introduced

(i) For $X,Y \in \Gamma_U A\Omega$, $Ad(\sigma)[X,Y] = [Ad(\sigma)X,Ad(\sigma)Y]$.

(ii) If $X \in \Gamma_U A\Omega$ and U is a flow neighbourhood for X, then $V = (\beta\circ\sigma)(U)$ is flow neighbourhood for $Ad(\sigma)X$ and, for $|t|$ sufficiently small,

$$\operatorname{Exp}tAd(\sigma)X = \tilde{I}_\sigma(\operatorname{Exp}tX)$$

where $\tilde{I}_\sigma(\operatorname{Exp}tX) \in \Gamma_V\Omega$ is

$$y \mapsto I_\sigma(\operatorname{Exp}tX((\beta\circ\sigma)^{-1}(y))).$$

(iii) If $X,Y \in \Gamma_U A\Omega$ and U is a flow neighbourhood for X, then

$$[X,Y] = -\frac{d}{dt} Ad(\operatorname{Exp}tX)(Y)\Big|_0.$$

Proof: (i) is the equation (5) for $\phi = I_\sigma$.

(ii) is a version of 4.3 for the case in which ϕ_o is a general diffeomorphism, and is easily established.

(iii) can be deduced easily from the formula for $[\vec{X},\vec{Y}]$ as a Lie derivative of $\vec{Y}$ with respect to the flow $L_{\text{Exp}tX}$ of $\vec{X}$, and we urge the reader to work through the details. //

The results 4.11 generalize well-known identities for the relationship between vector fields and their flows. For example, (ii) generalizes the following: If $\{\phi_t\}$ is a local flow for a vector field X on a manifold M and $\phi: M \to M$ is a diffeomorphism, then $\{\phi\circ\phi_t\circ\phi^{-1}\}$ is a local flow for $\phi_*(X)$. In turn, (ii) can be deduced from this result by applying it to the right-invariant vector field $\vec{X}$ corresponding to $X \in \Gamma_U A\Omega$.

Similarly we obtain the following formula for "canonical co-ordinates of the second kind" on a differentiable groupoid Ω.

Proposition 4.12. Let $X_1,\ldots,X_r$ be a local basis for $A\Omega$ on an open set $U \subseteq B$ and suppose that U is a flow neighbourhood for each X_i. Then the map

$$(t_1,\ldots,t_r) \mapsto (\text{Exp}t_1X_1 * \text{Exp}t_2X_2 * \cdots * \text{Exp}t_rX_r)(x),$$

where $x \in U$ is fixed, is a diffeomorphism of an open neighbourhood of $\underset{\sim}{0} \in \mathbb{R}^r$ onto an open neighbourhood of $\tilde{x}$ in Ω_x.

Proof: This follows immediately from the result: If $X_1,\ldots,X_r$ are linearly independent vector fields on an open subset U of a manifold M, and if for each i, $\{\phi^i_t\}$ is a local flow for X_i on U, then the map

$$(t_1,\ldots,t_r) \mapsto (\phi^1_{t_1}\circ\phi^2_{t_2}\circ\cdots\circ\phi^r_{t_r})(x),$$

where $x \in U$ is fixed, is a diffeomorphism of an open neighbourhood of $\underset{\sim}{0} \in \mathbb{R}^r$ onto an open neighbourhood of x in M. //

If U itself is the domain of a chart for B, we obtain coordinates $\mathbb{R}^n \times \mathbb{R}^r \cong U \times \mathbb{R}^r \rightsquigarrow \Omega_U$.

Remark: Compared to the corresponding result for Lie groups, the proof of this result is immediate. However the purpose of the more delicate analysis in the case of Lie groups is to prove the existence of analytic co-ordinates; such coordinates need not exist for differentiable (or Lie) groupoids.

Proposition 4.13. Let $X,Y \in \Gamma_U A\Omega$ where U is a flow neighbourhood for both X and Y. Then

(i) If $[X,Y] = 0$, then $\mathrm{Exp}t(X + Y) = \mathrm{Exp}tX * \mathrm{Exp}tY$ on U;

(ii) $\frac{d}{dt}(\mathrm{Exp}{-\sqrt{t}Y} * \mathrm{Exp}{-\sqrt{t}X} * \mathrm{Exp}\sqrt{t}Y * \mathrm{Exp}\sqrt{t}X)\big|_0 = [X,Y]$ on U.

Proof: (i) From $[X,Y] = 0$ it follows that $[\vec{X},\vec{Y}] = 0$. Hence the local flows $\phi_t(\xi) = \mathrm{Exp}tX(\beta\xi)\xi$ and $\psi_t(\xi) = \mathrm{Exp}tY(\beta\xi)\xi$ commute. It is now easy to check that $\theta_t = \phi_t \circ \psi_t$ is a local 1-parameter group of local transformations, and that $\frac{d}{dt}\theta_t(\xi)\big|_0 = \vec{X}(\xi) + \vec{Y}(\xi)$. Lastly, θ_t, being a composition of left-translations, is itself a left-translation and corresponds to $\mathrm{Exp}tX * \mathrm{Exp}tY$.

(ii) follows, in the same way as does (i), from the corresponding result for general vector fields (see, for example, Spivak (1979, I, pp. 220 e.s.)). //

From 4.11(iii) the following result is immediate.

Proposition 4.14. Let Ω be a Lie groupoid on B. Then

$$\mathrm{Ad}_* = \mathrm{ad}\colon A\Omega \to CDO[L\Omega]. \qquad //$$

For Lie groups, the formula in 4.11(iii) follows as a special case from 4.3. However in the groupoid setting, we cannot write $[X,Y] = \mathrm{ad}(X)(Y)$ for $X,Y \in \Gamma A\Omega$, and get a Lie algebroid representation and so this method is not available. Nonetheless it is possible to overcome this difficulty by lifting to the 1-jet prolongation groupoid Ω^1 (see Kumpera and Spencer (1972) or Kumpera (1975) for the definition): The adjoint map defined above 4.11 is well-defined as a map

$$\Omega^1 \to \Pi(A\Omega)$$

$$j^1_x(\sigma) \mapsto \left(\mathrm{Ad}(\sigma)\big|_x : A\Omega\big|_x \to A\Omega\big|_{(\beta\circ\sigma)(x)}\right),$$

and gives a smooth representation of Ω^1 on the vector bundle $A\Omega$, which we denote by Ad^1. Now, by Kumpera (1975, §18), $A(\Omega^1)$ is naturally isomorphic to the natural Lie algebroid structure on $J^1(A\Omega)$ and 4.11(iii) now states that the induced representation $(Ad^1)_*: J^1(A\Omega) \to CDO(A\Omega)$ is

$$j^1X \longmapsto (Y \longmapsto [X,Y]), \qquad X,Y \in \Gamma A\Omega.$$

Thus 4.11(iii) can now be written as $[X,Y] = (Ad^1)_*(j^1X)(Y)$.

Lastly, we note the following for future reference.

<u>Proposition 4.15</u>. Let E be a vector bundle on B. Then Ad: $\Pi(E) \to \Pi(End(E))$ is given by

$$Ad(\xi)(\phi) = \xi \circ \phi \circ \xi^{-1}$$

for $\xi \in \Pi(E)^y_x$, $\phi: E_x \to E_y$, $x,y \in B$.

<u>Proof</u>: $I_\xi: GL(E_x) \to GL(E_y)$ is the restriction to open sets of the linear map $\mathfrak{gl}(E_x) \to \mathfrak{gl}(E_y)$, $\phi \longmapsto \xi \circ \phi \circ \xi^{-1}$ and is therefore its own derivative. (The change of sign in the identification of $T(GL(V))_{id}$ with $\mathfrak{gl}(V)$ of course cancels out.) //

§5. Infinitesimal connection theory and the concept of transition form

Infinitesimal connection theory is now part of mainstream mathematics and thorough treatments of it exist (for example, Kobayashi and Nomizu (1963), Greub et al. (1973)). We present an account of connection theory in the first part of this section and in §7, partly because we need in Chapter IV sharper forms of some results than are available elsewhere and partly because infinitesimal connection theory is an integral part of the theory of (transitive) Lie algebroids and this fact has not been evidenced before. Primarily, however, we include this material in order to show that the Lie groupoid/Lie algebroid language contributes something to elementary connection theory itself. (See also Chapter V.) Traditionally the motivation for the constructions and results of general connection theory (i.e., the connection theory of principal bundles) have come from the special case of linear connections and, in particular, from Riemannian geometry. It will become clear in the course of this and the next two sections that the general theory of connections is essentially coextensive with the Lie theory of Lie groupoids and Lie algebroids and as such, granted the importance of the Lie theory of Lie groups and Lie algebras, has a natural algebraic interpretation and justification. This cannot be brought out in a treatment which uses only the principal bundle concept. We submit that the Lie groupoid/Lie algebroid formulation of connection theory offers a genuinely new insight into the theory, and that it is the first to do so since the foundational account of Kobayashi and Nomizu (1963).

This section treats those parts of connection theory which do not use the concept of path-lifting or holonomy. Infinitesimal connection theory in this sense can in fact be developed in the context of abstract transitive Lie algebroids and without reference to Lie groupoids; this can only be done however once some nontrivial results about transitive Lie algebroids have been established, and these results are proved in IV§1 by making essential use of connection theory. Once the results of IV§1 are established, however, the results and proofs of this section can be applied to the abstract situation without change.

It is also possible to extend the results of this section (and of Chapter IV) to a purely algebraic setting, in which manifolds are replaced by commutative and unitary rings and vector bundles by projective modules over such rings. This was done in Mackenzie (1979).

In the second part of this section we present the concept of transition form, introduced in Mackenzie (1979). The transition forms of a transitive Lie algebroid play a role analogous to that played by the transition functions of a Lie groupoid or principal bundle. Consider a Lie groupoid Ω. Given a section-atlas $\{\sigma_i : U_i \to \Omega_b\}$ there are local morphisms $\theta_i : U_i \times U_i \to \Omega$, induced by the σ_i, which differentiate to Lie algebroid morphisms $TU_i \to A\Omega|_{U_i}$. These may be considered to be local flat connections in $A\Omega$. On an overlap $U_{ij} \neq \emptyset$ two such connections differ by a tensor, which is essentially a Maurer-Cartan form on U_{ij} with values in $\mathfrak{g}$. These Maurer-Cartan forms are in fact the right-derivatives of the transition functions $\sigma_i^{-1}\sigma_j$ for Ω, and we call them the transition forms of $A\Omega$. In IV§4 we will prove that every abstract transitive Lie algebroid possesses a system of local flat connections, and hence a family of transition forms. Here we prove (5.15) that a transitive Lie algebroid can be constructed from a family of transition forms. Together with the results of IV§4, this will show that there is a complete classification of transitive Lie algebroids up to equivalence by families of transition forms. This classification will be central to the integrability results proved in Chapter V.

Definition 5.1 is due to Atiyah (1957) and Pradines (1967); 5.11 appears in Pradines (1967). The formalism of induced connections in associated vector bundles appears in, for example, Koszul (1960) and Pradines (1967), but the derivation of them from the Lie algebroid representations 4.8 induced by the Lie groupoid representations 1.25 is new and appears here for the first time. The concept of transition form and 5.15 are due to the author (Mackenzie (1979), (1980)). The local description of connections and their curvatures, and 5.19 and 5.20 are slightly sharper and more general versions of standard results (Kobayashi and Nomizu (1963, II 1.4 and II.9)).

* * * * * * * *

<u>Definition 5.1.</u>

(i) Let $L \overset{j}{\to} A \overset{q}{\to} TB$ be a transitive Lie algebroid. A <u>connection</u> in A is a morphism of vector bundles γ: TB $\to$ A over B such that $q \circ \gamma = \mathrm{id}_{TB}$. A <u>back-connection</u> in A is a morphism of vector bundles ω: A $\to$ L over B such that $\omega \circ j = \mathrm{id}_L$.

The <u>curvature</u> of a connection γ in A is the alternating vector bundle morphism $\bar{R}_\gamma$: TB $\oplus$ TB $\to$ L defined by

$$j(\bar{R}_\gamma(X,Y)) = \gamma[X,Y] - [\gamma X, \gamma Y]$$

for $X,Y \in \Gamma TB$.

A connection γ is <u>flat</u> if $\bar{R}_\gamma = 0$; that is, iff γ is a morphism of Lie algebroids over B.

The Lie algebroid A is <u>flat</u> if it has a flat connection.

(ii) Let Ω be a Lie groupoid on B. An <u>infinitesimal connection</u> in Ω is a connection γ in the Lie algebroid $A\Omega$. A <u>back-connection</u> in Ω is a back-connection in $A\Omega$.

Ω is <u>flat</u> if its Lie algebroid is flat. //

The term "back-connection" is chosen to avoid the term "connection form", which we wish to retain in its standard meaning.

If $E' \overset{\iota}{\rightarrowtail} E \overset{\pi}{\twoheadrightarrow} E''$ is an exact sequence of vector bundles over a fixed base, then π has right-inverse morphisms $\rho: E'' \to E$, necessarily injective, and ι has left-inverse morphisms $\lambda: E \to E'$, necessarily surjective. If either of ρ, λ is chosen then the other is determined uniquely by the equation $\iota\circ\lambda + \rho\circ\pi = id_E$; such a pair ρ, λ are said to be associated, or to correspond and $E'' \overset{\rho}{\rightarrowtail} E \overset{\lambda}{\twoheadrightarrow} E'$ is then an exact sequence.

Applying this to the setting of 5.1, there is a bijective correspondence between connections γ and back-connections ω, such that

$$j\circ\omega + \gamma\circ q = id_A \; . \tag{1}$$

<u>Example 5.2.</u> Let $P(B,G,\pi)$ be a principal bundle and consider the Atiyah sequence

$$\frac{P\times\mathfrak{g}}{G} \overset{j}{\rightarrowtail} \frac{TP}{G} \overset{\pi_*}{\twoheadrightarrow} TB$$

(see Appendix A). There is a bijective correspondence between connections $\gamma: TB \to \frac{TP}{G}$ and invariant horizontal distributions Q on P, the determining relationship being $im(\gamma) = \frac{Q}{G}$ (see A§4). There is a bijective correspondence between back-connections $\omega: \frac{TP}{G} \to \frac{P\times\mathfrak{g}}{G}$ and connection forms $\vec{\omega} \in A^1(P,\mathfrak{g})$; here ω is the quotient over G of $\vec{\omega}$ (see A§4). For a connection γ and its back-connection ω, the corresponding invariant horizontal distribution Q and connection-form $\vec{\omega}$ correspond in the standard sense of Kobayashi and Nomizu (1963, II.1).

The curvature $\bar{R}_\gamma: TB \oplus TB \to \frac{P\times\mathfrak{g}}{G}$ of a connection γ and the curvature form $\Omega \in A^2(P,\mathfrak{g})$ of the corresponding invariant horizontal distribution are related in the obvious fashion: for $X,Y \in \Gamma(\frac{TP}{G})$, $\bar{R}_\gamma(\pi_*X,\pi_*Y)$ is the section of $\frac{P\times\mathfrak{g}}{G}$

corresponding to the G-equivariant map $\Omega(\vec{X},\vec{Y}): P \to \mathfrak{g}$ (see A.4.16). //

Note that a Lie algebroid $q: A \to TB$ admits a connection iff it is transitive. In particular, a differentiable groupoid on a connected base B admits a connection iff it is Lie.

Example 5.3. Let E be a vector bundle on B. Then it is easy to see that a connection $\gamma: TB \to CDO(E)$ in $CDO(E)$ or, equivalently, an infinitesimal connection in $\Pi(E)$, is a Koszul connection ∇ in E, where

$$\nabla_X(\mu) = \gamma(X)(\mu).$$

In what follows we will use the standard notation $X \mapsto \nabla_X$ for connections in CDO(E).

The curvature of ∇ is $\bar{R}_\nabla: TB \oplus TB \to End(E)$ defined by

$$\bar{R}_\nabla(X,Y)(\mu) = \nabla_{[X,Y]}(\mu) - \nabla_X(\nabla_Y(\mu)) + \nabla_Y(\nabla_X(\mu)). \tag{2}$$

This is the negative of the usual definition, but it is shown in A§4 that the definition 5.1 of $\bar{R}_\gamma$ for a connection in an arbitrary Lie algebroid corresponds in a natural way and with the correct sign, to the standard curvature form $\Omega \in A^2(P,\mathfrak{g})$ for a connection in a principal bundle P(B,G) and so we accept the change of sign in the case of vector bundles. To do so does not of course oblige us to reverse the sign of the curvature for specific Riemannian (or other) manifolds; one simply changes the sign in the formulas by which sectional and scalar curvature are obtained from $\bar{R}_\nabla$.

There is of course no difference between our concept of connection in a vector bundle and the usual one (compare the equations defining ∇ in Kobayashi and Nomizu (1963, III.1) with the formulas in 4.5); the change of sign in (2) is a consequence only of the way we identify $T(GL(V))_I$ with $\mathfrak{gl}(V)$ for V a vector space (see B§1). See Kobayashi and Nomizu (1963, p. 134) for the point at which this identification determines the sign of $\bar{R}_\nabla$.

The definition (2) has occasionally been used, notably by Milnor (1963).

If $\langle\ ,\ \rangle$ is a Riemannian structure in E then a connection in $CDO\langle E\rangle$ is a Koszul connection ∇ in E such that

$$\langle\nabla_X(\mu),\nu\rangle + \langle\mu,\nabla_X(\nu)\rangle = X(\langle\mu,\nu\rangle), \qquad X \in \Gamma TB, \quad \mu,\nu \in \Gamma E.$$

Such a connection is called a Riemannian connection; 4.9 shows that such a connection always exists.

Let L be an LAB on B. Then a connection in CDO[L] is a Koszul connection ∇ in the vector bundle L such that

$$\nabla_X([V,W]) = [\nabla_X(V),W] + [V,\nabla_X(W)], \qquad X \in \Gamma TB, \quad V,W \in \Gamma L.$$

Such a connection is called a <u>Lie connection</u> in L. Again, 4.9 shows that a Lie connection always exists. //

<u>Example 5.4</u>. Let $TB \oplus (B \times \mathfrak{g})$ be a trivial Lie algebroid. Then $TB \to TB \oplus (B \times \mathfrak{g})$, $X \mapsto X \oplus 0$ is a flat connection, called the <u>standard flat connection</u> in $TB \oplus (B \times \mathfrak{g})$ and denoted by γ^o.

An arbitrary connection in $TB \oplus (B \times \mathfrak{g})$ has the form $X \mapsto X \oplus \omega(X)$ where $\omega: TB \to B \times \mathfrak{g}$ is a $\mathfrak{g}$-valued 1-form on B. The corresponding back-connection is $X \oplus V \mapsto V - \omega(X)$. The curvature is $-(\delta\omega + [\omega,\omega]) \in A^2(B,\mathfrak{g})$. //

<u>Definition 5.5</u>. Let $\phi: A \to A'$ be a morphism of transitive Lie algebroids over B and let γ be a connection in A. Then $\gamma' = \phi \circ \gamma$ is called the <u>produced connection</u> in A'. //

Clearly, then

$$\bar{R}_{\gamma'} = \phi^+ \circ \bar{R}_\gamma. \tag{3}$$

The terminology "produced" is an extension of the usage in II 2.22 and corresponds to that in II 7.10.

<u>Example 5.6</u>. Let E be a vector bundle on B, and let ∇ be a connection in E. Then ∇ induces connections in the vector bundles $\mathrm{Hom}^n(E; B \times \mathbf{R})$, $\mathrm{Hom}^n(E;E)$ and in the various tensor, exterior and symmetric algebra bundles built over E through the representations 4.8. For example, the produced connection $\tilde{\nabla}$ in $\mathrm{Hom}^n(E; B \times \mathbf{R})$ is

$$\tilde{\nabla}_X(\phi)(\mu_1,\dots,\mu_n) = X(\phi(\mu_1,\dots,\mu_n)) - \sum_{r=1}^{n} \phi(\mu_1,\dots,\nabla_X(\mu_i),\dots,\mu_n).$$

Similarly, if E' is a second vector bundle on B and ∇' a connection in E' then the representation of $\mathrm{CDO}(E) \oplus_{TB} \mathrm{CDO}(E')$ on Hom(E;E') given in 4.8(iii) induces the connection $\tilde{\nabla}$ in Hom(E;E') given by

$$\tilde{\nabla}_X(\phi)(\mu) = \nabla'_X(\phi(\mu)) - \phi(\nabla_X(\mu)). \qquad //$$

Definition 5.7. Let $\phi: E \to E'$ be a morphism of vector bundles over B and let ∇ and ∇' be connections in E and E' respectively. Then ϕ maps ∇ to ∇' if $\phi(\nabla_X(\mu)) = \nabla'_X(\phi(\mu))$ for all $X \in \Gamma TB$ and all $\mu \in \Gamma E$. //

If ϕ in 5.7 is an isomorphism, then it induces an isomorphism of Lie groupoids $\tilde{\phi}: \Pi(E) \to \Pi(E')$, $\xi \mapsto \phi_{\beta\xi} \circ \xi \circ (\phi_{\alpha\xi})^{-1}$. This differentiates to $\tilde{\phi}_*: CDO(E) \to CDO(E')$ where $\tilde{\phi}_*(D)(\mu) = (\phi \circ D \circ \phi^{-1})(\mu)$ for $D \in \Gamma CDO(E)$ and $\mu \in \Gamma E$ (see §4). Now, given a connection ∇ in E, the produced connection $\tilde{\phi}_*(\nabla)$ is the unique connection in E' which ϕ maps ∇ to.

The following case of a produced connection will be used repeatedly in what follows.

Definition 5.8. Let Ω be a Lie groupoid on B and let $\gamma: TB \to A\Omega$ be an infinitesimal connection. Then the produced connection $ad \circ \gamma: TB \to CDO[L\Omega]$ in $L\Omega$ will be denoted ∇^{γ} and called the adjoint connection of γ. //

Example 5.9. If ∇ is a connection in a vector bundle E then the adjoint connection in End(E) is the connection

$$\nabla^{\nabla}_X(\phi)(\mu) = \nabla_X(\phi(\mu)) - \phi(\nabla_X(\mu)),$$

and coincides with that induced from ∇ via the action of $\Pi(E)$ on End(E). (See 4.15.) //

Proposition 5.10. Let Ω and γ be as in 5.8. Then

(i) ∇^{γ} is a Lie connection in the LAB $L\Omega$; and

(ii) if $L\Omega$ is abelian then ∇^{γ} is independent of γ - that is, there is a single adjoint connection in $L\Omega$ - and it is flat.

Proof: (i) follows from 4.10.

(ii) Let $\gamma' = \gamma + j \circ \ell$ be a second connection in $A\Omega$. Then $j(\nabla^{\gamma'}_X(V)) = [\gamma'(X), j(V)] = [\gamma(X), j(V)] + j[\ell(X), V] = [\gamma(X), j(V)] = j(\nabla^{\gamma}_X(V))$. By (3), the curvature of ∇^{γ} is $\overset{+}{ad} \circ \overline{R}_{\gamma}$, and $\overset{+}{ad}: L\Omega \to Der(L\Omega)$ is zero if $L\Omega$ is abelian. //

Proposition 5.11. Let Ω and γ be as in 5.8. Then

$$\mathfrak{S}\{\nabla^{\gamma}_{X}(\bar{R}_{\gamma}(Y,Z)) - \bar{R}_{\gamma}([X,Y],Z)\} = 0$$

for all $X,Y,Z \in \Gamma TB$.

Here $\mathfrak{S}$ denotes the cyclic sum over X,Y,Z.

Proof: Apply j to the LHS. The result is

$$\mathfrak{S}\{[\gamma X,\gamma[Y,Z] - [\gamma Y,\gamma Z]] - \gamma[[X,Y],Z] + [\gamma[X,Y],\gamma Z]\}$$

$$= \mathfrak{S}\{[\gamma X,\gamma[Y,Z]] - [\gamma X,[\gamma Y,\gamma Z]] - \gamma[[X,Y],Z] + [\gamma[X,Y],\gamma Z]\}.$$

The first and the last terms cancel, once $\mathfrak{S}$ is applied. The second and the third terms vanish by the Jacobi identity. //

5.11 is of course the (second) Bianchi identity. It does not have much importance for this account of connection theory but has a central role to play in the cohomology theory of Chapter IV.

We come now to the algebraic formalism of covariant derivatives. This material will not be used in the remainder of this chapter, but will be drawn on repeatedly in Chapter IV. Let Ω be a Lie groupoid on B and let $\gamma\colon TB \to A\Omega$ be an infinitesimal connection.

For $n > 0$ denote the vector bundle $\mathrm{Alt}^n(TB;L\Omega)$ by $C^n(TB,L\Omega)$ (compare 3.26). Thus elements of $\Gamma C^n(TB;L\Omega)$ are alternating n-forms on B with values in the vector bundle $L\Omega$. Treat $C^0(TB;L\Omega)$ as $L\Omega$ itself.

Define differential operators $\nabla^{\gamma}\colon \Gamma C^n(TB,L\Omega) \to \Gamma C^{n+1}(TB,L\Omega)$ by

$$\nabla^{\gamma}(f)(X_1,\ldots,X_{n+1}) = \sum_{r=1}^{n+1} (-1)^{r+1}\, \nabla^{\gamma}_{X_r}(f(X_1,\ldots,\hat{\ },\ldots,X_{n+1}))$$

(4)

$$+ \sum_{r<s} (-1)^{r+s}\, f([X_r,X_s],X_1,\ldots\hat{\ }\ldots\hat{\ }\ldots,X_{n+1}) .$$

∇^{γ} is the <u>(exterior) covariant derivative</u> associated with the connection ∇^{γ}. For $V \in \Gamma L\Omega = \Gamma C^0(TB,L\Omega)$, the covariant derivative $X \mapsto \nabla^{\gamma}(V)(X)$ is the adjoint connection $X \mapsto \nabla^{\gamma}_{X}(V)$ itself.

Observe that the Bianchi identity 5.11 can now be written $\nabla^{\gamma}(\bar{R}_{\gamma}) = 0$.

Proposition 5.12. Let γ' be a second connection in $A\Omega$ and let $\ell\colon TB \to L\Omega$ be the map with $\gamma' = \gamma + j\circ\ell$. Then

$$\bar{R}_{\gamma'} - \bar{R}_{\gamma} = -\{\nabla^{\gamma}(\ell) + [\ell,\ell]\}.$$

Proof: For $X,Y \in \Gamma TB$,

$$\begin{aligned} R_{\gamma'}(X,Y) - R_{\gamma}(X,Y) &= \gamma'[X,Y] - \gamma[X,Y] - [\gamma'X,\gamma'Y] + [\gamma X,\gamma Y] \\ &= j\ell[X,Y] - [\gamma X, j\ell Y] - [j\ell X, \gamma Y] - [j\ell X, j\ell Y] \\ &= j\{\ell[X,Y] - \nabla^{\gamma}_X(\ell Y) + \nabla^{\gamma}_Y(\ell X) - [\ell X,\ell Y]\}, \end{aligned}$$

which gives the result. //

In particular, if two connections have the same curvature, then their difference map $\ell\colon TB \to A\Omega$ satisfies $\nabla^{\gamma}(\ell) + [\ell,\ell] = 0$, which is a Maurer-Cartan equation with respect to the adjoint connection ∇^{γ}.

We also need a covariant derivative for forms on $A\Omega$. With $C^n(A\Omega,L\Omega) = \mathrm{Alt}^n(A\Omega;L\Omega)$, define $D^{\gamma}\colon \Gamma C^n(A\Omega,L\Omega) \to \Gamma C^{n+1}(A\Omega,L\Omega)$ by

$$D^{\gamma}(f)(X_1,\ldots,X_{n+1}) = \sum_{r=1}^{n+1} (-1)^{r+1}\, \nabla^{\gamma}_{qX_r}(f(X_1,\ldots\hat{},\ldots,X_{n+1})) \tag{5}$$
$$+ \sum_{r<s} (-1)^{r+s} f([X_r,X_s],X_1,\ldots\hat{}\ldots\hat{}\ldots,X_{n+1}).$$

If Ω corresponds to a principal bundle $P(B,G)$ and $C^n(\frac{TP}{G}, \frac{P\times\mathfrak{g}}{G})$ is identified with $A^n(P,\mathfrak{g})^G$, the pseudo-tensorial n-forms on P of type $(\mathrm{ad},\mathfrak{g})$, then D^{γ} is $(n+1)$ times the exterior covariant differentiation of Kobayashi and Nomizu (1963, II 5.1). For the proof see A 4.15. We call D^{γ} the (exterior) covariant derivative associated to $\nabla^{\gamma}\circ q\colon A\Omega \to CDO[L\Omega]$.

Proposition 5.13. Let $\omega\colon A\Omega \to L\Omega$ be the back-connection corresponding to γ. Then for $X,Y \in \Gamma A\Omega$,

$$\bar{R}_{\gamma}(qX,qY) = (D^{\gamma}(\omega) + [\omega,\omega])(X,Y).$$

Proof: Using $j\circ\omega + \gamma\circ q = \mathrm{id}$, we expand $[X,Y]$ and get

$$[X,Y] = [j\omega X, j\omega Y] + [j\omega X, \gamma qY] + [\gamma qX, j\omega Y] + [\gamma qX, \gamma qY]$$

$$= j[\omega X, \omega Y] - j\nabla^{\gamma}_{qY}(\omega X) + j\nabla^{\gamma}_{qX}(\omega Y) + [\gamma qX, \gamma qY]$$

so

$$[X,Y] - [\gamma qX, \gamma qY] = j\{[\omega X, \omega Y] - \nabla^{\gamma}_{qY}(\omega X) + \nabla^{\gamma}_{qX}(\omega Y)\}$$

and therefore

$$R_{\gamma}(qX, qY) = \gamma[qX, qY] - [\gamma qX, \gamma qY]$$

$$= \gamma q[X,Y] - [\gamma qX, \gamma qY]$$

$$= [X,Y] - j\omega[X,Y] - [\gamma qX, \gamma qY]$$

$$= j\{-\omega[X,Y] + [\omega X, \omega Y] - \nabla^{\gamma}_{qY}(\omega X) + \nabla^{\gamma}_{qX}(\omega Y)\},$$

whence the result. //

In particular, if γ is flat then ω satisfies an equation of Maurer-Cartan type. (The resemblance between 5.12 and 5.13 is explained in IV 3.10.)

If $\rho\colon A\Omega \to CDO(E)$ is any representation of $A\Omega$ on a vector bundle E, then exterior covariant derivatives can be defined in both $C^*(TB,E)$ and $C^*(A\Omega,E)$, using the produced connection $\rho\circ\gamma$ in place of ∇^{γ} in equations (4) and (5).

In the connection theory of principal bundles, the definition of a connection as an invariant horizontal distribution is usually given pre-eminence, perhaps because one can visualize an invariant horizontal distribution more readily than one can the associated connection form. Nonetheless it is easier to compute with the connection form than with the associated distribution, and most computations are done in terms of forms, either global or local.

In the connection theory of Lie algebroids we will usually work with connections $\gamma\colon TB \to A$ rather than with back-connections $\omega\colon A \to L$. Here the reason is that γ is anchor-preserving ($q\circ\gamma = id$) whereas ω is not and for this reason γ fits into the algebraic formalism of Lie algebroids better than does ω. This point will be more evident after Chapter IV. For the present, note that the definition of curvature in terms of ω would need to be via 5.13 and would present curvature as the failure of ω to be a Maurer-Cartan form; the simple definition 5.1

for the curvature of γ is only meaningful because γ is anchor-preserving.

On the other hand, it is desirable to have the formalism of back-connections available, firstly to relate the global Lie algebroid formalism to the standard theory, and secondly because the back-connection formalism is needed in the cohomology theory of Chapter IV.

* * * * * * * *

We come now to the local description of transitive Lie algebroids and their connections.

Let Ω be a Lie groupoid and let $\{\sigma_i : U_i \to \Omega_b\}$ be a section-atlas for Ω. Each $\theta_i : U_i \times U_i \to \Omega_{U_i}^{U_i}$, $(y,x) \mapsto \sigma_i(y)\sigma_i(x)^{-1}$ is a morphism of Lie groupoids over U_i and is a local right-inverse to $[\beta,\alpha]$, so each $(\theta_i)_* : TB|_{U_i} \to A\Omega|_{U_i}$ is a morphism of Lie algebroids over U_i with $q \circ (\theta_i)_* = \mathrm{id}$, and may therefore be considered to be a flat connection in $A\Omega|_{U_i}$. We call the $(\theta_i)_*$ the <u>local flat connections</u> induced by the section-atlas $\{\sigma_i\}$.[1]

When $U_{ij} \neq \emptyset$, there are two connections in $A\Omega|_{U_{ij}}$ namely $(\theta_i)_*$ and $(\theta_j)_*$, and so there is a unique vector bundle morphism $\ell_{ij} : TB|_{U_{ij}} \to L\Omega|_{U_{ij}}$ such that

$$(\theta_j)_* = (\theta_i)_* + \ell_{ij}.$$

Let $\{\psi_i : U_i \times \mathfrak{g} \to L\Omega|_{U_i}\}$ denote the atlas for $L\Omega$ induced from $\{\sigma_i\}$; that is, $\psi_{i,x} = \mathrm{Ad}\,\sigma_i(x)$; and define $\mathfrak{g}$-valued 1-forms χ_{ij} on U_{ij} by

$$\chi_{ij} = \psi_i^{-1} \circ \ell_{ij} : TU_{ij} \to U_{ij} \times \mathfrak{g}.$$

<u>Proposition 5.14.</u> With the notation above,

$$\chi_{ij} = \Delta(s_{ij}),$$

where $\{s_{ij}\}$ is the cocycle corresponding to $\{\sigma_i\}$. In particular, each χ_{ij} is a Maurer-Cartan form.

<u>Proof:</u> For $X \in T(U_{ij})_x$,

$$(\theta_j)_*(X) = T(R_{\sigma_j(x)^{-1}})T(\sigma_j)_x(X).$$

Now use $\sigma_j = \sigma_i s_{ij}$.   //

We call the χ_{ij} the transition forms corresponding to the section-atlas $\{\sigma_i\}$, or to the cocycle $\{s_{ij}\}$. Since $\ell_{ij} + \ell_{jk} = \ell_{ik}$ whenever $U_{ijk} \neq \emptyset$, it follows that

(6) $$\chi_{ik} = \chi_{ij} + a_{ij}(\chi_{jk})$$

where $\{a_{ij}: U_{ij} \to \mathrm{Aut}(\mathfrak{g})\}$ is the cocycle for $L\Omega$ corresponding to the atlas $\{\psi_i = \mathrm{Ad}\sigma_i\}$. We call (6) the cocycle condition for the χ_{ij}; it also follows directly from $s_{ik} = s_{ij}s_{jk}$ by the product rule for right derivatives (see B§2, equation (1)).

Consider the Lie groupoid isomorphism

$$\Sigma_i: U_i \times G \times U_i \to \Omega_{U_i}^{U_i}$$

$$(y,g,x) \mapsto \sigma_i(y)g\sigma_i(x)^{-1}.$$

It is straightforward to see that

$$(\Sigma_i)_*: TU_i \oplus (U_i \times \mathfrak{g}) \to A\Omega|_{U_i}$$

is

(7) $$X \oplus V \mapsto (\theta_i)_*(X) + \psi_i(V)$$

and we call $(\Sigma_i)_*$ the Lie algebroid chart for $A\Omega$ induced by σ_i. From the bracket-preservation equation for $(\Sigma_i)_*$, it follows that

(8) $$[(\theta_i)_*(X), \psi_i(V)] = \psi_i(X(V))$$

for $X \in \Gamma TU_i$ and $V: U_i \to \mathfrak{g}$. Now (8) may be interpreted as the statement that $\psi_i: U_i \times \mathfrak{g} \to L\Omega|_{U_i}$ maps the standard flat connection $\nabla^0_X(V) = X(V)$ in the trivial bundle $U_i \times \mathfrak{g}$ to the adjoint connection induced in $L\Omega|_{U_i}$ by $(\theta_i)_*$ in $A\Omega|_{U_i}$. Similarly, (8) signifies that $(\Sigma_i)_*$ maps the standard flat connection in $B \times G \times B$ to $(\theta_i)_*$.

When $U_{ij} \neq \emptyset$ we obtain an overlap isomorphism

$$(S_{ij})_* = (\Sigma_i)_*^{-1} \circ (\Sigma_j)_*: TU_{ij} \oplus (U_{ij} \times \mathfrak{g}) \to TU_{ij} \oplus (U_{ij} \times \mathfrak{g})$$

$$(X \oplus V) \mapsto X \oplus (\chi_{ij}(X) + a_{ij}(V)).$$

Since this is a morphism of trivial Lie algebroids, the compatibility condition of 2.4 must be satisfied, that is, we must have

$$X(a_{ij}(V)) - a_{ij}(X(V)) + [\chi_{ij}(X), a_{ij}(V)] = 0 \tag{9}$$

for all X and V. (9) can be proved directly from the equations $a_{ij} = Ad \circ s_{ij}$ and $\chi_{ij} = \Delta(s_{ij})$ by using B 2.1, and can be written more succinctly as

$$\Delta(a_{ij}) = ad \circ \chi_{ij} \tag{9a}$$

where Δ is the right derivative for the Lie group $Aut(\mathfrak{g})$ (use equation (6a) in B§2). Written in this form, the equation is an immediate consequence of $a_{ij} = Ad \circ s_{ij}$.

It is to be expected that the Lie algebroid $A\Omega$ can be reconstructed from a system of transition forms χ_{ij}. The presence in the cocycle condition of the a_{ij} inevitably complicates the formulation of this result and the reader is urged to check the details in the proof of the following theorem.

<u>Theorem 5.15</u>. Let B be a manifold and let $\mathfrak{g}$ be a Lie algebra. Let $\{U_i\}$ be an open cover of B and let $\{a_{ij}: U_{ij} \to Aut(\mathfrak{g})\}$ be a cocycle. Let $\{\chi_{ij} \in A^1(U_{ij}, \mathfrak{g})\}$ be a set of local $\mathfrak{g}$-valued 1-forms and suppose that

(i) each χ_{ij} is a Maurer-Cartan form,

(ii) $\chi_{ik} = \chi_{ij} + a_{ij}(\chi_{jk})$ whenever $U_{ijk} \neq \emptyset$,

(iii) $\Delta(a_{ij}) = ad \circ \chi_{ij}$ for all i,j.

Then there is a transitive Lie algebroid $L \rightarrowtail A \twoheadrightarrow TB$ on B whose adjoint bundle L is the LAB corresponding to $\{a_{ij}\}$ and which admits local flat connections $\gamma_i: TB|_{U_i} \to A|_{U_i}$ such that

$$\gamma_j = \gamma_i + \psi_i \circ \chi_{ij}$$

where $\{\psi_i: U_i \times \mathfrak{g} \to L_{U_i}\}$ is an LAB atlas for L with $\{a_{ij}\}$ as cocycle.

Let $L \rightarrowtail A \twoheadrightarrow TB$ and $L' \rightarrowtail A' \twoheadrightarrow TB$ be transitive Lie algebroids on B for which there are, firstly, LAB atlases $\{\psi_i: U_i \times \mathfrak{g} \to L_{U_i}\}$ and $\{\psi'_i: U_i \times \mathfrak{g} \to L'_{U_i}\}$ (with the same fibre type $\mathfrak{g}$ and open cover $\{U_i\}$) for L and L' which both have $\{a_{ij}\}$ as cocycle and, secondly, local flat connections $\gamma_i: TB|_{U_i} \to A|_{U_i}$ and $\gamma': TB|_{U_i} \to A'|_{U_i}$ such that $\gamma_j = \gamma_i + \psi_i \circ \chi_{ij}$ and $\gamma'_j = \gamma'_i + \psi'_i \circ \chi_{ij}$ whenever $U_{ij} \neq \emptyset$. Then there is a unique Lie algebroid isomorphism $\phi: A \to A'$ such that $\phi^+ \circ \psi_i = \psi'_i$ and $\phi \circ \gamma_i = \gamma'_i$ for all i.

<u>Proof</u>: For each i, let A^i be the <u>set</u> $TU_i \oplus (U_i \times \mathfrak{g})$ and on the disjoint sum $\coprod_i A^i$ define an equivalence relation ~ by

$$(i, X \oplus V) \sim (j, Y \oplus W) \iff X = Y \quad \text{and} \quad W = \chi_{ji}(X) + a_{ji}(V).$$

Denote the quotient set by A and equivalence classes by $\langle i, X \oplus V\rangle$.

Define a map $p: A \to B$ by $p(\langle i, X \oplus V\rangle) = x$ where $X \in T(U_i)_x$. Then it is easy to see that

$$\bar{\psi}_i: TU_i \oplus (U_i \times \mathfrak{g}) \to p^{-1}(U_i)$$

$$X \oplus V \mapsto \langle i, X \oplus V\rangle$$

is a bijection. Give A the smooth structure induced from the manifolds $TU_i \oplus (U_i \times \mathfrak{g})$ via the $\bar{\psi}_i$.

Clearly (A,p,B) is now a vector bundle, and further, the map $q: A \to TB$, $\langle i, X \oplus V\rangle \mapsto X$, is well-defined and a surjective vector bundle morphism over B. Denote the kernel of q by L. The $\bar{\psi}_i$ restrict to charts

$$\psi_i: U_i \times \mathfrak{g} \to L_{U_i}, \quad V \mapsto \langle i, 0 \oplus V\rangle$$

for L and the atlas $\{\psi_i\}$ has $\{a_{ij}\}$ as cocycle.

Now we define a bracket in ΓA. For $\mu, \nu \in \Gamma A$ and $x \in B$, choose U_i containing x and write $\mu = \langle i, X \oplus V\rangle$, $\nu = \langle i, Y \oplus W\rangle$ where X and Y are vector fields on U_i and V and W are maps $U_i \to \mathfrak{g}$. Define

$$[\langle i, X \oplus V\rangle, \langle i, Y \oplus W\rangle] = \langle i, [X,Y] \oplus \{X(W) - Y(V) + [V,W]\}\rangle.$$

It is an instructive exercise to verify that this is well-defined and makes A a transitive Lie algebroid on B with adjoint bundle L.

The local flat connections γ_i are defined by $X \mapsto \langle i, X \oplus 0\rangle$. The remainder of the proof is straightforward. //

The reader is urged to work out the relationship of 5.15 and its proof to II 2.19.

Remark. It is proved in IV§4 that every transitive Lie algebroid admits a system of local flat connections and thus of transition forms. //

We postpone until IV§4 the concept of equivalence for systems of transition forms and the proof that equivalent systems generate, under 5.15, isomorphic Lie algebroids. Here we merely note that if $\{\sigma_i'\colon U_i \to \Omega_b\}$ is a second section-atlas with respect to the same open cover $\{U_i\}$ and reference point b, then there are maps $r_i\colon U_i \to G = \Omega_b^b$ such that $\sigma_i' = \sigma_i r_i$, and the following formulae, which are easily proved, relate the primed data to the unprimed.

(10a) $$\theta_i'(y,x) = \theta_i(y,x)I_{\sigma_i(x)}(r_i(y)r_i(x)^{-1}),$$

(10b) $$(\theta_i')_* = (\theta_i)_* + \psi_i \circ \Delta(r_i),$$

(10c) $$\chi_{ij}' = \mathrm{Ad}(r_i)^{-1}\{-\Delta(r_i) + \chi_{ij} + a_{ij}\Delta(r_j)\},$$

(10d) $$a_{ij}' = \mathrm{Ad}(r_i)^{-1}a_{ij}\mathrm{Ad}(r_j).$$

(10c) also follows from $s_{ij}' = r_i^{-1}s_{ij}r_j$ by using 5.14 and the formulas in B§2.

We come now to the local description of infinitesimal connections. Let Ω be a Lie groupoid on B, let $\gamma\colon TB \to A\Omega$ be an infinitesimal connection in Ω and let $\{\sigma_i : U_i \to \Omega_b\}$ be a section atlas. Continue the notations θ_i, ψ_i, χ_{ij}, a_{ij} used above.

For each i, define $\omega_i : TU_i \to U_i \times \mathfrak{g}$ by

(11) $$\psi_i(\omega_i) = \gamma|_{U_i} - (\theta_i)_*.$$

The $\omega_i \in A^1(U_i,\mathfrak{g})$ are called the <u>local connection forms</u> for γ with respect to the atlas $\{\sigma_i\}$. Clearly, on a U_{ij} which is nonvoid,

(12) $$\omega_i = a_{ij}(\omega_j) + \chi_{ij}$$

and, conversely, if $\{\omega_i \in A^1(U_i,\mathfrak{g})\}$ is a family of forms which satisfy (12), then they define by (11) a connection γ in $A\Omega$. (Compare, for example, Kobayashi and Nomizu (1963, II 1.4).)

Proposition 5.16. With the above notation, the adjoint connection ∇^γ is given locally by

$$\nabla^\gamma_X(\psi_i(V)) = \psi_i\{X(V) + [\omega_i(X),V]\}$$

where $X \in \Gamma TU_i$ and $V\colon U_i \to \mathfrak{g}$.

Proof: Use (11) and (8). //

Corollary 5.17. Let Ω be an abelian Lie groupoid on base B. Then the (unique) adjoint connection ∇ in $L\Omega$ is the image of the standard flat connection ∇^0 in $B \times \mathfrak{g}$ under the canonical chart $\psi\colon B \times \mathfrak{g} \to L\Omega$.

Proof: Since $\mathfrak{g}$ is abelian the equation in 5.16 reduces to

$$\nabla^\gamma_X(\psi_i(V)) = \psi_i(X(V)).$$

Now ∇^γ is equal to ∇^0 for any connection γ, and (because Ω is abelian) ψ_i is equal to ψ for any i (see remark following 3.19). //

Define $R^i_\gamma \in A^2(U_i,\mathfrak{g})$ by

(13) $$R^i_\gamma = \psi_i^{-1}\circ \bar{R}_\gamma.$$

The R^i_γ are the local curvature forms for γ with respect to $\{\sigma_i\}$. Clearly $R^j_\gamma = a_{ji}(R^i_\gamma)$ when $U_{ij} \neq \emptyset$.

Proposition 5.18. Continuing the above notation,

$$R^i_\gamma = -(\delta\omega_i + [\omega_i,\omega_i]).$$

Proof: Follows easily from (11) and (8). //

Proposition 5.19. Let Ω be a Lie groupoid on B, and continue the above notations.

(i) If $\{\sigma_i\}$ is a section-atlas for Ω whose cocycle $\{s_{ij}\}$ consists of constant maps, then

$$\gamma|_{U_i} = (\theta_i)_*$$

is a well-defined and flat (global) infinitesimal connection in Ω.

(ii) If γ is a flat infinitesimal connection in Ω then there is a section-atlas $\{\sigma_i\}$ for Ω whose cocycle consists of constant maps and for which

$$\gamma|_{U_i} = (\theta_i)_*$$

on all U_i.

Proof: (i) Since s_{ij} is constant, it follows that $\chi_{ij} = 0$ and so (12) admits the solution $\omega_i = 0$ for all i.

(ii) Let $\{\sigma_i : U_i \to \Omega_b\}$ be a section-atlas in which each U_i is connected and simply-connected. Since γ if flat, the ω_i are Maurer-Cartan forms and so there exist maps $f_i : U_i \to G$ such that

$$\Delta(f_i) = \omega_i.$$

Define $\tau_i : U_i \to \Omega_b$ by $\tau_i = \sigma_i f_i$. It is easy to verify that $\Delta(t_{ij}) = 0$, where $\{t_{ij}\}$ is the cocycle for $\{\tau_i\}$, and so t_{ij} is constant.

Denote by θ_i' the local morphism induced by τ_i. Then, using

$$(\theta_i')_*(X) = T(R_{\tau_i(x)^{-1}})T(\tau_i)(X),$$

for $X \in T(U_i)_x$, it is easy to verify that

$$(\theta_i')_* = (\theta_i)_* + \psi_i(\Delta(f_i)).$$

Hence $(\theta_i')_* = \gamma|_{U_i}$ as claimed. //

The following version of 5.19 is used several times in the sequel.

Proposition 5.20. Let B × G × B be a trivial Lie groupoid with B connected and simply-connected. Then for any flat connection γ, there is an automorphism ϕ

of $B \times G \times B$ over B such that $\phi_*\bullet\gamma^o = \gamma$, where γ^o is the standard flat connection.

Proof: γ is of the form $X \mapsto X \oplus \omega(X)$ where $\omega \in A^1(B,\mathfrak{g})$ is a Maurer-Cartan form (see 5.4). Let $f: B \to G$ be such that $\Delta(f) = \omega$; then $\phi(y,g,x) = (y,f(y)gf(x)^{-1},x)$ has the required property. //

Ever since Spivak (1979, Volume II), any new account of connection theory has to address the following result.

Proposition 5.21. Let B be a manifold and $\langle\ ,\ \rangle$ a Riemannian structure in the vector bundle TB. Let ∇ be a Riemannian connection such that

$$\nabla_X(Y) - \nabla_Y(X) - [X,Y] = 0 \qquad X,Y \in \Gamma TB. \tag{14}$$

Then if ∇ is flat, B is locally isometric to Euclidean space.

Proof: Let $\phi: \mathbf{R}^n \to U$ be a chart for B. Pull the Riemannian structure and the connection on $TB|_U$ back to $T(\mathbf{R}^n)$. Continue to use the notations $\langle\ ,\ \rangle$ and ∇. By 1.26 there is a neighbourhood W of $0 \in \mathbf{R}^n$, which we may assume to be connected and simply-connected, and a decomposing section $\sigma: W \to \Pi\langle T\mathbf{R}^n\rangle_o$. This σ defines an automorphism of the vector bundle TW which maps the given Riemannian structure to another, still denoted $\langle\ ,\ \rangle$, for which $\Pi\langle TW\rangle$ is $W \times \mathcal{O}(n) \times W$; the value of σ at 0 can be chosen so that $\langle\ ,\ \rangle_0$ is the standard metric on $\mathbf{R}^n$. We transport ∇ under this automorphism also; ∇ is still a Riemannian connection in TW and still satisfies (14), and still is flat.

By 5.20 there is a map $f: W \to \mathcal{O}(n)$ such that $F: W \times \mathbf{R}^n \to W \times \mathbf{R}^n$, $(x,X) \mapsto (x,f(x)(X))$ maps the standard flat connection ∇^o to ∇; that is,

$$\nabla_X(Y) = F(X(F^{-1}(Y)));$$

we can also require that $f(0) = I \in \mathcal{O}(n)$. Let $\{\frac{\partial}{\partial x_i}\}$ be the standard vector fields on $\mathbf{R}^n$ and define $X_i = F\frac{\partial}{\partial x_i}$. Then for any vector field X,

$$\nabla_X(X_i) = F\left(X\left(\frac{\partial}{\partial x_i}\right)\right) = 0 \tag{15}$$

since $\frac{\partial}{\partial x_i}$ is constant as a map $W \to \mathbf{R}^n$. Hence from (14) it follows that

$[X_i, X_j] = 0$ for all i,j and so there is a local coordinate system $\{y_1, \ldots, y_n\}$ around 0 in W such that

$$\frac{\partial}{\partial y_i} = X_i \quad \text{for all } i$$

(see, for example, Spivak (1979, I.5.14)).

Now in this coordinate system the metric is canonical, for

$$X\left(\langle \frac{\partial}{\partial y_i}, \frac{\partial}{\partial y_j} \rangle\right) = \langle \nabla_X(X_i), X_j \rangle + \langle X_i, \nabla_X(X_j) \rangle$$

$$= 0 \quad \text{by (15)}$$

and so $\langle \frac{\partial}{\partial y_i}, \frac{\partial}{\partial y_j} \rangle$ is constant and we arranged the value at 0 to be δ_{ij}.
//

§6. The Lie theory of Lie groupoids over a fixed base

This section treats the correspondence between α-connected reductions of a Lie groupoid and reductions of its Lie algebroid, and the correspondence between local base-preserving morphisms of Lie groupoids and base-preserving morphisms of their Lie algebroids.

In 6.1 we prove that if Ω is a Lie groupoid and A' is a reduction of $A\Omega$, then there is a unique α-connected reduction Ω' of Ω such that $A\Omega' = A'$. This result is closely related to the Ambrose-Singer theorem of connection theory; our proof of 6.1 is essentially a groupoid formulation of the main idea of Kobayashi and Nomizu's proof of the latter result. Conversely, in §7 we deduce a strong form of Ambrose-Singer as an immediate corollary of 6.1 and the correspondence between infinitesimal and path connections. At the same time, the proof of 6.1 follows closely the outline of the proof of the corresponding result for Lie groups and Lie algebras.

In 6.5 we prove that if Ω and Ω' are Lie groupoids on the same base B and $\phi\colon A\Omega \to A\Omega'$ a morphism of their Lie algebroids over B, then if Ω is α-connected and α-simply connected, ϕ can be integrated to a global morphism $f\colon \Omega \to \Omega'$. This follows from 6.1 in a manner similar to the case of Lie groups. The local integrability of ϕ, in the case where Ω is an arbitrary Lie groupoid on B, is then deduced (6.7) from 6.5 via the results of II§6 on the monodromy groupoid of the α-identity component subgroupoid of Ω.

A second proof of 6.7, using connection theory, is given in §7.

There are generalizations of both 6.1 and 6.7 applicable to arbitrary differentiable groupoids and not-necessarily-base-preserving morphisms, stated in Pradines (1966, 1967) and proved in Almeida (1980) and Almeida and Kumpera (1981). The proofs of these generalizations are largely disjoint from the proofs of 6.1 and 6.7. In particular, the generalization of 6.1, or rather the recovery of 6.1 from it, depends on the very subtle construction of the holonomy groupoid of a microdifferentiable groupoid. This subject is essentially a generalization of foliation theory; a very brief discussion is included here, following 6.3. In the same way that our proof of 6.7 depends on 6.1, the general result on the local integrability of morphisms depends on the generalized subgroupoid-subalgebroid correspondence; here the main problem is to give a correct definition of the general concept of morphism of Lie algebroids. We have preferred to omit these substantial considerations and give instead proofs of the special cases which suffice for Chapter IV.

In category-theoretic terms, the main results of this section, taken together with the results of II§6, show that the Lie functor is both full and faithful, regarded as a functor from the category of germs of local base-preserving morphisms between Lie groupoids on a given base B, to the category of transitive Lie algebroids and base-preserving morphisms over B.

The section ends with a demonstration that some smaller parts of the Lie theory of Lie groups and Lie algebras do not generalize to Lie groupoids and Lie algebroids.

<u>Theorem 6.1</u>. Let Ω be a Lie groupoid on a a connected base B and let $L' \rightarrowtail A' \twoheadrightarrow TB$ be a reduction of $L\Omega \rightarrowtail A\Omega \twoheadrightarrow TB$. Then there is a unique α-connected Lie subgroupoid Φ of Ω such that $A\Phi = A'$.

<u>Proof</u>: The proof is modelled on the proof for Lie groups as given, for example, by Warner (1971), and we deal only with the features that are new.

Denote by Δ the inverse image bundle β^*A' on Ω. Since

$$\begin{array}{ccc} T^{\alpha}\Omega & \longrightarrow & A\Omega \\ \downarrow & & \downarrow \\ \Omega & \xrightarrow{\beta} & B \end{array}$$

is a pullback, it follows that there is a unique injective vector bundle morphism $\Delta \xrightarrow{i} T^{\alpha}\Omega$ over Ω such that

$$\begin{array}{ccc} T^{\alpha}\Omega & \xrightarrow{\mathcal{R}} & A\Omega \\ {\scriptstyle i}\uparrow & & \uparrow \\ \Delta & \longrightarrow & A' \end{array}$$

commutes. By a standard result (see C.4) it follows from $\Delta = \beta^*A'$ that $\Gamma\Delta = C(\Omega) \otimes_{C(B)} \Gamma A'$ and using this one can mimic the proof for the case of groups and show that Δ is an involutive distribution on Ω.

For each $x \in B$, let Δ_x be the restriction of Δ to Ω_x; Δ_x is an involutive distribution on Ω_x. Let Φ_x be the (connected) integral manifold of Δ_x which contains $\tilde{x}$ and let Φ be the union of the sets Φ_x, $x \in B$. For any $x,y \in B$ and $\xi \in \Omega_x^y$ it follows from the definition of Δ that $T(R_{\xi^{-1}})(\Delta_x) = \Delta_y$, and therefore if $\xi \in \Phi_x^y$ it follows that $R_{\xi^{-1}}(\Phi_y) = \Phi_x$. Thus Φ is a subgroupoid of Ω in the algebraic sense.

Write α',β' for the restrictions of α,β to Φ. Because $q'\colon A' \to TB$ is assumed to be surjective, it follows that $\beta'_x\colon \Phi_x \to B$, $x \in B$, is a submersion, for $T(\beta'_x)_\xi$, $\xi \in \Phi_x$, is the composite

$$T(\Phi_x)_\xi = \Delta_\xi \xrightarrow{\ \cong\ } A'_{\beta\xi} \xrightarrow{\ q'_\xi\ } T(B)_{\beta\xi}$$

where the middle map is the pullback. Hence each $\mathrm{im}(\beta'_x)$, $x \in B$, is open in B. Since Φ is a groupoid in the algebraic sense, the $\mathrm{im}(\beta'_x)$, $x \in B$, partition B and since B is connected it follows that each β'_x is onto B. Thus Φ is transitive.

We now give Φ a differentiable structure. Choose $b \in B$ and write $H = \Phi_b^b$. Since β'_b is a surjective submersion, H is a closed embedded sub-manifold of Φ_b. Also, β'_b has a family of local right-inverses $\sigma_i\colon U_i \to \Phi_b$, where the U_i cover B. Let

$$\Sigma_i^j\colon U_j \times H \times U_i \to \Phi_{U_i}^{U_j}$$

be the bijections defined using $\{\sigma_i\}$; to show that the overlap maps are smooth, it suffices to show that the transition functions $s_{ij}\colon U_{ij} \to H$ for $\{\sigma_i\}$ are smooth. Now, again using the fact that β'_b is a surjective submersion,

$$\Phi_b * \Phi_b = \{(\eta,\xi) \in \Phi_b \times \Phi_b \mid \beta'(\eta) = \beta'(\xi)\}$$

is a submanifold of $\Phi_b \times \Phi_b$, and hence of $\Omega_b \times \Omega_b$; consider the restriction of

$$\delta'\colon \Omega_b * \Omega_b \to \Omega_b \qquad (\eta,\xi) \mapsto \eta^{-1}\xi$$

to $\Phi_b * \Phi_b$. Since Φ_b is a leaf of the distribution Δ_b on Ω_b, it follows that $\Phi_b * \Phi_b \to \Omega_b$ is smooth as a map into Φ_b, and since H is an embedded submanifold of Φ_b, it follows that $\Phi_b * \Phi_b \to \Phi_b$ is smooth as a map into H. Hence $s_{ij}\colon U_{ij} \to H$, $x \mapsto \sigma_i(x)^{-1}\sigma_j(x)$ is smooth, as required. Also, it follows that $H \times H \to H$, $(h,h') \mapsto h^{-1}h'$ is smooth and so H is a Lie subgroup of Ω_b^b. The maps Σ_i^j now define a differentiable structure on Φ with respect to which it is a Lie groupoid on B. Because the σ_i are smooth into Φ_b with its leaf differentiable structure, it follows that each Φ_x, $x \in B$, inherits from Φ the differentiable structure it was originally given by Δ, and because the inclusion $\Phi \subseteq \Omega$ is locally (with respect to the Σ_i^j) represented by $id_{U_j} \times (H \subseteq \Omega_b^b) \times id_{U_i}$, it follows that Φ is a Lie subgroupoid of Ω.

The inclusion $\Phi \subseteq \Omega$ induces an injective Lie algebroid morphism $A\Phi \to A\Omega$ over B; since $A\Phi|_x = T(\Phi_x)_{\tilde{x}} = \Delta_{\tilde{x}} = A'_x$, $\forall x \in B$, it follows that $A\Phi \to A\Omega$ is a Lie algebroid isomorphism onto A'.

Suppose now that $\psi\colon \Psi \to \Omega$ is a Lie subgroupoid of Ω with $\psi_*(A\Psi) = A'$. From the diagram

$$\begin{array}{ccccc} T^\alpha(\Psi) & & \Delta & \subseteq & T^\alpha(\Omega) \\ \downarrow & & \downarrow & & \downarrow \\ A\Psi & \xrightarrow{\psi^*} & A' & \subseteq & A\Omega \end{array}$$

in which each vertical arrow is a pullback, it follows that $T^\alpha(\psi)$ is onto Δ and hence for each $x \in B$, $\psi_x(\Psi_x)$ is an integral manifold for Δ through $\tilde{x}$. So there exists a smooth map $\phi_x\colon \Psi_x \to \Phi_x$ such that

$$(1) \qquad \begin{array}{ccc} \Psi_x & \xrightarrow{\phi_x} & \Phi_x \\ & {\scriptstyle\psi_x}\searrow \quad \swarrow{\scriptstyle\cong} & \\ & \Omega_x & \end{array}$$

commutes. It is easy to see that $\phi_y \circ R_\xi = R_{\phi_x(\xi)} \circ \phi_x$ for each $\xi \in \Psi_x^y$, $x,y \in B$ and therefore $\phi = \bigcup_x \phi_x$ is a morphism of algebraic groupoids $\Psi \to \Phi$ over B. Since Ψ and Φ are Lie and ϕ_x is smooth, it follows that ϕ is smooth. From (1) it follows that ϕ_x is an injective immersion and so ϕ is an injective immersion. Lastly, from $\psi_*(A\Psi) = A' = A\Phi$, it follows that Ψ and Φ have the same dimension, so ϕ is étale, and therefore, by 3.14, ϕ is onto Φ. //

A brief outline of this proof was given in Bowshell (1971).

Corollary 6.2. Let $\phi: \Omega \to \Omega'$, $\phi_o: B \to B'$ be a morphism of Lie groupoids and let Φ be a reduction of Ω'. Then if ϕ takes values in Φ, it is smooth as a map $\Omega \to \Phi$.

Proof: First assume that Ω is α-connected. Let Ψ be the α-identity component subgroupoid of Φ. Then for each $x \in B$, the submanifold $\Psi_{\phi_o(x)}$ of Ω' is a leaf of the foliation on Ω' defined by $A\Phi$. So $\phi_x: \Omega_x \to \Psi_{\phi_o(x)}$ is smooth. Hence, by the smooth version of II 1.21(i), $\phi: \Omega \to \Psi$ is smooth; since Ψ is open in Φ it follows that $\phi: \Omega \to \Phi$ is smooth.

The case where Ω is not α-connected now follows from 1.3 and II 1.21(ii).

//

The assumption that A' is transitive is essential to the possibility of transferring the differentiable structures on the Φ_x globally to the groupoid Φ. If A' is not transitive one obtains in general only the algebraic groupoid Φ and a differentiable structure on a subset $W \subseteq \Phi$ which contains $\tilde{B}$ and generates Φ. This constitutes what was called by Pradines (1966) "un morceau différentiable de groupoide"; we propose to call it a local differentiable groupoid structure. The precise definition follows.

Definition 6.3. (Pradines (1966))

Let Φ be a groupoid in the algebraic sense on a manifold B. Then a local differentiable groupoid structure on Φ is a subset W of Φ together with a differentiable structure on W such that

(i) $\tilde{x} \in W$, $\forall x \in B$ and W generates Φ;

(ii) $\alpha|_W: W \to B$, $\beta|_W: W \to B$ are smooth submersions and $\varepsilon: B \to W$ is smooth;

(iii) $(W \underset{\alpha}{\times} W) \cap \delta^{-1}(W)$ is an open subset of $W \underset{\alpha}{\times} W$ and the restriction of δ to $(W \underset{\alpha}{\times} W) \cap \delta^{-1}(W) \to W$ is smooth.

A locally differentiable groupoid is a pair (Φ, W), where Φ is a groupoid in the algebraic sense on a manifold B and W is a local differentiable groupoid structure on Φ. Two local differentiable groupoid structures W and W' on a groupoid Φ are equivalent if for all $x \in B$ there is a set $\tilde{x} \in U \subseteq W \cap W'$ such that U is open in both W and W'. An equivalence class of local differentiable groupoid structures on a groupoid Φ is a micro differentiable groupoid structure on Φ, and Φ together with this structure is a microdifferentiable groupoid. //

A smooth foliation $\mathcal{F}$ on a connected manifold B defines a microdifferentiable groupoid structure on $X \subseteq B \times B$, the groupoid corresponding to the equivalence relation defined by $\mathcal{F}$. Conversely a microdifferentiable structure on a wide subgroupoid of $B \times B$ defines a foliation on B. (Pradines (1966).)

Almeida (1980) proves the following generalization of 6.1:

<u>Theorem.</u> Let Ω be a differentiable groupoid on B and A' a Lie subalgebroid of $A\Omega$ on a submanifold B' of B. Then there is a unique microdifferentiable subgroupoid Ω' of Ω such that $A\Omega' = A'$. //

The proof of this result is not difficult, though the correct definition of the general concept of Lie subalgebroid is not obvious and requires care. However the <u>deduction</u> of 6.1 from this theorem depends on the construction of the holonomy groupoid of a microdifferentiable groupoid, and this is a very subtle and substantial theory. We have preferred here to give a direct proof for the locally trivial case.

* * * * * *

We come now to the correspondence between local morphisms of Lie groupoids over a fixed base and morphisms of their Lie algebroids.

It is easy to extend the construction of the Lie functor to the case of local smooth morphisms $\phi: \Omega \rightsquigarrow \Omega'$, $\phi_o: B \to B'$ of differentiable groupoids. Denote the domain of ϕ by $\mathcal{U}$. Then $T(\phi): T\Omega|_{\mathcal{U}} \to T\Omega'$ restricts to $T^\alpha(\phi): T^\alpha\Omega|_{\mathcal{U}} \to T^\alpha\Omega'$ and one can form the composition

$$\begin{array}{ccccc} A\Omega & \longrightarrow & T^\alpha\Omega|_{\mathcal{U}} & \longrightarrow & T^\alpha\Omega' \\ \downarrow & & \downarrow & & \downarrow \\ B & \xrightarrow{\ \varepsilon\ } & \mathcal{U} & \xrightarrow{\ \phi\ } & \Omega' \end{array}$$

and proceed as in §4. The version of equation (3) of §4 needed is that $\vec{X}|_{\mathcal{U}} \overset{\phi}{\sim} \vec{X}'$ iff $\phi_* \circ X = X' \circ \phi_o$, and equation (4) holds without change. The following unicity result can now be proved in the most general setting.

<u>Proposition 6.4.</u> Let $\phi, \psi: \Omega \rightsquigarrow \Omega'$ be local morphisms of differentiable groupoids.

If $\phi_o = \psi_o \colon B \to B'$ and $\phi_* = \psi_* \colon A\Omega \to A\Omega'$, then ϕ and ψ are germ-equivalent. If ϕ and ψ are global morphisms and Ω is α-connected then $\phi_o = \psi_o$, $\phi_* = \psi_*$ imply $\phi = \psi$.

<u>Proof</u>: Germ-equivalence is defined in II 6.7. Let $\mathcal{U}$ be the intersection of the domains of ϕ and ψ. From $\phi_* = \psi_*$ it follows immediately that $T^\alpha(\phi)$ and $T^\alpha(\psi)$ coincide on $T^\alpha\Omega|_{\mathcal{U}}$. Now the diagram

$$\begin{array}{ccccc}
T(\Omega_{\alpha\xi})_\xi & \rightarrowtail & T(\Omega)_\xi & \xrightarrow{T(\alpha)} & T(B)_{\alpha\xi} \\
{\scriptstyle T^\alpha(\phi)_\xi}\downarrow & & {\scriptstyle T(\phi)_\xi}\downarrow & & \downarrow{\scriptstyle T(\phi_o)_{\alpha\xi}} \\
T(\Omega'_{\phi_o(\alpha\xi)})_{\phi(\xi)} & \rightarrowtail & T(\Omega')_{\phi(\xi)} & \xrightarrow{T(\alpha')} & T(B')_{\phi_o(\alpha\xi)}
\end{array}$$

is valid for ϕ and ψ, and $\phi_o = \psi_o \colon B \to B'$, so it follows that $T(\phi)$ and $T(\psi)$ coincide on $T(\Omega)_{\mathcal{U}}$.

ϕ and ψ are now two maps $\Omega \rightsquigarrow \Omega'$ which coincide on the closed embedded submanifold B of Ω and whose tangent maps coincide on an open neighbourhood of B; it follows that ϕ and ψ themselves coincide on an open neighbourhood of B.

The second assertion follows from II 3.11. //

We now address the construction of a local morphism of Lie groupoids corresponding to a base-preserving morphism of their Lie algebroids.

<u>Theorem 6.5</u>. Let Ω and Ω' be Lie groupoids on base B with Ω α-connected and α-simply connected, and let $\phi \colon A\Omega \to A\Omega'$ be a morphism of Lie algebroids over B. Then there is a morphism $f \colon \Omega \to \Omega'$ of Lie groupoids over B such that $f_* = \phi$.

<u>Proof</u>: Define $\bar{\phi} \colon A\Omega \to A\Omega \underset{TB}{\oplus} A\Omega'$ by $X \mapsto X \oplus \phi(X)$. Then $\bar{\phi}$ is an injective vector bundle morphism over B, and so its image, $\operatorname{im}(\bar{\phi})$, is a sub vector bundle of $A\Omega \underset{TB}{\oplus} A\Omega'$. It is easy to see that $\operatorname{im}(\bar{\phi})$ is a transitive Lie subalgebroid of $A\Omega \underset{TB}{\oplus} A\Omega'$. Therefore, by 6.1, there is a unique α-connected Lie subgroupoid Φ of $\Omega \underset{B\times B}{\times} \Omega'$ such that $A\Phi = \operatorname{im}(\bar{\phi})$. Let π denote the restriction of the projection $\Omega \underset{B\times B}{\times} \Omega' \to \Omega$ to Φ. Then $\pi_* \colon A\Phi \to A\Omega$ is $X \oplus \phi(X) \mapsto X$ and is evidently an isomorphism of Lie algebroids over B.

From 6.6 below it follows that each $\pi_x: \Phi_x \to \Omega_x$ is a covering and since Φ and Ω are α-connected and Ω is α-simply connected, it follows that each π_x is a diffeomorphism. So, by 3.13, π itself is a diffeomorphism and it is easy to check now that $f = \pi_2 \circ \pi^{-1}: \Omega \to \Omega'$ has the required properties. //

Compare the proof for Lie groups; for example, Varadarajan (1974, 2.7.5).

Theorem 6.6. Let $\phi: \Omega \to \Omega'$ be a morphism of Lie groupoids over B.

(i) If ϕ is a surjective submersion, then for each $b \in B$, $\Omega_b(\Omega'_b, \ker(\phi_b^b), \phi_b)$ is a principal bundle.

(ii) If $\phi_*: A\Omega \to A\Omega'$ is an isomorphism and Ω and Ω' are α-connected, then each $\phi_b: \Omega_b \to \Omega'_b$ is a covering. (In particular, ϕ is a surjective submersion.)

Proof: Use the notation $P = \Omega_b$, $G = \Omega_b^b$, $Q = \Omega'_b$, $H = \Omega'^b_b$, $K = \ker(\phi_b^b)$, $\pi = \beta_b$, $\pi' = \beta'_b$.

(i) From 3.13 it follows that $\phi_b: P \to Q$ is a surjective submersion. The algebraic requirements are easily verified, and it only remains to prove that $P(Q,K,\phi_b)$ admits local charts.

Given $\xi_o \in P$, let $\sigma: U \to P$ be a local section with $\sigma(x_o) = \xi_o$, where $x_o = \pi(\xi_o)$. Then, under the chart $U \times G \to \pi^{-1}(U)$ induced by σ, the point $(x_o, 1)$ corresponds to ξ_o. Let $U \times H \to \pi'^{-1}(U)$ be the chart for $Q(B,H)$ induced by $\phi_b \circ \sigma: U \to Q$. Then ϕ_b is locally $\mathrm{id} \times \phi_b^b: U \times G \to U \times H$.

Now let $\tau: W \to G_o$ be a local section of $\phi_b^b: G_o \to H_o$, where G_o, H_o are the identity components of G and H. Then $\mathrm{id} \times \tau: U \times W \to U \times G_o$ defines a principal bundle chart for $P(Q,K,\phi_b^b)$.

(ii) From 3.13 it follows that ϕ is étale and so $\phi(\Omega)$ is open in Ω'. By II 3.11, it follows that ϕ is onto, and so, by (i), $P(Q,K,\phi_b^b)$ is a principal bundle.

Now $\phi_*: A\Omega \to A\Omega'$ is an isomorphism and so, by 2.8, $(\phi_b^b)_*: \mathfrak{g} \to \mathfrak{g}'$ is an isomorphism and so $K = \ker(\phi_b^b) < G$ is discrete. Since P and Q are connected, it follows that ϕ_b is a covering (see, for example, Hu (1959, pp. 104-105)). //

We will need the full generality of 6.6 in Chapter IV.

Theorem 6.7. Let Ω and Ω' be Lie groupoids on B and $\phi: A\Omega \to A\Omega'$ a morphism of Lie algebroids over B. Then there is a local morphism $F: \Omega \rightsquigarrow \Omega'$ of Lie groupoids over B, such that $F_* = \phi$.

Proof: It is no loss of generality to assume that Ω is α-connected. Let $\psi: M\Omega \to \Omega$ be the projection of the monodromy groupoid, and apply 6.5 to $\phi\circ\psi_*: AM\Omega \to A\Omega'$. There is then a morphism $f: M\Omega \to \Omega'$ over B such that $f_* = \phi\circ\psi_*$. Now, by II 6.11, ψ has a local right-inverse morphism $\chi: \Omega \rightsquigarrow M\Omega$, and $f\circ\chi: \Omega \rightsquigarrow \Omega'$ is now a local morphism over B with $(f\circ\chi)_* = \phi\circ\psi_*\circ\chi_* = \phi$. //

Corollary 6.8. Let $\phi: \Omega \rightsquigarrow \Omega'$ be a local morphism of Lie groupoids over B. Then ϕ is a local isomorphism iff $\phi_*: A\Omega \to A\Omega'$ is an isomorphism.

Proof: Follows from 6.7 and 6.4. //

Two instances of 6.7 need to be noticed.

Example 6.9. Let Ω be a Lie groupoid on a connected base B. A flat infinitesimal conection $\gamma: TB \to A\Omega$ is a morphism of Lie algebroids and so, by 6.7, integrates to a local morphism of Lie groupoids $\theta: B \times B \rightsquigarrow \Omega$. By II 6.8, local morphisms $B \times B \rightsquigarrow \Omega$ are equivalent to section-atlases whose transition functions are constant, and this argument therefore gives an alternative proof of 5.19(ii). The reader may like to trace through in detail the relationship between the two proofs.

Since $A\pi(B) \cong TB$, γ may also be integrated to a global morphism $h^\gamma: \pi(B) \to \Omega$, called the holonomy morphism of γ. Conversely, any Lie groupoid morphism $\pi(B) \to \Omega$ differentiates to a flat infinitesimal connection in Ω.

See also 7.29. The holonomy of general connections is treated in §7. //

Example 6.10. Let Ω be an α-connected Lie groupoid on B and $\rho: A\Omega \to CDO(E)$ a representation of $A\Omega$ in a vector bundle E on B. Then, by 6.5, there is a representation $P: M\Omega \to \Pi(E)$ of the monodromy groupoid of Ω in E with $P_* = \rho$. In the case where Ω is $B \times B$, the monodromy groupoid of Ω is $\pi(B)$ and a representation of $\pi(B)$ in E is precisely a local system of coefficients on B with values in E (see, for example, Hu (1959, IV.15)). In analogy with this case we call a representation $M\Omega \to \Pi(E)$ a local system of coefficients on Ω with values in E.

Note in particular that a flat connection in a vector bundle E on B constitutes a local system of coefficients on B with values in E. //

Lastly, we note for reference the following result, whose proof will now be evident.

Theorem 6.11 Let Ω be a Lie groupoid on base B. Then the covering projection $\psi: M\Omega \to \Omega$ induces an isomorphism of Lie algebroids $\psi_*: AM\Omega \to A\Omega$. //

* * * * * * *

Two items in the Lie theory of Lie groups and Lie algebras which do not generalize to Lie groupoids and transitive Lie algebroids are the correspondence between connected normal subgroups and ideals, and the result that a connected Lie group with abelian Lie algebra is abelian.

The concept of ideal of a transitive Lie algebroid is not defined until IV§1, but assume that we have some concept of ideal which satisfies the minimal requirement that every transitive Lie algebroid is an ideal of itself. Then the α-connected Lie subgroupoid of Ω corresponding to $A\Omega$ itself is the α-identity component subgroupoid Ψ of Ω and II 3.7 shows that Ψ need not be normal in Ω.

If Ω is an α-connected Lie groupoid with abelian Lie algebroid then the adjoint bundle $L\Omega$ is abelian and so the identity components of the vertex groups of Ω are abelian. If the base B is simply-connected then the vertex groups of Ω must be connected (by the long exact homotopy sequence for the vertex bundles (see the proof of II 6.6)), and so Ω is abelian. However if B is not simply connected, then the vertex groups of Ω need not be abelian, and $I: \Omega \to \Pi(G\Omega)$ need not quotient to $B \times B \to \Pi(G\Omega)$: consider, for example, the fundamental groupoid of any manifold, such as the Klein bottle, which has a nonabelian fundamental group. (Compare II 3.7.) It is however true that if $A\Omega$ is abelian, then $Ad: \Omega \to \Pi[L\Omega]$ quotients through a map $h: \pi(B) \to \Pi[L\Omega]$ with $ad = h_* \circ q$.

There remains one major topic in the Lie theory of Lie groupoids and transitive Lie algebroids: the integrability of transitive Lie algebroids. This topic is treated in Chapter V.

§7. Path connections in Lie groupoids

§7 is concerned with the relationship between the action of an infinitesimal connection and the action of its holonomy groupoid.

The first part of this section formalizes the concept of C^∞-path connection in a Lie groupoid Ω and establishes its correspondence with connections in the Lie algebroid $A\Omega$. The path lifting associated with an infinitesimal connection has usually been treated as a subsidiary concept; though passing references to an independent concept of path connection have been made in the literature (for example, Bishop and Crittenden (1964, 5.2), Singer and Thorpe (1967, 7.1)) no full discussion seems to have appeared. Our purpose in treating this concept here is to keep clear the distinction between the infinitesimal aspect of connection theory, which may be developed in the context of abstract transitive Lie algebroids, and those parts of connection theory - the concept of path lifting and holonomy - which require the Lie algebroid to be realized as the Lie algebroid of a specific Lie groupoid. The situation is exactly parallel to that existing with Lie groups and Lie algebras: the one transitive Lie algebroid may arise from several distinct Lie groupoids, which are only locally isomorphic, and although the curvature, for example, depends only on the Lie algebroid, the holonomy, and its associated concepts, depend on the Lie groupoid. This point does not need to be made in the standard treatments of connection theory, because there a connection is regarded as existing on a specific Lie groupoid or principal bundle, but we have argued elsewhere in this book the need to regard abstract Lie algebroids as mathematical structures in their own right.

In 7.11 we prove a very general result, crucial to the developments of Chapter IV, concerning structures on vector bundles defined by tensor fields. 1.20 may be reformulated to state that such a structure is locally trivial iff the structures defined on the fibres of the vector bundle are pairwise isomorphic; in 7.11 we prove that this is so iff the bundle admits a connection compatible with the structure. From this it follows, for example, that (7.13) a morphism of vector bundles $\phi: E^1 \to E^2$ over a base B is of locally constant rank iff ϕ maps some connection ∇^1 in E^1 to a connection ∇^2 in E^2. The proof of 7.11 is a concatenation of results already established. 7.11 is a slight generalization of a result of Greub et al (1973, Chapter VIII); the proof given here is new.

In 7.25 to 7.28 we give a strong, Lie groupoid form of the Ambrose-Singer theorem. 7.25 and 7.26 give an abstract construction of the Lie algebroid of the holonomy groupoid of a connection which is easily seen (IV§1) to hold in any transitive Lie algebroid.

7.30 is a connection-theoretic analysis of morphisms of trivial Lie algebroids over a fixed base. In I 2.13 it is pointed out that a morphism of trivial groupoids $\phi: B \times G \times B \to B \times H \times B$ can be constructed from any morphism $f = \phi^b_b: G \to H$ and any map $\theta: B \to H$; however 7.30 shows that an arbitrary Maurer-Cartan form $\omega = \Delta(\theta) \in A^1(B,\mathfrak{h})$ and Lie algebra morphism $\mathfrak{g} \to \mathfrak{h}$ need to satisfy a further compatibility condition before they define a morphism of trivial Lie algebroids $TB \oplus (B \times \mathfrak{g}) \to TB \oplus (B \times \mathfrak{h})$. This difference in behaviour turns out to be typical. Using 7.30, we obtain a second proof of the local integrability of base-preserving morphisms of transitive Lie algebroids.

Throughout this section we assume that B is a connected manifold. Until we reach 7.9, we work with a fixed Lie groupoid Ω on B.

We modify the notations $C(I,B)$, $P^\alpha = P^\alpha(\Omega)$ and $P^\alpha_o = P^\alpha_o(\Omega)$ of II§6: each now denotes the corresponding set of continuous and piecewise-smooth paths. No topology is required on these sets.

Definition 7.1. A C^∞-path connection in Ω is a map $\Gamma: C(I,B) \to P^\alpha_o(\Omega)$, usually written $c \mapsto \bar{c}$, satisfying the conditions (i) and (ii) of II 7.1, and consequently (iii)-(v) of II 7.4, and in addition,

(vi) If $c \in C(I,B)$ is differentiable at $t_o \in I$ then $\bar{c}$ is also differentiable at t_o;

(vii) If $c_1,c_2 \in C(I,B)$ have $\frac{dc_1}{dt}(t_o) = \frac{dc_2}{dt}(t_o)$ for some $t_o \in I$, then $\frac{d\bar{c}_1}{dt}(t_o) = \frac{d\bar{c}_2}{dt}(t_o)$;

(viii) If $c_1,c_2,c_3 \in C(I,B)$ are such that $\frac{dc_1}{dt}(t_o) + \frac{dc_2}{dt}(t_o) = \frac{dc_3}{dt}(t_o)$ for some $t_o \in I$, then $\frac{d\bar{c}_1}{dt}(t_o) + \frac{d\bar{c}_2}{dt}(t_o) = \frac{d\bar{c}_3}{dt}(t_o)$. //

We refer to (vi) and (vii) as the tangency conditions and to (viii) as the additivity condition. All three are clearly necessary, if it is to be possible to differentiate Γ to an infinitesimal connection.

Proposition 7.2. Let $\Gamma: c \mapsto \bar{c}$ be a C^∞-path connection in Ω. Then

(ix) If $c_1,c_2 \in C(I,B)$ are such that $\frac{dc_1}{dt}(t_o) = k\frac{dc_2}{dt}(t_o)$ for some $t_o \in I$ and $k \in \mathbf{R}$, then $\frac{d\bar{c}_1}{dt}(t_o) = k\,\frac{d\bar{c}_2}{dt}(t_o)$.

(x) If $\phi_t: U \times (-\varepsilon,\varepsilon) \to B$ is a local 1-parameter group of local transformations on B, then the map $\bar{\phi}_t: \Omega^U \times (-\varepsilon,\varepsilon) \to \Omega$ constructed as in II 7.3 is a local 1-parameter group of local transformations on Ω, and

$$\beta\circ\bar{\phi}_t = \phi_t\circ\beta$$

for all $t \in (-\varepsilon,\varepsilon)$.

Proof: (ix) follows from the reparametrization condition on Γ and (x) is proved exactly as for (the local form of) II 7.3. //

Theorem 7.3. There is a bijective correspondence between C^∞-path connections $\Gamma: c \mapsto \bar{c}$ in Ω and infinitesimal connections $\gamma: TB \to A\Omega$ in $A\Omega$, such that a corresponding Γ and γ are related by

$$(1) \qquad \frac{d}{dt}\bar{c}(t_o) = T(R_{\bar{c}(t_o)})\left(\gamma\left(\frac{d}{dt}c(t_o)\right)\right), \qquad c \in C(I,B),\ \ t_o \in I\ .$$

Remark: Note that there is no continuity condition on the map Γ.

Proof: Suppose given a C^∞-path connection Γ. For $X \in T(B)_x$ take any $c \in C(I,B)$ with $c(t_o) = x$ and $\frac{dc}{dt}(t_o) = X$ for some $t_o \in I$, and define

$$\gamma(X) = T\left(R_{\bar{c}(t_o)^{-1}}\right)\left(\frac{d}{dt}\bar{c}(t_o)\right)\ .$$

Since $\alpha\circ\bar{c}$ is constant, the RHS is defined, and lies in $T(\Omega_x)_{\tilde{x}}$. By the tangency conditions, $\gamma(X)$ is well-defined with respect to the choice of c. By (viii) and (ix), $\gamma: T(B)_x \to A\Omega|_x$ is $\mathbf{R}$-linear.

Now let X be a vector field on B. We prove that $\gamma(X)$ is a smooth section of $A\Omega$. Let $\phi_t: U \times (-\varepsilon,\varepsilon) \to B$ be a local flow for X and $\bar{\phi}_t: \Omega^U \times (-\varepsilon,\varepsilon) \to \Omega$ the Γ-lift of $\{\phi_t\}$. Let X* be the (local) vector-field on Ω derived from $\{\bar{\phi}_t\}$. Then, from the definition of $\bar{\phi}_t$ in II 7.3, it is clear that X* is right-invariant. From the definition of γ above, it is clear that X* is the (local) vector field on Ω associated to $\gamma(X)$; in the notation of 3.10, $\tilde{X}^* = \gamma(X)|_U$. Hence, by 3.10, $\gamma(X)$ is smooth. Therefore $\gamma: TB \to A\Omega$ is smooth.

Conversely, suppose given an infinitesimal connection $\gamma: TB \to A\Omega$, and a path $c \in C(I,B)$. Let $\sigma: U \to \Omega_b$ be a local decomposing section of Ω with $c(0) \in U$, and let $\Sigma: U \times G \times U \to \Omega_U^U$ be the corresponding chart. Let $t_1 > 0$ be such that $c([0,t]) \subseteq U$ and c is smooth on $[0,t_1]$ and write $\bar{c}$ on $[0,t_1]$ as $\Sigma\circ(c,a,c(0))$, where $a: [0,t_1] \to G$ has $a(0) = 1$. Then (1) becomes

$$\frac{da}{dt} = T(R_{a(t)})\left(\omega\left(\frac{dc}{dt}\right)\right), \tag{2}$$

where $\omega \in A^1(U,\mathfrak{g})$ is the local connection form of γ with respect to σ. In terms of the right-derivative of a this can be rewritten as

$$\Delta(a) = c^*\omega, \tag{2a}$$

where $c^*\omega \in A^1([0,t_1],\mathfrak{g})$ is the pullback of ω. Now $c^*\omega$ is a Maurer-Cartan form, since $[0,t_1]$ is 1-dimensional, and so there is a unique smooth solution a to (2a) on $[0,t_1]$ with $a(0) = 1$. Since c has only a finite number of points where it is not smooth, and since $c(I)$ is covered by a finite number of domains of decomposing sections $\sigma_i: U_i \to \Omega_b$, this process yields a curve $\bar{c} \in P_0^\alpha$ satisfying (1) and with properties (i), (vi), (vii) and (viii). The remaining property, (ii), is easily seen from the form of (2) and the uniquenss of its solutions. //

The second part of this proof is a reformulation of that of Kobayashi and Nomizu (1963, II 3.1).

<u>Corollary 7.4</u>. (of the proof) Let $\gamma: TB \to A\Omega$ be an infinitesimal connection in Ω and let Γ be the corresponding path connection. Then, for all $X \in \Gamma TB$,

$$\operatorname{Exp} t\gamma(X)(x) = \Gamma(\phi,x)(t)$$

where $\{\phi_t\}$ is a local flow for X near x, and $\Gamma(\phi,x): \mathbb{R} \rightsquigarrow \Omega_x$ is the lift of $t \mapsto \phi_t(x)$.

<u>Proof</u>: Follows from the definition of γ in terms of Γ in 7.3 and the construction of $\bar{\phi}$ in II 7.3. //

<u>Corollary 7.5</u>. Let $\phi: \Omega \to \Omega'$ be a morphism of Lie groupoids over B, let γ be an

infinitesimal connection in Ω, Γ the corresponding path-connection, and γ' the produced connection, $\gamma' = \phi_* \circ \gamma$. Then the C^∞-path connection Γ' associated to γ' is the produced C^∞-path connection, $\Gamma' = \phi \circ \Gamma$, and $\phi(\Psi) = \Psi'$.

<u>Proof</u>: Take $c \in C(I,B)$. Then, since $\bar{c}$ satisfies equation (1) for γ it immediately follows that $\phi \circ \bar{c}$ satisfies equation (1) for $\phi_* \circ \gamma$. Thus $\Gamma' = \phi \circ \Gamma$.

That $\phi(\Psi) = \Psi'$ follows immediately, as in II 7.12. //

<u>Corollary 7.6</u>. If Ω' is a reduction of Ω, and γ: $TB \to A\Omega$ takes values in $A\Omega'$, then $\Psi \leqslant \Omega'$.

<u>Proof</u>: This is a particular case of 7.5. //

For the proof of the following crucial theorem, we refer the reader to Kobayashi and Nomizu (1963, II 7.1, II 4.2).

<u>Theorem 7.7</u>. Let Γ be a C^∞-path connection in Ω. Then the holonomy groupoid Ψ of Γ is a Lie subgroupoid of Ω. //

Using the correspondence between principal bundles and groupoids set up in II 1.19, and the particular account for the holonomy groupoid in II 7.14, a translation of the proof of Kobayashi and Nomizu into Lie groupoid terms is immediate, and need not be given here.

It would be interesting to obtain a proof of 7.7 which works directly with the groupoids, rather than via the holonomy group and holonomy bundle. For example, one may ask for conditions under which a "C^∞ Yamabe theorem" for Lie groupoids is true!

<u>Problem</u>. Let Ω' be a wide, transitive subgroupoid of the Lie groupoid Ω, and suppose that each $\xi \in \Omega'$ may be joined to $\widetilde{\alpha\xi}$ by an element of $P_0^\alpha(\Omega)$ which lies entirely in Ω'. Find conditions under which Ω' is a Lie subgroupoid of Ω. //

It seems likely that a further, rather strong, condition will be needed, to guarantee the transitivity of the associated Lie algebroid.

Still with reference to 7.7, note that no continuity or smoothness condition on the map Γ is needed to guarantee that Ψ is a Lie subgroupoid of Ω (compare II 7.7). However, Ψ need not have the relative topology from Ω: for examples, see

Hano and Ozeki (1956) or Kobayashi and Nomizu (1963, p. 290).

Lastly, we recall that the identity components of the holonomy groups Ψ_x^x are the restricted holonomy groups $H_x^o = \{\hat{\ell} \mid \ell$ is a piecewise smooth loop at x, contractible in B to x$\}$. See Kobayashi and Nomizu (1963, II 4.2) for the proof.

Corollary 7.8. Continuing the notation of 7.7, for each $X \in \Gamma TB$ and all t sufficiently near 0,

$$\gamma(X) \in \Gamma A\Psi \quad \text{and} \quad \text{Exp}\, t\gamma(X) \in \Psi.$$

Proof: These are reformulations of 7.4. //

Henceforth we will call a C^∞-path connection simply a path connection, unless it is necessary to emphasize the differentiability of the paths.

Later in this section we will calculate the Lie algebroid of the holonomy groupoid Ψ and in so doing will establish a form of the Ambrose-Singer theorem. We give now however a series of applications of the concept of holonomy, which are fundamental to all the developments in Chapter IV.

Theorem 7.9. Let E be a vector bundle on (a connected base) B, and let ∇ be a connection in E. Let $\Psi = \Psi(\nabla)$ be the holonomy groupoid of ∇.

Write

$$(\Gamma E)^\nabla = \{\mu \in \Gamma E \mid \nabla(\mu) = 0\}$$

and recall (II 4.14) that

$$(\Gamma E)^\Psi = \{\mu \in \Gamma E \mid \xi\mu(\alpha\xi) = \mu(\beta\xi),\ \forall \xi \in \Psi\} .$$

Then

$$(\Gamma E)^\nabla = (\Gamma E)^\Psi .$$

Remark: Equivalently, $\nabla(\mu) = 0 \iff \Psi(\nabla) < \Phi_\mu$. A section μ of E satisfying $\nabla(\mu) = 0$ is said to be parallel with respect to ∇.

Proof: ($\supseteq$) Take $\mu \in (\Gamma E)^\Psi$. Then μ is Ψ-deformable, and so Ω-deformable. Hence, by 1.20, the isotropy subgroupoid $\Phi = \Phi_\mu$ is a closed reduction of $\Pi(E)$. Now, by 4.7,

$$\Gamma A\Phi = \{D \in \Gamma CDO(E) \mid D(\mu) = 0\}$$

and since $\Psi \leq \Phi$ by assumption, it follows that $D(\mu) = 0$ for all $D \in \Gamma A\Psi$. But Ψ is the holonomy groupoid for ∇ and therefore, by 7.8, $\nabla_X \in \Gamma A\Psi$, $\forall X \in \Gamma TB$. Hence $\nabla(\mu) = 0$.

($\supseteq$) Assume $\nabla(\mu) = 0$. We are to prove that $\xi\mu(\alpha\xi) = \mu(\beta\xi)$ for all $\xi \in \Psi$. Since Ψ is α-connected it is sufficient (by II 3.11) to establish the equation for $\xi \in \Psi_U^U$, U the domain of a decomposing section $U \to \Pi(E)_b$ for E.

So we can assume that E is a trivial vector bundle $U \times V$. Now $\nabla\colon TU \to TU \oplus (U \times \mathfrak{gl}(V))$ has the form $\nabla(X) = X \oplus \omega(X)$, where $\omega \in A^1(U, \mathfrak{gl}(V))$ is the local connection form of ∇ with respect to $x \mapsto (x, id_V, x)$. As in 7.3, the lift of any $c\colon I \to U$ is $t \mapsto (c(t), a(t), c(0))$, where $\Delta(a) = c^*\omega$ and $a(0) = id_V$.

We need to show that $a(t)\mu(c(0)) = \mu(c(t))$ for all $t \in I$, where μ is regarded as a map $U \to V$. We show that $\frac{d}{dt}\left(a(t)^{-1}\mu(c(t))\right)$ is identically zero.

Write $f = \mu \circ c\colon I \to V$. From B§2, equation (6a), we obtain

$$\frac{d}{dt}(a^{-1}(f)) = -\Delta(a^{-1})\left(\frac{d}{dt}\right)(a^{-1}(f)) + a^{-1}\left(\frac{d}{dt} f\right)$$

and from B§2, equation (2), we have

$$\Delta(a^{-1}) = -Ad(a^{-1})(\Delta(a)).$$

Putting these together we get

$$\frac{d}{dt}(a^{-1}(f)) = Ad(a^{-1})\left(\Delta(a)\left(\frac{d}{dt}\right)(a^{-1}(f))\right) + a^{-1}\left(\frac{d}{dt} f\right)$$

$$= a^{-1}\left(\Delta(a)\left(\frac{d}{dt}\right)(f)\right) + a^{-1}\left(\frac{d}{dt} f\right)$$

$$= a^{-1}\left(\omega\left(\frac{dc}{dt}\right)(\mu) + \frac{dc}{dt}(\mu)\right).$$

Now the hypothesis $\nabla(\mu) = 0$ is exactly that $X(\mu) + \omega(X)(\mu) = 0$ for all $X \in T(U)$, and so, putting $X = \frac{dc}{dt}$, the result follows. //

Remark: If, in the second part of the proof of 7.9, one knew in advance that μ was $\Pi(E)$-deformable, then the result would follow directly from 4.7 and 7.6. //

Corollary 7.10. Continue the notation of 7.9 and let $b \in B$ be any reference point. There is an isomorphism of vector spaces

$$(\Gamma E)^{\nabla} \to V^{H}, \quad \mu \mapsto \mu(b)$$

where $V = E_b$ and $H = \Psi_b^b$.

Proof: Apply II 4.15 to $\Psi * E \to E$. //

Theorem 7.11. Let Ω be a Lie groupoid on B and let $\rho: \Omega * E \to E$ be a smooth linar action of Ω on a vector bundle E. For $\mu \in \Gamma E$, the following four conditions are equivalent

(i) μ is Ω-deformable;

(ii) the isotropy groupoid Φ of μ is a Lie subgroupoid of Ω;

(iii) Ω possesses a section-atlas $\{\sigma_i : U_i \to \Omega_b\}$ such that $\rho(\sigma_i(x)^{-1})\mu(x)$ is a constant map $U_i \to E_b$;

(iv) Ω possesses an infinitesimal connection γ such that $(\rho_* \circ \gamma)(\mu) = 0$.

Proof: (i) => (ii) is 1.20. (ii) => (iii) is immediate; (iii) => (i) follows from the connectivity of B.

(ii) => (iv) follows from 4.7.

(iv) => (i) Let Ψ be the holonomy groupoid of γ, and let $\Psi' < \Pi(E)$ be the holonomy groupoid of $\rho_* \circ \gamma$. Then $\rho(\Psi) = \Psi'$ by 7.5. Now $(\rho_* \circ \gamma)(\mu) = 0$ so, by 7.9, $\mu \in (\Gamma E)^{\Psi'}$ and now $\Psi' = \rho(\Psi)$ shows that μ is Ψ-deformable. In particular, μ is Ω-deformable. //

In (iii), the various constant maps $U_i \to E_b$ may be chosen so as to have the same value.

7.11 includes Theorems 1 and 2 of Greub et al (1973, Chapter VIII). In the following series of applications of 7.11, the first, 7.12, may be deduced equally well from Greub et al (1973, loc. cit.).

Theorem 7.12. Let L be a vector bundle on B and [,] a field of Lie algebra brackets on L. Then the following three conditions are equivalent:

(i) The fibres of L are pairwise isomorphic as Lie algebras;

(ii) L admits a connection ∇ such that $\nabla_X([V,W]) = [\nabla_X(V),W] + [V,\nabla_X(W)]$ for all $X \in \Gamma TB$ and $V,W \in \Gamma L$;

(iii) L is an LAB.

Proof: Let $\rho: \Pi(L) * Alt^2(L;L) \to Alt^2(L;L)$ denote the action 1.25(ii). Then (i) is the condition that [,] is $\Pi(L)$-deformable, and (iii) is the condition that $\Pi(L)$ admits a section atlas $\{\sigma_i\}$ such that the corresponding charts for $Alt^2(L;L)$ via ρ map $[\ ,\ ] \in \Gamma Alt^2(L;L)$ to constant maps $U_i \to Alt^2(L_b;L_b)$. So (i) and (iii) are equivalent by the equivalence (i) <=> (iii) of 7.11.

From 4.8(ii) it follows that $\rho_*: CDO(L) \to CDO(Alt^2(L;L))$ is $\rho_*(D)(\phi)(V,W) = D(\phi(V,W)) - \phi(D(V),W) - \phi(V,D(W))$. Therefore (ii) is the condition that L admits a connection ∇ such that $(\rho_* \circ \nabla)([\ ,\ ]) = 0$. Hence (i) <=> (ii) follows from the equivalence (i) <=> (iv) of 7.11. //

Recall from 5.3 that a connection in L satisfying (ii) is called a Lie connection in L.

Theorem 7.13. Let E and E' be vector bundles on B and let $\phi: E \to E'$ be a morphism over B. Then the following three conditions are equivalent:

(i) $x \mapsto rk(\phi_x)$ is constant;

(ii) there exist connections ∇ in E and ∇' in E' such that $\nabla'_X(\phi(\mu)) = \phi(\nabla_X(\mu))$ for all $\mu \in \Gamma E$ and $X \in \Gamma TB$;

(iii) there exist atlases $\{\psi_i: U_i \times V \to E_{U_i}\}$ and $\{\psi'_i: U_i \times V' \to E'_{U_i}\}$ for E and E' such that each $\phi: E_{U_i} \to E'_{U_i}$ is of the form $\phi(x,v) = (x,f_i(v))$, where $f_i: V \to V'$ is a linear map depending only on i.

Proof: This follows from 7.11 in the same way as does 7.12, using now 1.25(iii), 4.8(iii) and the following lemma. //

Lemma 7.14. Let $\phi_1: V \to V'$ and $\phi_2: W \to W'$ be morphisms of vector spaces with $\dim V = \dim W$, $\dim V' = \dim W'$ and $rk(\phi_1) = rk(\phi_2)$. Then there are isomorphisms $\alpha: V \to W$, $\alpha': V' \to W'$ such that $\alpha' \circ \phi_1 = \phi_2 \circ \alpha$. //

In order to facilitate reference to 7.13, we use the following terminology:

Definition 7.15. Let M be any manifold, not necessarily connected, and let $\phi: E \to E'$ be a morphism of vector bundles over M. Then ϕ is of locally constant rank if $x \mapsto rk(\phi_x)$, $M \to \mathbb{Z}$, is locally constant and ϕ is a locally constant morphism if it satisfies condition (iii) of 7.13. //

Then 7.13 may be paraphrased for morphisms $\phi: E \to E'$ of vector bundles over any base as follows: ϕ is of locally constant rank iff it is locally constant and iff there are connections ∇, ∇' in E, E' such that ϕ maps ∇ to ∇'. In 7.13 itself, where the base is connected, the maps f_i may be arranged to be identical.

We will also need the following LAB version of 7.13.

Theorem 7.16. Let L and L' be LAB's on B and let $\phi: L \to L'$ be a morphism of LAB's over B. Then the following three conditions are equivalent:

(i) For each x and y in B, there are Lie algebra isomorphisms $\alpha: L_x \to L_y$ and $\alpha': L'_x \to L'_y$ such that $\phi_y \circ \alpha = \alpha' \circ \phi_x$;

(ii) L and L' possess Lie connections ∇ and ∇' such that $\phi(\nabla_X(V)) = \nabla'_X(\phi(V))$ for all $V \in \Gamma L$ and $X \in \Gamma TB$;

(iii) there exist LAB atlases $\{\psi_i: U_i \times \mathfrak{g} \to L_{U_i}\}$ and $\{\psi'_i: U_i \times \mathfrak{g}' \to L'_{U_i}\}$ for L and L' such that each $\phi: L_{U_i} \to L'_{U_i}$ is of the form $\phi(x,A) = (x, f_i(A))$, where $f_i: \mathfrak{g} \to \mathfrak{g}'$ is a Lie algebra morphism depending only on i. //

The proof is similar to that of 7.13. We refer to a morphism $\phi: L \to L'$ of LAB's over any (not necessarily connected) base M which satisfies (iii) of 7.16 as a locally constant morphism of LAB's. When the base is connected, f_i may be chosen to be independent of i as well.

Further applications of 7.11 are made in IV§1.

The following result is similar in spirit to 7.12, 7.13 and 7.16.

Theorem 7.17. Let E be a vector bundle on B, and let E^1 and E^2 be sub vector bundles of E. Then $E^1 \cap E^2$ is a sub vector bundle of E if there is a connection ∇ in E such that $\nabla(\Gamma E^1) \subseteq \Gamma E^1$ and $\nabla(\Gamma E^2) \subseteq \Gamma E^2$. //

Here $\nabla(\Gamma E') \subseteq \Gamma E'$ is an abbreviation for "$\mu' \in \Gamma E'$, $X \in \Gamma TB \Rightarrow \nabla_X(\mu') \in \Gamma E'$". Though this result is related to 7.11, it requires a more circuitous proof and will only be completed after 7.23.

Proposition 7.18. Let E be a vector bundle on B and let E' be a subbundle. Let $\rho: \Omega * E \to E$ be a smooth linear action of a Lie groupoid Ω on E. Denote by Φ the subgroupoid $\{\xi \in \Omega \mid \xi(E'_{\alpha\xi}) = E'_{\beta\xi}\}$ of Ω.

Then if Φ is a transitive subgroupoid of Ω, it is a closed embedded reduction of Ω.

Proof: Let q be the rank of E' and let $G_q(E) \to B$ be the fibre bundle with $G_q(E)_x$, for $x \in B$, the Grassmannian of q-dimensional subspaces of E_x and charts induced from the charts of E in the natural fashion. Then E' is a smooth Ω-deformable section of $G_q(E)$. Now apply 1.20. //

Definition 7.19. Let E be a vector bundle on B and E' a sub vector bundle of E. Then $\Pi(E,E')$ denotes the Lie groupoid $\{\phi \in \Pi(E) \mid \phi(E'_{\alpha\phi}) = E'_{\beta\phi}\}$. //

Proposition 7.20. With the notation of 7.19, the isomorphism $\mathcal{D}: A\Pi(E) \to CDO(E)$ of 4.5 maps $\Gamma A\Pi(E,E')$ isomorphically onto $\{D \in \Gamma CDO(E) \mid D(\Gamma E') \subseteq \Gamma E'\}$.

The proof of 7.20 is completed after 7.22. Although 7.20 resembles 4.7, the method of 4.7 cannot be used in a general fibre bundle (such as a Grassmannian) and we are obliged to give a different proof.

Lemma 7.21. E admits a connection ∇ such that $\nabla(\Gamma E') \subseteq \Gamma E'$.

Proof: Let < , > be a Riemannian structure on E, and let E" be the orthogonal complement to E' in E. Let ∇' and ∇'' be connections in E' and E' and define ∇ in $E = E' \oplus E''$ by $\nabla_X(\mu' \oplus \mu'') = \nabla'_X(\mu') \oplus \nabla''_X(\mu'')$. //

Proposition 7.22. There is a transitive sub Lie algebroid CDO(E,E') of CDO(E) which has the property

$$\Gamma CDO(E,E') = \{D \in \Gamma CDO(E) \mid D(\Gamma E') \subseteq \Gamma E'\}.$$

Proof: Let End(E,E') be the sub LAB of End(E) defined by

$$\Gamma End(E,E') = \{\phi \in \Gamma End(E) \mid \phi(\Gamma E') \subseteq \Gamma E'\};$$

it is easy to prove (see IV 1.1 and the subsequent discussion) that a unique such LAB exists. Let ∇ be a connection in E such that $\nabla(\Gamma E') \subseteq \Gamma E'$ and define

$$i: TB \oplus End(E,E') \to CDO(E)$$

by $i(X \oplus \phi) = \nabla_X + \phi$. Then i is an injection since $End(E,E') \to End(E)$ is an injection, and so im(i) is a sub vector bundle of CDO(E). It is easily verified that $[\nabla_X, \nabla_Y] = \nabla_X \circ \nabla_Y - \nabla_Y \circ \nabla_X$ maps $\Gamma E'$ into $\Gamma E'$ for all $X,Y \in \Gamma TB$, and that $[\nabla_X, \psi] = \nabla_X \circ \psi - \psi \circ \nabla_X$ does likewise, for $X \in \Gamma TB$, $\psi \in \Gamma End(E,E')$. It now follows that im(i) is closed under the bracket on CDO(E), and hence im(i) is a reduction of CDO(E).

If $\tilde{\nabla}$ is a second connection in E such that $\tilde{\nabla}(\Gamma E') \subseteq \Gamma E'$, then $\tilde{\nabla} - \nabla$ takes values in End(E,E') and so it is easily seen that $im(\tilde{i}) = im(i)$. Denote this common image by CDO(E,E'); if $D \in \Gamma CDO(E)$ has the property that $D(\Gamma E') \subseteq \Gamma E'$ then $D = \nabla_{q(D)} + (D - \nabla_{q(D)})$ shows that $D \in \Gamma CDO(E,E')$. //

A similar construction may be carried out with any suitable family of connections on E.

Proof of 7.20: If $X \in \Gamma A\Pi(E,E')$ then Exp tX takes values in $\Pi(E,E')$ for all t and so, by the definition of $\mathfrak{D}$ in 4.5, $\mathfrak{D}(X) \in \Gamma CDO(E,E')$. Thus $\mathfrak{D}$ maps $A\Pi(E,E')$ into CDO(E,E').

To prove that $\mathfrak{D}(A\Pi(E,E')) = CDO(E,E')$, is suffices (by 2.8) to prove that $\mathfrak{D}^+(L\Pi(E,E')) = End(E,E')$. Fibrewise, $\mathfrak{D}^+$ is $T(GL(V,V'))_I \to \mathfrak{gl}(V,V')$, and is an isomorphism by the following (classical) lemma. //

Lemma 7.23. Let V be a vector space and V' a subspace. Let $GL(V,V') = \{A \in GL(V) \mid A(V') = V'\}$ and let $\mathfrak{gl}(V,V') = \{X \in \mathfrak{gl}(V) \mid X(V') \subseteq V'\}$. Then $\mathfrak{gl}(V,V')$ is the Lie algebra of GL(V,V').

Proof: Take $X \in \mathfrak{gl}(V,V')$. Then $X^n \in \mathfrak{gl}(V,V')$ for all integers $n > 0$ and therefore, since V' is closed, $\exp tX$ maps V' into V' for all t. Thus $\exp tX \in GL(V,V')$. //

Proof of 7.17. Let $\Psi < \Pi(E)$ be the holonomy groupoid of ∇. Now $\nabla_X \in \Gamma CDO(E,E^1) = \Gamma A\Pi(E,E^1)$ so, by 7.6, $\Psi < \Pi(E,E^1)$. Similarly $\Psi < \Pi(E,E^2)$. So every element ξ of Ψ maps $E^1_{\alpha\xi} \cap E^2_{\alpha\xi}$ to $E^1_{\beta\xi} \cap E^2_{\beta\xi}$.

Let $\{\phi_i : U_i \to \Psi_b\}$ be a section-atlas for Ψ. Then the associated charts $\psi_i: U_i \times E_b \to E_{U_i}$, $(x,v) \mapsto \sigma_i(x)(v)$, restrict to charts for $E^1 \cap E^2$. //

We also need an LAB version of 7.17; the proof is exactly analogous.

Theorem 7.24. Let L be an LAB on B, and let L^1 and L^2 be sub LAB's of L. Then $L^1 \cap L^2$ is a sub LAB of L if there is a Lie connection ∇ in L such that $\nabla(\Gamma L^1) \subseteq \nabla(\Gamma L^1)$ and $\nabla(\Gamma L^2) \subseteq \Gamma L^2$. //

We arrive now at the Ambrose-Singer theorem. First we show that, given an infinitesimal connection γ in a Lie algebroid $A\Omega$, there is a least reduction, denoted $(A\Omega)^\gamma$, of $A\Omega$ which contains γ. As we will see in IV§1, this construction may be carried out in any transitive Lie algebroid. It then follows immediately that this reduction $(A\Omega)^\gamma$ is the Lie algebroid of the holonomy groupoid of γ.

Until we reach 7.27, let Ω be a Lie groupoid on B and let γ be an infinitesimal connection in Ω.

Proposition 7.25. Let L' be a sub LAB of $L\Omega$ such that

(i) $\bar{R}_\gamma(X,Y) \in L'$ $\quad \forall X,Y \in T(B)$, and

(ii) $\nabla^\gamma(\Gamma L') \subseteq \Gamma L'$.

Then there is a reduction $A' < A\Omega$ defined by

$$\Gamma A' = \{X \in \Gamma A \mid X - \gamma q(X) \in \Gamma L'\}$$

which has L' as adjoint bundle and is such that $\gamma(X) \in A'$ for all $X \in TB$.

Proof: Define $\phi: TB \oplus L' \to A\Omega$ by $\phi(X \oplus V') = \gamma(X) + V'$. Then $\mathrm{im}(\phi) = A'$ and, applying the 5-lemma to

$$\begin{array}{ccccc} L' & \rightarrowtail & TB \oplus L' & \twoheadrightarrow & TB \\ \downarrow & & \downarrow & & \| \\ L\Omega & \rightarrowtail & A\Omega & \twoheadrightarrow & TB \end{array}$$

as in 2.8, it follows that A' is a sub vector bundle of $A\Omega$.

Clearly $\gamma(X) \in A'$ for $X \in TB$ and so the restriction of q to A' is surjective. Clearly $\ker(q|_{A'}) = L'$.

To prove that $\Gamma A'$ is closed under the bracket on ΓA, take $X,Y \in \Gamma A'$ and write $X = \gamma q X + V'$, $Y = \gamma q Y + W'$, where $V',W' \in \Gamma L'$. Then

$$[X,Y] = \gamma[qX,qY] - \bar{R}_\gamma(qX,qY) + \nabla^\gamma_{qX}(W')$$
$$- \nabla^\gamma_{qY}(V') + [V',W']$$

and the last four terms are in $\Gamma L'$ by (i) and (ii). //

7.25 resembles 7.22; we give a general statement of this procedure in IV 3.20.

Proposition 7.26. There is a least sub LAB, denoted $(L\Omega)^\gamma$, of $L\Omega$ which has the properties (i) and (ii) of 7.25.

Proof: It suffices to prove that if L^1 and L^2 both satisfy (i) and (ii), then $L^1 \cap L^2$ does also. The only point that is not clear is that $L^1 \cap L^2$ is a sub LAB, and since ∇^γ is a Lie connection this is established by 7.24. //

The corresponding reduction of $A\Omega$ is denoted by $(A\Omega)^\gamma$ and called the γ-curvature reduction of $A\Omega$.

Theorem 7.27. (Ambrose-Singer) Let Ω be a Lie groupoid on B and let $\gamma: TB \to A\Omega$ be an infinitesimal connection. Denote the associated C^∞-path connection by Γ and the holonomy groupoid of Γ by Ψ.

Then $A\Psi = (A\Omega)^\gamma$.

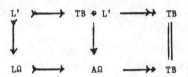

Proof: By 7.8, γ takes values in $A\Psi$. Hence $L\Psi$ satisfies the conditions of 7.25 and therefore $L\Psi \geq (L\Omega)^{\gamma}$ and $A\Psi \geq (A\Omega)^{\gamma}$. On the other hand, γ takes values in $(A\Omega)^{\gamma}$ and $A\Psi \leq (A\Omega)^{\gamma}$. //

Since Ψ is α-connected, Ψ is determined by $(A\Omega)^{\gamma}$ via 6.1. Though 7.27 is considerably stronger than the standard statement of the Ambrose-Singer theorem, it should be pointed out that everything required to prove 7.27 is implicit in the proofs of Kobayashi and Nomizu (1963, II 7.1, II 8.1); what the language of Lie groupoids and Lie algebroids has provided is the means to formulate these results with their full force.

To complete this account of the Ambrose-Singer theorem, we indicate how the standard formulation may be obtained from 7.27.

Example 7.28. Let $P(B,G)$ be a principal bundle and let $\omega \in A^1(P,\mathfrak{g})$ and $\Omega \in A^2(P,\mathfrak{g})$ be a connection 1-form and its curvature 2-form. (For the relationship between ω and $\gamma: TB \to \frac{TP}{G}$ and between Ω and $\bar{R}_{\gamma}: TB \oplus TB \to \frac{P \times \mathfrak{g}}{G}$, see A§4.) Let $\mathfrak{h}$ be the Lie subalgebra of $\mathfrak{g}$ generated by $\{\Omega(X,Y) \mid X,Y \in T(P)\}$. Since $\mathrm{Ad}g\Omega(X,Y) = \Omega(T(R_{g^{-1}})X, T(R_{g^{-1}})Y)$, $g \in G$, it follows that $\mathfrak{h}$ is stable under $\mathrm{Ad}G$.

Let $\{\sigma_i: U_i \to P\}$ be a section-atlas for P and let $\psi_i: U_i \times \mathfrak{g} \to \frac{P \times \mathfrak{g}}{G}\Big|_{U_i}$ be the associated charts for $\frac{P \times \mathfrak{g}}{G}$. Since the transition functions $\psi_i^{-1}\psi_j$ take values in $\mathrm{Ad}G \leq \mathrm{Aut}(\mathfrak{g})$, it follows that $\mathfrak{h}$ translates into a well-defined sub LAB K of $\frac{P \times \mathfrak{g}}{G}$.

From A 4.16 it follows that $\bar{R}_{\gamma}(X,Y) \in K$, $\forall X,Y \in TB$. To show that condition (ii) of 7.25 holds, note that $\nabla^{\gamma}_X(\psi_i(V)) = \psi_i(X(V) + [\omega_i(X),V])$, where $V: U_i \to \mathfrak{g}$, and the $\omega_i \in A^1(U_i,\mathfrak{g})$ are the local connection forms of γ with respect to $\{\sigma_i\}$ (see 5.16). From this it follows easily that $\nabla^{\gamma}(\Gamma K) \subseteq \Gamma K$.

Hence $(L\Omega)^{\gamma} \leq K$. Now $\mathfrak{h}$ is the least Lie subalgebra of $\mathfrak{g}$ which contains all the values of Ω so, by following through the relationships between $\bar{R}_{\gamma}$ and Ω, and K and $\mathfrak{h}$, it follows that any sub LAB of $L\Omega$ which satisfies (i) of 7.25, also contains K. Hence $(L\Omega)^{\gamma} \geq K$.

Note that $\mathfrak{h}$ is an ideal of $\mathfrak{g}$, since it is stable under $\mathrm{Ad}G$. //

The account of flat connections given in §6 can now be made more precise.

Proposition 7.29. Let Ω be a Lie groupoid on B and let γ: TB → AΩ be a flat connection. Then the holonomy groupoid Ψ of γ is a quotient of π(B).

Proof: Since $\bar{R}_\gamma = 0$ we can take L' = B × {0} in 7.25. So $L\Psi = (L\Omega)^\gamma = B \times \{0\}$ and γ is an isomorphism TB → AΨ. Now, as in 6.9, γ integrates to a morphism $h^\gamma: \pi(B) \to \Psi$. Since both π(B) and Ψ are α-connected, it follows from 6.6(ii) that h^γ is a surjective submersion and fibrewise a covering. //

Thus the locally constant transition functions found in 6.9 form a cocycle for Ψ, and the path connection for γ in Ω is the image under h^γ of the unique path connection in π(B).

In the case of flat connections in a vector bundle E, h^γ gives a smooth action of π(B) on E.

We close this section with a more detailed analysis of morphisms of transitive Lie algebroids, based on 7.11-7.16.

Example 7.30. Let ϕ: TB ⊕ (B × $\mathfrak{g}$) → TB ⊕ (B × $\mathfrak{h}$) be a morphism of trivial Lie algebroids. By 2.4, ϕ is $\phi(X \oplus V) = X \oplus (\omega(X) + \phi^+(V))$ where $\omega \in A^1(B,\mathfrak{h})$ is a Maurer-Cartan form, $\phi^+: B \times \mathfrak{g} \to B \times \mathfrak{h}$ is the induced morphism of LAB's, and ϕ^+ and ω satisfy the compatibility equation

(3) $$X(\phi^+(V)) - \phi^+(X(V)) + [\omega(X),\phi^+(V)] = 0.$$

Let ∇^0 be the standard flat connection in B × $\mathfrak{g}$, and define a connection ∇^1 in B × $\mathfrak{h}$ by

$$\nabla^1_X(W) = X(W) + [\omega(X),W].$$

Since ω is a Maurer-Cartan form, ∇^1 is flat, and it is easily seen to be Lie. In terms of 7.16, (3) now asserts that $\phi^+(\nabla^0_X(V)) = \nabla^1_X(\phi^+(V))$, for V: B → $\mathfrak{g}$, X ε ΓTB. Hence ϕ^+ is a locally constant morphism of LAB's.

Further, ∇^0 and ∇^1 together induce, by 4.8(iii), a connection ∇ in Hom(B × $\mathfrak{g}$, B × $\mathfrak{h}$) = B × Hom($\mathfrak{g}$,$\mathfrak{h}$), the vector bundle whose sections are the vector bundle morphisms B × $\mathfrak{g}$ → B × $\mathfrak{h}$. ∇ is

$$\nabla_X(\psi)(V) = \nabla^1_X(\psi(V)) - \psi(\nabla^0_X(V))$$

$$= X(\psi(V)) - \psi(X(V)) + [\omega(X),\psi(V)].$$

Since ∇^0 and ∇^1 are both flat, ∇ is flat.

(3) now asserts that $\nabla(\phi^+) = 0$ and, applying 7.10, ϕ^+ is determined by its restriction to any single fibre, $f = \phi^+|_b : \mathfrak{g} \to \mathfrak{h}$. By analogy with the situation for morphisms of trivial groupoids, one would expect this. However 7.10 also shows that f cannot be an arbitrary Lie algebra morphism, but must lie in $\mathrm{Hom}(\mathfrak{g},\mathfrak{h})^H$, where H is the holonomy group of ∇. Since ∇ is induced from ∇^0 and ∇^1 via 4.8(iii), it follows (from 7.5) that H is the direct product $H^0 \times H^1$ of the holonomy groups of ∇^0 and ∇^1 and acts on $\mathrm{Hom}(\mathfrak{g},\mathfrak{h})$ by $(\alpha^0,\alpha^1)(\psi) = \alpha^1 \circ \psi \circ (\alpha^0)^{-1}$. Since ∇^0 has trivial holonomy, we conclude that f must lie in $\mathrm{Hom}(\mathfrak{g}, \mathfrak{h}^{\pi_1 B})$ where $\mathfrak{h}^{\pi_1 B}$ is the vector space of elements of $\mathfrak{h}$ invariant under the ∇^1-holonomy action of $\pi_1 B$ (see 7.29). A direct calculation with (2) of 7.3 shows that $\pi_1 B$ must act by elements of $\mathrm{Int}(\mathfrak{h})$.

This is a real restriction; there are certainly Lie algebras with nontrivial discrete groups of inner automorphisms. (An example is the 3-dimensional Lie algebra with $[e_1,e_2] = e_3$, $[e_2,e_3] = [e_3,e_1] = 0$; see Helgason (1978, p. 130).)

Thus if $\pi_1 B \neq 0$, if $\omega \neq 0$, if $\mathfrak{h}$ is not abelian, and if the action of $\pi_1 B$ on $\mathfrak{h}$ induced by ω is nontrivial, then there will be Lie algebra morphisms $\mathfrak{g} \to \mathfrak{h}$ which are not the restriction of a Lie algebroid morphism which induces ω. //

This analysis also yields an alternative proof of 6.7.

<u>Proposition 7.31</u>. Let B be a simply-connected manifold and let $\phi\colon TB \oplus (B \times \mathfrak{g}) \to TB \oplus (B \times \mathfrak{h})$ be a morphism of trivial Lie algebroids over B. If G and H are Lie groups with Lie algebras $\mathfrak{g}$ and $\mathfrak{h}$, then there is a local morphism M of trivial Lie groupoids $B \times G \times B \rightsquigarrow B \times H \times B$ such that $M_* = \phi$, and M is unique up to germ-equivalence.

<u>Proof:</u> For convenience we assume that G is connected and simply-connected; the general case is only notationally more complicated.

With ϕ^+ and ω as in 7.30, choose $b \in B$ and define $\theta\colon B \to H$ to be the solution to $\Delta(\theta) = \omega$, $\theta(b) = 1$, and $F\colon G \to H$ to be the Lie group morphism with $F_* = \phi^+|_b$. Define $M\colon B \times G \times B \to B \times H \times B$ by $M(y,g,x) = (y,\theta(y)F(g)\theta(x)^{-1},x)$. Then, by 3.21, M_* is $X + V \mapsto X + \{\Delta(\theta)(X) + \mathrm{Ad}(\theta)F_*(V)\}$; thus the Maurer-Cartan form for M_* is $\Delta(\theta) = \omega$ and $M_*^+ = Ad(\theta)F_*$. We need to show that $M_*^+ = \phi^+$.

Since both M_*^+ and ϕ^+ are LAB morphisms compatible with ω, they are both parallel with respect to ∇. Also, $M_*^+|_b = (M_b^b)_* = F_* = \phi^+|_b$, since $\theta(b) = 1$. So, as in 7.30, it follows that $M_*^+ = \phi^+$.

That $M_*^+ = Ad(\theta)F_*$ is parallel with respect to ∇ also follows from B 2.1.

To prove uniqueness, let $M'\colon B \times G \times B \to B \times H \times B$ be any other morphism with $M'_* = \phi$. Then, by I 2.13, $M'(y,g,x) = (y,\theta'(y)F'(g)\theta'(x)^{-1},x)$ where $\theta'(b) = 1$. It easily follows that $\theta' = \theta$ and $F' = F$. //

<u>Theorem 7.32.</u> Let Ω and Ω' be Lie groupoids on B and let $\phi\colon A\Omega \to A\Omega'$ be a morphism of Lie algebroids over B. Then there is a local morphism $M\colon \Omega \rightsquigarrow \Omega'$ of Lie groupoids over B such that $M_* = \phi$, and M is unique up to germ-equivalence.

<u>Proof</u>: Let $\{U_i\}$ be a simple cover of B, and let $\{\sigma_i\colon U_i \to \Omega_b\}$ and $\{\sigma'_i\colon U_i \to \Omega'_b\}$ be section-atlases for Ω and Ω' over $\{U_i\}$. Let Σ_i denote the isomorphism $U_i \times G \times U_i \to \Omega_{U_i}^{U_i}$, $(y,g,x) \mapsto (y,\sigma_i(y)g\sigma_i(x)^{-1},x)$ and $(\Sigma_i)_*\colon TU_i \oplus (U_i \times \mathfrak{g}) \to A\Omega|_{U_i}$ its derivative. Let $S_{ij} = \Sigma_i^{-1}\circ\Sigma_j$ for $U_{ij} \neq \emptyset$. Similarly with Σ'_i and S'_{ij}.

Define $\phi^i = (\Sigma'_i)_*\circ\phi\circ(\Sigma_i^{-1})_*$; by 7.31, ϕ^i integrates to a well-defined local morphism $M^i\colon U_i \times G \times U_i \rightsquigarrow U_i \times G' \times U_i$. In order to show that the $\Sigma'_i\circ M^i\circ\Sigma_i^{-1}$ stick together into a well-defined local morphism $\Omega \rightsquigarrow \Omega'$, it is sufficient to prove that $(S'_{ij})^{-1}\circ M^i\circ S_{ij} = M^j$ whenever $U_{ij} \neq \emptyset$. By the uniqueness result in 7.31, it suffices to prove that $(S'_{ij})_*^{-1}\circ M_*^i\circ(S_{ij})_* = M_*^j$, and this follows from the definition of ϕ^i, ϕ^j.

Likewise, the uniqueness statement may be deduced from the uniqueness result in 7.31. //

The idea of this proof of 6.7 is due to Ngô van Quê (1968). Although part of the analysis in 7.30 relies on 6.7, it would be possible to develop connection theory sufficiently to prove 7.32 without making use of 6.4 to 6.10. Thus this connection-theoretic proof of 7.32 is independent of the results in §6 on the local integrability of morphisms of Lie algebroids.

CHAPTER IV THE COHOMOLOGY OF LIE ALGEBROIDS

It is one of the principal theses of this book that Lie algebroids deserve to be recognized as mathematical objects in their own right. It will already be apparent from III§5 that infinitesimal connection theory - that part of standard connection theory which does not depend on the concepts of path-lifting or holonomy - can be developed entirely within the context of transitive Lie algebroids, and the final step in demonstrating this is accomplished early in §1. The cohomology theory which is developed in this chapter includes the equivariant de Rham cohomology $H^*_{deRh}(P)^G$ of a principal bundle P(B,G) (with coefficients in vector bundles associated to P(B,G) now allowed) and incidentally shows that this cohomology, which has been the subject of an enormous body of work (see Greub et al (1976)) is strictly an infinitesimal invariant: principal bundles which are locally isomorphic will have the same equivariant de Rham cohomology although they need not be isomorphic.

The advantages of this point of view, both for connection theory and cohomology theory, are immense. One should try to imagine a situation in which Lie group theory was actively pursued without any use being made of the Lie algebra. The situation for Lie groupoid and principal bundle theory is exactly comparable: it is for example well-known that curvature is a more accessible, but less subtle, invariant than holonomy; from the results of III§7 it is clear that two locally isomorphic bundles will admit the same infinitesimal connections and that corresponding connections will have the same curvature, but the holonomy of these connections will depend upon the connectivity properties of the underlying bundle.

The situation with the equivariant de Rham cohomology is similar. In the case of cohomology theory however, there are further benefits arising from the use of the Lie algebroid concept. Firstly the Lie algebroid cohomology produces results for the equivariant de Rham cohomology $H^*_{deRh}(P,V)^G$ which, in the case of compact groups G and trivial coefficients reduces to $H^*_{deRh}(P) \otimes V$ itself; it is reasonable to expect that the vast body of work done on the cohomology structure of principal bundles with compact groups will generalize to the equivariant (or Lie algebroid) de Rham cohomology. Some beginnings on this programme are made here; their development will be continued elsewhere.

The concept of Lie algebroid also allows de Rham cohomology to be treated as a cohomology theory of algebraic type, comparable to the cohomology theories of Lie algebras or discrete groups. Together with the fact that coefficients in general vector bundles are now allowed, this enables the enormous body of results and techniques established for discrete group and Lie algebra cohomology to be applied to principal bundle cohomology. Again, a beginning on this programme is made in this chapter; its development will be continued elsewhere.

The concept of Lie algebroid is due to Pradines (1967); the closely related concept of Lie pseudo-algebra has been found (under a variety of names) by many authors - see III§2. The cohomology of Lie pseudo-algebras has likewise been defined by a number of authors - see §2 in this chapter. However we believe that this is the first occasion on which the technical problems peculiar to Lie algebroid cohomology - namely, the problems involved in staying within the category of smooth vector bundles - have been dealt with, and we believe this account goes considerably further than any previous account.

The central results of this chapter are two classifications of transitive Lie algebroids. Firstly, the results of §3 give a global classification of transitive Lie algebroids in terms of curvature forms and adjoint connections. Secondly, 4.1 allows the classification of transitive Lie algebroids by transition forms, which was begun in III§5, to be extended to all transitive Lie algebroids. Both these classifications are cohomological in nature, and it should be noted that both are classifications up to equivalence, not up to isomorphism.

The results of this chapter are from Chapter III of Mackenzie (1979), but have been substantially revised in the account given here. The proof of 4.1 is a revision and the detailed classification (including III 5.15) appear here for the first time. Most of the proofs in §3 have been rewritten for clarity. As well, the proofs of 1.6 and 1.16 now use the results of III§7 instead of the concept of local flat connection (as in 4.7). This seems preferable, since the earlier approach was in danger of appearing circular.

The theory of transitive Lie algebroids may be developed without reference to the theory of Lie groupoids. A reader familiar with principal bundles may read this chapter, together with Appendix A, III§2 and III§5, independently of the rest of the book, although they will miss some explanatory material by so doing.

We now give a brief description of the sections. §1 proves several technical results about transitive Lie algebroids, which enable a proper algebraic theory to be developed. These results establish that transitive Lie algebroids do in many ways behave like Lie groupoids. §2 defines the cohomology of an (arbitrary) Lie algebroid and gives interpretations in degrees 2, 1 and 0. §3 deals with the theory of general (non-abelian) extensions of Lie algebroids and their cohomological classification. This section embodies a re-interpretation of infinitesimal connection theory; in particular, we obtain a necessary and sufficient condition for an LAB-valued 2-form to be the curvature of a connection in a transitive Lie algebroid. §4 gives the proof that every transitive Lie algebroid is locally isomorphic to a trivial Lie algebroid; this result is central to Chapter V and will, we believe, be central to any further development of the theory of transitive Lie algebroids. §5 constructs a spectral sequence for the cohomology of a transitive Lie algebroid in terms of the cohomology of its base and its adjoint bundle, and uses algebraic methods to calculate a few of the higher-order differentials. The relationship between extensions of a transitive Lie algebroid and extensions of its adjoint bundle is explicated.

§1. The abstract theory of transitive Lie algebroids

This section uses the results of III§7 to prove several algebraic results of basic importance about transitive Lie algebroids. In 1.4 we prove that the adjoint bundle of a transitive Lie algebroid is an LAB; in 1.6 we prove that if $\phi: A \to A'$ is a base-preserving morphism of transitive Lie algebroids then $\phi^+: L \to L'$ is a locally constant morphism of LAB's; it follows that such morphisms have well-defined kernels and images. In 1.16 we prove that if ρ is a representation of a transitive Lie algebroid A on a vector bundle E, then E^L is a flat vector bundle with a natural flat connection; in 1.19 we prove that (under the same hypotheses) $(\Gamma E)^A$ is naturally isomorphic to $(V^{\mathfrak{g}})^{\pi_1 B}$, where V and $\mathfrak{g}$ are the fibre types of E and L and $\pi_1 B$ acts via the holonomy of the natural flat connection in E^L. These results are fundamental to the cohomology theory developed in the following sections, and to any development of the algebraic theory of transitive Lie algebroids.

Each of these results is an infinitesimal version of results established for locally trivial groupoids in Chapters II and III; the groupoid results are comparatively elementary.

The results of this section are due to the author. 1.4, 1.6, 1.15 and several of the subsidiary results appeared (with different proofs in the case of 1.6 and 1.15) in Mackenzie (1979).

We begin with some necessary observations about LAB's. The definition, recall, was given in III 2.3. Throughout this section, except in 1.19, B is a fixed arbitrary manifold.

Many constructions in the category of Lie algebras carry over to LAB's. Examples we will need include the centre LAB ZL of an LAB L, the derived LAB [L,L], the LAB of derivations Der(L), and the adjoint LAB ad(L). In the first two cases, the following construction principle applies.

Proposition 1.1. Let L be an LAB on B with fibre type $\mathfrak{g}$. Let $\mathfrak{h}$ be a characteristic subalgebra of $\mathfrak{g}$; that is, $\phi(\mathfrak{h}) = \mathfrak{h}$ for all $\phi \in \mathrm{Aut}(\mathfrak{g})$. Then there is a well-defined sub LAB K of L such that any LAB chart $\psi: U \times \mathfrak{g} \to L_U$ for L restricts to an LAB chart $U \times \mathfrak{h} \to K_U$ for K.

Proof: Immediate. //

Taking $\mathfrak{h} = \mathfrak{z}$, the centre of $\mathfrak{g}$, the resulting LAB, denoted ZL, clearly has fibres $ZL|_x$, $x \in B$, which are the centres of the fibres L_x of L. Further, for any open $U \subseteq B$, the (infinite-dimensional) $\mathbb{R}$-Lie algebra $\Gamma_U(ZL)$ is the centre of $\Gamma_U L$. The LAB ZL is called the centre of L. The derived sub LAB [L,L] is obtained in the same way.

For Der(L), consider first a vector bundle E on B. The vector bundle End(E) is the unique vector bundle with fibres $\mathrm{End}(E)_x = \mathrm{End}(E_x)$, $x \in B$, and charts $\bar{\psi}: U \times \mathfrak{gl}(V) \to \mathrm{End}(E)_U$ induced from charts $\psi: U \times V \to E_U$ for E by $\bar{\psi}_x(A) = \psi_x \circ A \circ \psi_x^{-1}$. Here V is the fibre type of E and ψ_x is the isomorphism $V \to E_x$ obtained by restricting ψ. It follows that End(E) is an LAB with respect to these charts.

Now given an LAB L, with fibre type $\mathfrak{g}$, observe that the Lie subalgebra $\mathrm{Der}(\mathfrak{g})$ of $\mathfrak{gl}(\mathfrak{g})$ is invariant under automorphisms of $\mathfrak{gl}(\mathfrak{g})$ of the form $A \mapsto s \circ A \circ s^{-1}$, where $s: \mathfrak{g} \to \mathfrak{g}$ is a Lie algebra automorphism. Applying the same method of proof as for 1.1, it follows that Der(L) is a sub LAB of End(L). It is called the LAB of derivations of L. (Der(L) was introduced in III§4 by a different construction; the equivalence of the two definitions follows from III 3.17.)

Proposition 1.2. Let L be an LAB on B. Then the LAB morphism ad: $L \to \mathrm{Der}(L)$, defined as being fibrewise the adjoint map $\mathrm{ad}_x: L_x \to \mathrm{Der}(L_x)$ of the fibres of L, is locally constant as a morphism of LAB's, in the sense of III 7.16(iii).

Proof: To prove that ad is smooth, note that $[V,W] = \mathrm{ad}(V)(W)$ is smooth whenever V and W are smooth sections of L. To prove that it is locally constant, note that ad is locally $id_U \times ad : U \times \mathfrak{g} \to U \times Der(\mathfrak{g})$, with respect to an LAB chart of L and the corresponding chart for Der(L). //

It follows that ad(L), the image of ad, is a sub LAB of Der(L), called the adjoint LAB of L. It is also an ideal of Der(L), in the sense of the following definition, and is also called the ideal of inner derivations of L.

Definition 1.3. Let L be an LAB on B and let K be a sub LAB of L. Then K is an ideal of L, denoted $K \triangleleft L$, if K_x is an ideal of L_x, for all $x \in B$. //

Given K an ideal of L, a quotient LAB L/K can be constructed in an obvious fashion. Its elements will be written V + K or $\bar{V}$.

The terminology of Lie algebra theory will be taken over without comment: an LAB is reductive, semisimple, nilpotent, abelian, etc., if each of its fibres has the corresponding property. Many deep results of the structure theory of Lie algebras generalize without effort: for example, a reductive LAB L is the direct sum $ZL \oplus [L,L]$.

With these preliminaries established, we turn to the abstract theory of transitive Lie algebroids. The first result shows the relevance of the preliminary discussion.

Theorem 1.4. Let $L \overset{j}{\rightarrowtail} A \overset{q}{\twoheadrightarrow} TB$ be a transitive Lie algebroid on base B. Then L is an LAB with respect to the bracket structure on ΓL induced from the bracket on ΓA.

Proof: Recall the adjoint representation ad: $A \to CDO(L)$ of A on the vector bundle L (III 2.11). Let γ be a connection in A and consider the produced connection $\nabla = ad\circ\gamma$: $TB \to CDO(L)$ in the vector bundle L. A calculation with the Jacobi identity for ΓA, similar to that for III 5.10(i), shows that ∇ satisfies the condition (ii) of III 7.12 with respect to the field of Lie algebra brackets on L induced from the bracket on ΓA (see III§2). Hence, by III 7.12, L is an LAB. //

The following result, like III 4.10, is an immediate consequence of the Jacobi identity. It is only by virtue of 1.4, however, that it is possible to formulate it.

Proposition 1.5. Let $L \rightarrowtail A \twoheadrightarrow TB$ be a transitive Lie algebroid on B. Then the adjoint representation of A on L, ad: $A \to CDO(L)$, takes values in CDO[L]. //

For a connection γ in A, the produced connection $ad\circ\gamma$: TB → CDO[L] will be called, as in III 5.8, the <u>adjoint connection</u> of γ, and denoted ∇^{γ}. It is a Lie connection in L.

From now on we will call L the <u>adjoint LAB</u> of A. This should not be confused with the expression "adjoint LAB of L", which refers to ad(L).

The following theorem is a Lie algebroid analogue of III 1.31(i).

<u>Theorem 1.6</u>. Let ϕ: A → A' be a morphism of transitive Lie algebroids over B. Then ϕ^{+} : L → L' is a locally constant morphism of LAB's and, as a morphism of vector bundles over B, ϕ: A → A' is of locally constant rank.

<u>Proof</u>: Let γ be a connection in A and let $\gamma' = \phi\circ\gamma$ be the produced connection in A'. Let ∇^{γ} and $\nabla^{\gamma'}$ be the corresponding adjoint connections in L and L'. Then it is easily checked that ϕ^{+}: L → L' maps ∇^{γ} to $\nabla^{\gamma'}$, that is, ϕ^{+} satisfies (ii) of III 7.16. Now III 7.16 establishes that ϕ^{+} is a locally constant morphism of LAB's.

That ϕ itself is of locally constant rank follows from applying the 5-lemma to ϕ^{+} and ϕ, as in the proof of III 2.8. //

Thanks to 1.6, there is a significant algebraic theory of transitive Lie algebroids. For example, it is only by virtue of 1.6 that the following definition is usable.

<u>Definition 1.7</u>. Let ϕ: A → A' be a morphism of transitive Lie algebroids over B. Then the <u>kernel</u> of ϕ, denoted ker(ϕ), is the sub bundle ker(ϕ^{+}) of L. The <u>image</u> of ϕ, denoted im(ϕ), is the transitive Lie algebroid im(ϕ^{+}) ↣ im(ϕ) ↠ TB. //

<u>Proposition 1.8</u>. Let ϕ: A → A' be a morphism of transitive Lie algebroids over B. Then ker(ϕ) is a sub LAB of L.

<u>Proof</u>: Let γ be a connection in A and ∇^{γ} the corresponding adjoint connection in L. Write K = ker(ϕ). Then $\nabla^{\gamma}(\Gamma K) \subseteq \Gamma K$, for if V ε ΓK and X ε ΓTB, then $\phi^{+}(\nabla^{\gamma}_{X}(V)) = \phi([\gamma(X),V]) = [\phi\gamma(X),\phi^{+}(V)] = 0$. So K is a vector bundle with a field of Lie algebra brackets which admits a connection - the restriction of ∇^{γ} - which satisfies III 7.12(ii). //

We leave to the reader the proof that im(ϕ) is actually a reduction of A', as implied in 1.7.

<u>Definition 1.9</u>. Let $L \rightarrowtail A \twoheadrightarrow TB$ be a transitive Lie algebroid on B. An <u>ideal</u> of A is a sub LAB K of L such that

$$X \in \Gamma A,\ V \in \Gamma K \Rightarrow [X,V] \in \Gamma K.$$

That K is an ideal of A is denoted $K \triangleleft A$.

An <u>ideal reduction</u> of A is a reduction $L' \rightarrowtail A' \twoheadrightarrow TB$ of A such that L' is an ideal of A. //

Clearly the kernel of a morphism of transitive Lie algebroids over B is an ideal of its domain Lie algebroid. Other examples of ideals of a transitive Lie algebroid $L \rightarrowtail A \twoheadrightarrow TB$ include ZL and [L,L].

<u>Example 1.10</u>. If $\mathfrak{g}'$ is an ideal of a Lie algebra $\mathfrak{g}$ then $B \times \mathfrak{g}'$ is an ideal of the trivial Lie algebroid $A = TB \oplus (B \times \mathfrak{g})$, and $A' = TB \oplus (B \times \mathfrak{g}')$ is an ideal reduction of A. But note that, for $X \in \Gamma A$ and $Y' \in \Gamma A'$, it is not necessarily true that $[X,Y'] \in \Gamma A'$. //

<u>Proposition 1.11</u>. Let $L \overset{j}{\rightarrowtail} A \overset{q}{\twoheadrightarrow} TB$ be a transitive Lie algebroid on B and L' an ideal of A. Let $\bar{A}$ and $\bar{L}$ be the quotient vector bundles $A/j(L')$ and L/L' and let $\bar{q}: \bar{A} \to TB$ and $\bar{j}: \bar{L} \to \bar{A}$ be the vector bundle morphisms induced by q and j. Define a bracket on $\Gamma(\bar{A})$ by

$$[X + \Gamma L', Y + \Gamma L'] = [X,Y] + \Gamma L'$$

for $X,Y \in \Gamma A$. Then $\bar{L} \overset{\bar{j}}{\rightarrowtail} \bar{A} \overset{\bar{q}}{\twoheadrightarrow} TB$ is a transitive Lie algebroid on B and the natural projection $\natural: A \twoheadrightarrow \bar{A}$, $X \mapsto X + L$, is a surjective submersion of Lie algebroids over B, and has kernel L'.

If $\phi: A \to A''$ is any surjective submersion of transitive Lie algebroids over B, and $K \triangleleft A$ its kernel, then there is a unique isomorphism $\bar{\phi}: A/j(K) \to A''$ of Lie algebroids over B, such that $\phi = \bar{\phi} \circ \natural$.

<u>Proof</u>: Straightforward. //

$\bar{A} = A/j(L')$ is the <u>quotient transitive Lie algebroid</u> of A over the ideal L'. We usually denote $A/j(L')$ by A/L'.

Proposition 1.12. Let $\phi: \Omega \twoheadrightarrow \Omega'$ be a surjective submersive morphism of Lie groupoids over B, and let M denote its kernel. (See III 1.32.) Then $AM = M_*$ is an ideal of $A\Omega$ and $\phi_*: A\Omega \twoheadrightarrow A\Omega'$ induces an isomorphism $\overline{\phi_*}: A\Omega/M_* \cong A\Omega'$.

Proof: This follows by putting together III 1.32, III 3.15 and 1.11 above. //

The following remark extends III 7.30.

Remark 1.13. Let $\phi: A \to A'$ be a morphism of transitive Lie algebroids over B. As in 1.6, let γ be a connection in A, let $\gamma' = \phi \circ \gamma$, and let ∇^{γ} and $\nabla^{\gamma'}$ be the corresponding adjoint connections. Then the condition that ϕ^+ maps ∇^{γ} to $\nabla^{\gamma'}$ is equivalent to $\tilde{\nabla}(\phi^+) = 0$, where $\tilde{\nabla}$ is the connection in Hom(L,L') induced from ∇^{γ} and $\nabla^{\gamma'}$.

$\tilde{\nabla}(\phi^+) = 0$ may be paraphrased roughly as the statement that the rate-of-change of $\phi^+: B \to \mathrm{Hom}(L,L')$ is zero in every direction within B; the morphisms $\phi^+_x: L_x \to L'_x$ are in this sense constant with respect to x. When ϕ is equal to $f_*: A\Omega \to A\Omega'$ for a morphism of Lie groupoids $f: \Omega \to \Omega'$ over B, we have $\phi^+_y = \mathrm{Ad}(\phi(\xi))^{-1} \circ \phi^+_x \circ \mathrm{Ad}(\xi)$, for every $\xi \in \Omega^y_x$ (this follows from $f^y_y = I^{-1}_{\phi(\xi)} \circ f^x_x \circ I_{\xi}$). The condition $\phi^+(\nabla^{\gamma}) = \nabla^{\gamma'}$ is an infinitesimal version of this equation. It is remarkable that the structure of a transitive Lie algebroid is sufficiently tight to impose this local constancy on ϕ^+.

In §4 we will give a second proof of 1.6, which sheds further light on the structure of ϕ. //

The following generalization of 1.6 is proved by the same method.

Theorem 1.14. Let A^1 and A^2 be transitive Lie algebroids on B, let $\phi: A^1 \to A^2$ be a morphism of Lie algebroids over B, let ρ^1 and ρ^2 be representations of A^1 and A^2 on vector bundles E^1 and E^2, and let $\psi: E^1 \to E^2$ be a ϕ-equivariant morphism of vector bundles over B, as defined in III 2.9. Then ψ is of locally constant rank. //

1.6 is actually a special case of 1.14, for if $\phi: A \to A'$ is a morphism of transitive Lie algebroids over B, then there is the representation ad of A on L and the representation $X \mapsto (V' \mapsto \mathrm{ad}(\phi(X))(V'))$ of A on L', and ϕ^+ is id_A-equivariant with respect to them.

For representations of transitive Lie algebroids, a refinement of 1.6 is necessary.

Theorem 1.15. Let A be a transitive Lie algebroid on B, and let $\rho\colon A \to CDO(E)$ be a representation of A on a vector bundle E. Then there exist an LAB atlas $\{\psi_i\colon U_i \times \mathfrak{g} \to L_{U_i}\}$ for L and an atlas $\{\phi_i\colon U_i \times V \to E_{U_i}\}$ for E, and representations $f_i\colon \mathfrak{g} \to \mathfrak{gl}(V)$ of $\mathfrak{g}$ on V, such that

$$\begin{array}{ccc} L_{U_i} & \xrightarrow{\rho^+} & End(E)_{U_i} \\ \uparrow{\scriptstyle \psi_i} & & \uparrow{\scriptstyle \overline{\phi_i}} \\ U_i \times \mathfrak{g} & \xrightarrow{id \times f_i} & U_i \times \mathfrak{gl}(V) \end{array}$$

commutes, where $\{\overline{\phi_i}\}$ is the atlas for End(E) induced from the atlas $\{\phi_i\}$ for E (see the discussion preceding 1.2).

Proof: Let γ be a connection in A, let ∇^γ be the adjoint connection in L, let $\rho\circ\gamma$ be the produced connection in E, and let $\overline{\nabla}$ be the connection in End(E) induced by $\rho\circ\gamma$.

Let $\Omega = \Pi[L] \underset{B\times B}{\times} \Pi(E)$ and consider the action of Ω on Hom(L,End(E)) which is constructed from a double application of III 1.25(iii). Then $A\Omega = CDO[L] \underset{TB}{\oplus} CDO(E)$; let $\tilde{\nabla}$ be the connection in $A\Omega$ defined by ∇^γ and $\rho\circ\gamma$. Then it is easy to check that $\tilde{\nabla}(\rho^+) = 0$.

Now apply III 7.11 to Ω, $\overline{\nabla}$ and $\rho^+ \in \Gamma Hom(L,End(E))$. It is easy to see that a section-atlas for Ω, with respect to which ρ^+ is locally constant, is composed of an LAB atlas $\{\psi_i\}$ and a vector bundle atlas $\{\phi_i\}$ with the required property. //

When B is connected, a single representation f of $\mathfrak{g}$ on V may be used for all i.

Corollary 1.16. With the assumptions of 1.15, let $E^L|_x$, where $x \in B$, denote $E_x^{L_x} = \{u \in E_x \mid \rho_x^+(W)(u) = 0,\ \forall\ W \in L_x\}$. Then $E^L = \bigcup_{x\in B} E^L|_x$ is a sub vector bundle of E.

Proof: From the diagram it follows that

$$\rho^+(\psi_i(x,W))(\phi_i(x,u)) = \phi_i(x,f_i(W)(u))$$

for $W \in \mathfrak{g}$, $u \in V$. From this it is easy to see that each $\phi_{i,x}: V \to E_x$ restricts to an isomorphism $V^{\mathfrak{g}} \to E_x^{L_x}$. //

Continue the notation of 1.15. The representation ρ of A on E restricts to a representation, denoted $\bar{\rho}$, of A on E^L, for we have

$$\rho^+(W)(\rho(X)(\mu)) = [\rho^+(W),\rho(X)](\mu) + \rho(X)(\rho^+(W)(\mu))$$

for all $W \in \Gamma L$, $X \in \Gamma TB$, $\mu \in \Gamma E$, and so if $\mu \in \Gamma(E^L)$, then the second term obviously vanishes, and the first term vanishes because $[W,X]$ is in (the image in ΓA of) ΓL.

For $\bar{\rho}: A \to CDO(E^L)$, the representation $(\bar{\rho})^+: L \to End(E^L)$ is of course zero.

Let γ be any connection in A, and consider the produced connection $\bar{\rho}\circ\gamma$ in E^L. Since $\bar{R}_{\bar{\rho}\circ\gamma} = (\bar{\rho})^+\circ \bar{R}_\gamma$ (by equation (3) of III§5) and since $(\bar{\rho})^+: L \to End(E^L)$ is zero, $\bar{\rho}\circ\gamma$ is flat. Further, if γ' is a second connection in A, say $\gamma' = \gamma + j\circ\ell$, $\ell: TB \to L$, then for $\mu \in \Gamma(E^L)$ and $X \in \Gamma TB$,

$$\begin{aligned}(\bar{\rho}\circ\gamma')(X)(\mu) &= (\bar{\rho}\circ\gamma)(X)(\mu) + (\bar{\rho})^+(\ell(X))(\mu)\\ &= (\bar{\rho}\circ\gamma)(X)(\mu).\end{aligned}$$

Thus the representation ρ of A on E induces a unique flat connection in E^L, which we will denote by ∇^ρ. We summarize all this for reference.

Proposition 1.17. Let $L \rightarrowtail A \twoheadrightarrow TB$ be a transitive Lie algebroid, and ρ a representation of A on a vector bundle E. Then ρ restricts to a representation of A on E^L, denoted $\bar{\rho}$, and $\bar{\rho}$ maps every connection in A to a single flat connection, ∇^ρ, in E^L. //

This phenomenon, apparently differential-geometric, is well-known in algebra: if $N \rightarrowtail G \twoheadrightarrow Q$ is an exact sequence of (discrete) groups, then every representation of G on a vector space V induces a representation of Q on V^N.

<u>Definition 1.18.</u> Let A be a Lie algebroid on B and ρ a representation of A on a vector bundle E. Then

$$(\Gamma E)^A = \{\mu \in \Gamma E \mid \rho(X)(\mu) = 0 \quad \forall\, X \in \Gamma A\}$$

is called the <u>space of A-parallel sections</u> of E. //

If A is totally intransitive, then $(\Gamma E)^A$ is a $C(B)$-submodule of ΓE. Otherwise $(\Gamma E)^A$ is merely an $\mathbb{R}$-vector subspace of ΓE. If A is totally intransitive, $(\Gamma E)^A$ need not correspond to a sub vector bundle of E, even if A is an LAB.

<u>Theorem 1.19.</u> Let A be a transitive Lie algebroid on a connected base B, and let ρ be a representation of A on a vector bundle E. Choose $b \in B$ and write $\mathfrak{g} = L_b$, $V = E_b$, $\pi_1 B = \pi_1(B,b)$. Then the evaluation map $\Gamma E \to V$ restricts to an isomorphism of vector spaces

$$(\Gamma E)^A \xrightarrow{\cong} (V^{\mathfrak{g}})^{\pi_1 B}$$

where $\pi_1 B$ acts on $V^{\mathfrak{g}}$ via the holonomy morphism of ∇^ρ.

<u>Proof:</u> Clearly $(\Gamma E)^A = (\Gamma(E^L))^{\nabla^\rho}$, in the notation of III 7.9. The result now follows from III 7.10 and III 7.29. //

By way of comparison, if ρ is a representation of a Lie groupoid Ω on a vector bundle E, then $E^{G\Omega}$ is a trivializable sub bundle of E, isomorphic to $B \times V^G$ (see II 4.18), and $(\Gamma E)^\Omega$ is isomorphic to V^G (see II 4.15). One may say that the sections of $E^{G\Omega}$ invariant under the action of $B \times B$ are the constant sections, that is, they are the elements of V^G.

1.19 may be regarded as the calculation of the Lie algebroid cohomology of A with coefficients in E, at degree zero. In §5 we will extend this to arbitrary degrees.

* * * * * * *

With the results of this section established, one can now develop a full theory of abstract transitive Lie algebroids. The greater part of the present chapter is devoted to the development of their cohomology theory and connection theory, two subjects which are inextricably linked.

A number of the concepts and constructions of III§5 and III§7 carry over to the abstract setting immediately. For instance, the construction in III 7.25 to 7.26 of the γ-curvature reduction of a Lie algebroid $A\Omega$ derived from a Lie groupoid Ω and corresponding to a connection γ in $A\Omega$, may be extended to abstract transitive Lie algebroids without difficulty. The only point in III 7.25-7.26 where Ω is used is to ensure that $L\Omega$ is an LAB.

Likewise, the definitions of the exterior covariant derivatives in III§5 and propositions III 5.10 and III 5.11 are valid without change, in any transitive Lie algebroid.

§2. The cohomology of Lie algebroids

This section constructs the cohomology of an arbitrary Lie algebroid with coefficients in an arbitrary representation and gives interpretations in degrees 2, 1 and 0.

The definition 2.1 is in terms of a standard resolution of de Rham, or Chevalley-Eilenberg, type. This definition, as well as being the simplest, is the closest to the geometric applications. We prove in 2.5 and 2.6 that when A is a transitive Lie algebroid with adjoint bundle L, the cohomology spaces $\mathscr{H}^*(L,\rho^+,E)$ are the modules of sections of certain flat vector bundles $H^*(L,\rho^+,E)$. This is a generalization to all degrees of 1.16 and is proved using a generalization of the calculus of differential forms on a manifold, and the results of III§7. In 2.7 and 2.8 we calculate $\mathscr{H}^*(A\Omega,\rho,E)$ for a Lie groupoid Ω and any representation ρ of $A\Omega$, in terms of the equivariant de Rham cohomology of an associated principal bundle.

In the second part of the section we interpret $\mathscr{H}^2(A,\rho,E)$ in terms of equivalence classes of operator extensions of A by E. This is a straightforward generalization of the corresponding extension theory of Lie algebras, but we have given at least sketch proofs of most results, since there is no readily available account of the Lie algebra theory in the detail which is required for the geometric applications. In Lie algebra cohomology, as in other cohomology theories of algebraic type, there is little interest in specific cocycles or in specific transversals for extensions: one is there only interested in cohomological invariants. In Lie algebroid cohomology, however, transversals are (at least in the applications to geometry) infinitesimal connections and cocycles are, in degree two, curvature forms, and, in degree three, the left-hand sides of Bianchi identities, and the focus of geometric interest is usually on specific transversals or cocycles. It is for this reason that the explicit definition 2.1 is the best for our purposes.

In this section we treat only extensions by abelian (totally intransitive) Lie algebroids; that is, by vector bundles. The general case, which includes the most important applications, is treated in §3. If P(B,G) is a principal bundle with abelian structure group then

$$B \times \mathfrak{g} \cong \frac{P \times \mathfrak{g}}{G} \rightarrowtail \frac{TP}{G} \twoheadrightarrow TB$$

is an extension of TB by the trivial vector bundle $B \times \mathfrak{g}$; if G is compact and B is simply-connected then the cohomology class of $\frac{TP}{G}$ in $\mathscr{H}^2(TB, B \times \mathfrak{g}) \cong$

$H^2_{deRh}(B,\mathfrak{g})$ is the sum of the Chern classes of the component SO(2)-bundles of P. In the case where G is not compact and $\pi_1 B$ is arbitrary, the class of $\frac{TP}{G}$ in $\mathcal{H}^2(TB, B\times\mathfrak{g})^{\pi_1 B} \cong H^2_{deRh}(\tilde{B},\mathfrak{g})$ may still be regarded as a characteristic class of P(B,G). The question of the relationship of the results of this section and of §3 to the extension theory of Lie groupoids and principal bundles will be taken up elsewhere.

This section closes with a proof that the cohomology $\mathcal{H}^*(A,\rho,E)$ coincides with that defined, on the level of the modules of sections, by G.S. Rinehart (1963).

The definition of the cohomology of Lie algebroids has been given many times previously, under a variety of names. See, for example, Palais (1961b), Hermann (1967), Nelson (1967), and N. Teleman (1972). However, much of this work was done at the level of the module of sections and was only concerned with the algebraic formalism. The first major result in this area was the Poincaré-Birkhoff-Witt theorem of Rinehart (1963), which enabled Rinehart to define Lie algebroid cohomology as derived functors of $\mathrm{Hom}_{U(A)}(B\times\mathbb{R}, -)$. Rinehart's results were sheafified by Kamber and Tondeur (1971).

The results and constructions of this section from 2.4 on, in particular their establishment within the geometric context of smooth vector bundles are due to the author, and first appeared in Mackenzie (1979).

* * * * * * * * *

Until 2.4, let A be an arbitrary Lie algebroid on a base B. It is not assumed that A is transitive. Let $\rho\colon A \to CDO(E)$ be a representation of A on a vector bundle E.

Definition 2.1. The standard complex associated with the vector bundle E and the representation ρ of A is the sequence $C^n(A,E)$, $n \geq 0$, where $C^n(A,E)$ denotes the vector bundle $\mathrm{Alt}^n(A;E)$, and the sequence of differential operators $d^n\colon \Gamma C^n(A,E) \to \Gamma C^{n+1}(A,E)$ which is defined by

$$df(X_1,\ldots,X_{n+1}) = \sum_{r=1}^{n+1} (-1)^{r+1}\rho(X_r)(f(X_1,\ldots\hat{}\ldots,X_{n+1}))$$

$$+ \sum_{r<s} (-1)^{r+s} f([X_r,X_s],X_1,\ldots\hat{}\ldots\hat{}\ldots,X_{n+1})$$

for $f \in \Gamma C^n(A,E)$ and $X_1,\ldots,X_{n+1} \in \Gamma A$.

The <u>cohomology spaces</u> $\mathscr{H}^n(A,\rho,E)$, or $\mathscr{H}^n(A,E)$, are the cohomology spaces of this complex, namely $\mathscr{H}^n(A,E) = \mathscr{Z}^n(A,E)/\mathscr{B}^n(A,E)$, where $\mathscr{Z}^n(A,E) = \ker d\colon \Gamma C^n(A,E) \to \Gamma C^{n+1}(A,E)$ and $\mathscr{B}^n(A,E) = \operatorname{im} d\colon \Gamma C^{n-1}(A,E) \to \Gamma C^n(A,E)$ for $n > 1$, $\mathscr{B}^o(A,E) = (0)$. //

It is routine to verify that $d^2 = 0$. When A is totally intransitive (so that it is a vector bundle together with a field of Lie algebra brackets), the operators d are C(B)-linear. This is easy to check. Hence in this case, each d^n induces a vector bundle morphism $C^n(A,E) \to C^{n+1}(A,E)$, also denoted d^n. It is not true, however, that in this case the d^n must be of locally constant rank, and so the images and kernels may not be sub bundles. This is so even if A is an LAB. Examples are easy to construct, using the same device as in III 1.28.

For a general Lie algebroid A, with $q \neq 0$, the maps $d^n\colon \Gamma C^n(A,E) \to \Gamma C^{n+1}(A,E)$ are first-order differential operators, and do not induce morphisms of the underlying vector bundles.

Thus the $\mathscr{H}^n(A,E)$ are quotients of infinite-dimensional real vector spaces, and are at this stage rather formless. However for a transitive Lie algebroid A, we will show in §5 that the $\mathscr{H}^n(A,E)$ are computable.

When B is a point and A a finite-dimensional real Lie algebra, the $\mathscr{H}^n(A,E)$ clearly reduce to the Chevalley-Eilenberg cohomology spaces (see, for example, Cartan and Eilenberg (1956, XIII§8)). When A is the tangent bundle TB and ρ is the trivial representation of TB in a product vector bundle $B \times V$ (see III 2.10), $\mathscr{H}^n(TB, B \times V)$ is clearly the real de Rham cohomology space $H^n_{deRh}(B,V)$. In 2.7 we will calculate $\mathscr{H}^n(A\Omega,\rho_*,E)$ for a Lie groupoid Ω and a representation ρ_* induced from a groupoid representation $\rho\colon \Omega \to \Pi(E)$.

<u>Definition 2.2.</u> (i) The <u>Lie derivative</u> $\theta_X\colon \Gamma C^n(A,E) \to \Gamma C^n(A,E)$ for $X \in \Gamma A$, is defined by

$$\theta_X(f)(X_1,\ldots,X_n) = \rho(X)(f(X_1,\ldots,X_n)) - \sum_{r=1}^{n} f(X_1,\ldots,[X,X_r],\ldots,X_n).$$

Here $f \in \Gamma C^n(A,E)$ and $X_r \in \Gamma A$, $1 \leqslant r \leqslant n$.

(ii) The <u>interior multiplication</u> $\iota_X\colon \Gamma C^{n+1}(A,E) \to \Gamma C^n(A,E)$, for $X \in \Gamma A$, $n \geqslant 0$, is defined by

$$\iota_X(f)(X_1,\ldots,X_n) = f(X,X_1,\ldots,X_n),$$

for $f \in \Gamma C^n(A,E)$, $X_r \in \Gamma A$, $1 \leq r \leq n$. //

θ_X, ι_X and d satisfy a set of formulas identical in form to those which hold in the calculus of vector-valued forms on a manifold. Those which are used in the sequel follow.

Proposition 2.3: (i) $\iota_X(uf) = u\iota_X(f)$, $\iota_{uX}(f) = u\iota_X(f)$ and $\iota_X \circ \iota_Y = -\iota_Y \circ \iota_X$, for $X,Y \in \Gamma A$, $u \in C(B)$, $f \in \Gamma C^n(A,E)$;

(ii) $\theta_X(uf) = u\theta_X(f) + q(X)(u)f$ for $X \in \Gamma A$, $u \in C(B)$, $f \in \Gamma C^n(A,E)$;

(iii) $$\theta_{uX}(f)(X_1,\ldots,X_n) = u\theta_X(f)(X_1,\ldots,X_n) + \sum_{r=1}^{n} (-1)^{r-1} q(X_r)(u)\,\iota_X(f)(X_1,\ldots,\hat{\ },\ldots,X_n)$$

for $X,X_1,\ldots,X_n \in \Gamma A$, $u \in C(B)$, $f \in \Gamma C^n(A,E)$;

(iv) $\theta_{[X,Y]} = \theta_X \circ \theta_Y - \theta_Y \circ \theta_X$ for $X,Y \in \Gamma A$;

(v) $\theta_X = \iota_X \circ d + d \circ \iota_X$ for $X \in \Gamma A$;

(vi) $\theta_X \circ d = d \circ \theta_X$ for $X \in \Gamma A$.

Proof: Standard. //

With the definition of wedge-product given in §5, (iii) can be written

$$\theta_{uX}(f) = u\theta_X(f) + du \wedge \iota_X(f),$$

where du is taken with respect to the representation q of A on $B \times \mathbb{R}$.

Proposition 2.4. Let $L \overset{j}{\rightarrowtail} A \overset{q}{\twoheadrightarrow} TB$ be a transitive Lie algebroid on B. Then for $X \in \Gamma A$ and $f \in \Gamma C^n(L,E)$, the Lie derivative $\theta_X(f)$ is in $\Gamma C^n(L,E)$, the map $\theta_X\colon \Gamma C^n(L,E) \to \Gamma C^n(L,E)$ is in $\Gamma CDO(C^n(L,E))$ and $X \mapsto \theta_X$ defines a representation of A on $C^n(L,E)$.

Proof: For the first statement, note that

$$\theta_X(f)(V_1,\ldots,V_n) = \rho(X)(f(V_1,\ldots,V_n)) - \sum_{r=1}^{n} f(V_1,\ldots,\mathrm{ad}(X)(V_r),\ldots,V_n)$$

for $V_r \in \Gamma L$, $X \in \Gamma A$, $f \in \Gamma C^n(L,E)$. The remaining assertions follow from 2.3 (ii),(iii),(iv). //

In the cohomology theory of Lie algebras, θ is a representation of a Lie algebra $\mathfrak{g}$ on the associated spaces $C^n(\mathfrak{g},V)$. In the context of general Lie algebroids however, the map $\theta: \Gamma A \to \Gamma CDO(C^n(A,E))$ is not $C(B)$-linear, and therefore cannot be said to be a representation. As in the case of $\mathrm{ad}: X \mapsto (Y \mapsto [X,Y])$, $\Gamma A \to \Gamma CDO(A)$, which θ of course generalizes, this lack of $C(B)$-linearity can be avoided by lifting θ to the 1-jet prolongation of A. However the action of A on $C^n(L,E)$ suffices for our purposes.

Although $\theta_X = \iota_X \circ d + d \circ \iota_X$ is meaningless for the θ_X of 2.4, its consequence, $d \circ \theta_X = \theta_X \circ d$, continues to be valid. This follows from 2.3(vi). From this formula we have the following crucial result.

Theorem 2.5. Let $L \xrightarrow{j} A \xrightarrow{q} TB$ be a transitive Lie algebroid on B. Then the coboundaries

$$d^n: C^n(L,E) \to C^{n+1}(L,E)$$

are of locally constant rank, and consequently there are well-defined vector bundles $Z^n(L,\rho,E) = \ker d^n$, $B^n(L,\rho,E) = \mathrm{im}\, d^{n-1}$ and $H^n(L,\rho,E) = Z^n(L,\rho,E)/B^n(L,\rho,E)$ such that $\Gamma H^n(L,\rho,E) = \mathcal{H}^n(L,\rho^+,E)$.

Further, $\theta: A \to CDO(C^n(L,E))$ induces a well-defined representation, also denoted θ, of A on $H^n(L,E)$.

Proof: Let γ be a connection in A, and let γ^n be the connection $\theta \circ \gamma$ in $C^n(L,E)$. Then the equation $\theta_X \circ d^n = d^{n+1} \circ \theta_X$ for $X \in \Gamma A$ implies in particular that $\theta_{\gamma(X)}(d(f)) = d(\theta_{\gamma(X)}(f))$ for $X \in \Gamma TB$, $f \in \Gamma C^n(L,E)$; that is, that $\gamma^{n+1}(X)(d(f)) = d(\gamma^n(X)(f))$. Thus d maps γ^n to γ^{n+1} and so, by III 7.13, d is of locally constant rank.

That the representation θ of A on $C^n(L,E)$ induces a well-defined representation of A on $H^n(L,E)$ follows from this same equation $\theta_X \circ d = d \circ \theta_X$, $X \in \Gamma A$. //

Theorem 2.6. Continuing the notation of 2.5, for the representation θ of A on $H^n(L,E)$ we have

$$H^n(L,E)^L = H^n(L,E).$$

In particular, $H^n(L,E)$ is a flat vector bundle, and θ maps every connection in A to a single flat connection, denoted $\nabla^{\rho,n}$, in $H^n(L,E)$.

Proof: Applying 2.3 to the Lie algebroid L and the representation ρ^+ we have

$$\theta_V = d\circ\iota_V + \iota_V\circ d, \quad \text{for} \quad V \in \Gamma L.$$

Now take $f \in C^n(L,E)$ with $df = 0$. We have $\theta_V(f) = d(\iota_V(f)) + 0$ and so

$$\theta_V([f]) = [\theta_V(f)] = [d(\iota_V(f))] = 0.$$

Thus $[f] \in H^n(L,E)^L$.

The remaining statements follow from applying 1.17 to $\theta\colon A \to CDO(H^n(L,E))$. //

We hope that the proofs of 2.5 and 2.6 give the reader some amusement. Thanks to 2.5 and 2.6 it will be possible, in §5, to consider the cohomology of TB with coefficients in $H^*(L,E)$, and to relate the spaces $\mathcal{H}^*(TB,H^*(L,E))$ to the cohomology of A.

In the case of the transitive Lie algebroid of a Lie groupoid, the Lie algebroid cohomology is the equivariant de Rham cohomology of the corresponding principal bundle.

Theorem 2.7. Let Ω be a Lie groupoid on B and let $\rho\colon \Omega \to \Pi(E)$ be a representation of Ω on a vector bundle E. Then there are natural isomorphisms

$$\mathcal{H}^n(A\Omega,\rho_*,E) \cong H^n_{deRh}(P,V)^G$$

where $P = \Omega_b$, $G = \Omega_b^b$ and $V = E_b$ for some chosen $b \in B$.

Proof: This is an immediate consequence of A 4.13. //

Corollary 2.8. If Ω is an α-connected Lie groupoid on B and $\rho: A\Omega \to CDO(E)$ is any representation of $A\Omega$ on a vector bundle E, then there are natural isomorphisms

$$\mathscr{H}^n(A\Omega,\rho,E) \cong H^n_{deRh}(\tilde{P},V)^H$$

where H is the structure group of the monodromy bundle $\tilde{P}(B,H)$ of P(B,G). //

In particular, for a flat vector bundle E on B, and a flat connection ∇ in E,

$$\mathscr{H}^n(TB,\nabla,E) \cong H^n_{deRh}(\tilde{B},V)^{\pi_1 B},$$

the equivariant de Rham cohomology of the universal cover of B, constructed from forms $\omega \in A^*(\tilde{B},V)$ which are equivariant with respect to the holonomy action of $\pi_1(B)$ on the fibre type V. In this way Lie algebroid cohomology may be regarded as a generalization of de Rham cohomology in which coefficients in local systems of vector spaces are permitted.

Thus in the case of the Lie algebroids of Lie groupoids, the Lie algebroid cohomology is a known invariant, though one which has only been extensively studied in the case where the structure group is compact (see, for instance, Greub et al (1976)). One of the strengths of the Lie algebroid formulation, however, is that it is a cohomology theory of algebraic type, comparable to the cohomology theories of Lie algebras and of discrete groups, and that what is significant from the point of view of the algebraic cohomology theory is also significant geometrically. We will spend much of the next three sections justifying this observation and developing its consequences. The first step in this process is the interpretation of $\mathscr{H}^2(A,E)$ in terms of equivalence classes of extensions of A by E, and for this we need the following general concept of curvature.

Proposition 2.9. Let A and A' be Lie algebroids on B, not necessarily transitive, and let $\phi: A \to A'$ be a morphism of vector bundles over B such that $q'\circ\phi = q$. Then

$$R_\phi(X,Y) = \phi([X,Y]) - [\phi(X),\phi(Y)] \tag{1}$$

defines a map $\Gamma A \times \Gamma A \to \Gamma A'$ which is alternating and $C(B)$-bilinear, and thus defines a section of $Alt^2(A;A')$, called the curvature of ϕ.

For $X,Y,Z \in \Gamma A$ we have

(2) $$\mathfrak{S}\{[\phi(X),R_\phi(Y,Z)] - R_\phi([X,Y],Z)\} = 0,$$

where $\mathfrak{S}$ is the cyclic sum.

<u>Proof:</u> In the first paragraph only the C(B)-bilinearity is not clear, and it follows by calculation. Formula (2) follows from the Jacobi identity in $\Gamma A'$. //

The curvature map $\bar{R}_\gamma$: TB $\oplus$ TB $\to$ L defined in III 5.1 for a connection γ: TB $\to$ A in a transitive Lie algebroid is related to this R_γ by $R_\gamma = j\circ\bar{R}_\gamma$. Equation (2) generalizes III 5.11, and may be called an abstract Bianchi identity. Vector bundle morphisms ϕ: A $\to$ A' with $q'\circ\phi = q$ will be called <u>anchor-preserving maps</u>.

<u>Proposition 2.10</u>. Let ϕ: A $\to$ A' and ψ: A' $\to$ A" be anchor-preserving maps of Lie algebroids. Then

$$R_{\psi\circ\phi} = R_\psi\circ(\phi \times \phi) + \psi\circ R_\phi.$$

<u>Proof:</u> Calculation. //

We return now to the conventions made at the start of this section: A is a Lie algebroid on base B, not necessarily transitive, and ρ is a representation of A on a vector bundle B.

<u>Definition 2.11</u>. An <u>extension</u> of A by the vector bundle E is an exact sequence

(3) $$E \overset{\iota}{\rightarrowtail} A' \overset{\pi}{\twoheadrightarrow} A$$

of Lie algebroids over B, where E is considered to be an abelian Lie algebroid.

A <u>transversal</u> in the extension (3) is a vector bundle morphism χ: A $\to$ A' such that $\pi\circ\chi = id_A$. A <u>back-transversal</u> in (3) is a vector bundle morphism λ: A' $\to$ E such that $\lambda\circ\iota = id_E$.

A transversal is <u>flat</u> if it is a morphism of Lie algebroids; equivalently, if it has zero curvature. The extension (3) is <u>flat</u> if it has a flat transversal. //

The definition of an exact sequence of Lie algebroids is given in III 2.14.

As with connections (see the discussion following III 5.1) there is a bijective correspondence between transversals χ and back-transversals λ, given by

$$\iota\circ\lambda + \chi\circ\pi = \mathrm{id}_{A'}, \tag{4}$$

and a corresponding pair satisfy $\lambda\circ\chi = 0$. Since π is a surjective submersion and a morphism of vector bundles over B, transversals always exist. Any choice of transversal determines an isomorphism of vector bundles $A' \cong A \oplus E$.

From $\pi\circ\chi = \mathrm{id}_A$ it follows that $q'\circ\chi = q$, so a transversal is automatically anchor-preserving. From this and $q\circ\pi = q'$ it follows that $\mathrm{im}(q) = \mathrm{im}(q')$. Hence A' is transitive iff A is transitive, and A' is totally intransitive iff A is totally intransitive.

For transversals $\chi\colon A \to A'$ of (3), we will normally use as curvature the map $\bar{R}_\chi\colon A \oplus A \to E$ with $\iota\circ\bar{R}_\chi = R_\chi$. Note that $\bar{R}_\chi \in \Gamma C^2(A,E)$.

Recall from III 2.16 that an extension such as (3) induces a representation $\rho^{A'}$ of A on E, which can be written as

$$\iota(\rho^{A'}(X)(\mu)) = [\chi(X),\iota(\mu)],$$

$X \in \Gamma A$, $\mu \in \Gamma E$, for any transversal χ.

<u>Definition 2.12.</u> Given the Lie algebroid A and the action ρ of A on E, the extension (3) is an <u>operator extension</u> of A by E if $\rho^{A'} = \rho$.

Two operator extensions $E \overset{\iota_s}{\dashrightarrow} A^s \overset{\pi_s}{\dashrightarrow} A$, $s = 1,2$ are <u>equivalent</u> if there is a morphism of Lie algebroids $\phi\colon A^1 \to A^2$ over B (necessarily an isomorphism) such that $\phi\circ\iota_1 = \iota_2$ and $\pi_2\circ\phi = \pi_1$. //

The set of equivalence classes of operator extensions is denoted by $\mathcal{O}\mathrm{pext}(A,\rho,E)$, or by $\mathcal{O}\mathrm{pext}(A,E)$ if ρ is understood. There is a natural bijection between $\mathcal{H}^2(A,\rho,E)$ and $\mathcal{O}\mathrm{pext}(A,\rho,E)$ described in the following proposition.

<u>Proposition 2.13.</u> (i) Let $f \in \Gamma C^2(A,E)$ be a cocycle, that is, let

$$\mathfrak{S}\{\rho(X)(f(Y,Z)) - f([X,Y],Z)\} = 0$$

for all $X,Y,Z \in \Gamma A$. Denote by A^f the vector bundle $A \oplus E$ equipped with the anchor $q^f(X \oplus \mu) = q(X)$, and the bracket

$$[X \oplus \mu, Y \oplus \nu] = [X,Y] \oplus \{\rho(X)(\nu) - \rho(Y)(\mu) - f(X,Y)\}$$

on ΓA^f. Then A^f is a Lie algebroid on B and $E \overset{\iota_2}{\rightarrowtail} A^f \overset{\pi_1}{\twoheadrightarrow} A$ is an operator extension of A by E. (Here $\iota_2: E \to A \oplus E$ and $\pi_1: A \oplus E \to A$ are the natural maps.) The transversal $\iota_1: A \to A^f$ has curvature $\bar{R}_{\iota_1} = f$.

(ii) Conversely, let $E \overset{\iota}{\rightarrowtail} A' \overset{\pi}{\twoheadrightarrow} A$ be an operator extension of A by E and let χ be a transversal. Then $\bar{R}_\chi$ is a cocycle, that is, $d(\bar{R}_\chi) = 0$, with respect to the coboundary induced by ρ. Further

$$\mathcal{E}_\chi : A^{\bar{R}_\chi} \to A', \quad X \oplus \mu \mapsto \chi(X) + \iota(\mu)$$

is an equivalence of $E \rightarrowtail A^{\bar{R}_\chi} \twoheadrightarrow A$ with $E \rightarrowtail A' \twoheadrightarrow A$, and $\mathcal{E}_\chi$ maps ι_1 to χ.

(iii) Let $g \in \Gamma C^1(A,E)$; recall that

$$dg(X,Y) = \rho(X)(g(Y)) - \rho(Y)(g(X)) - g([X,Y])$$

for $X,Y \in \Gamma A$.

Given any $f \in \mathcal{Z}^2(A,E)$, the map

$$\varepsilon_g: A^f \to A^{f+dg}, \quad X \oplus \mu \mapsto X \oplus (\mu + g(X))$$

is an equivalence of extensions.

(iv) Conversely, if $E \overset{\iota_s}{\rightarrowtail} A^s \overset{\pi_s}{\twoheadrightarrow} A$, $s = 1,2$, are operator extensions, and $\phi: A^1 \to A^2$ is an equivalence, then for any pair of transversals χ_1, χ_2 for A^1, A^2, there is a unique $g \in \Gamma C^1(A,E)$ such that

$$\phi = \mathcal{E}_{\chi_2} \circ \varepsilon_g \circ \mathcal{E}_{\chi_1}^{-1},$$

namely $g = \lambda_2 \circ \phi \circ \chi_1$, and $dg = f_2 - f_1$.

Indeed each cochain $g \in \Gamma C^1(A,E)$ induces a permutation $\chi \mapsto \chi^g$ of the transversals in any operator extension $E \overset{\iota}{\rightarrowtail} A' \overset{\pi}{\twoheadrightarrow} A$. If χ is a transversal, then $\chi^g = \chi + \iota \circ g$ is another, and $\bar{R}_{\chi^g} = \bar{R}_\chi - dg$.

<u>Proof</u>: A straightforward manipulation. //

The linear structure on $\mathcal{H}^2(A,E)$ may now be transferred to $\mathcal{O}$pext(A,E). Recall that if A is totally intransitive, $\mathcal{H}^2(A,E)$ is a C(B)-module and is otherwise merely an $\mathbb{R}$-vector space. The multiplicative structure is given by the following construction.

<u>Proposition 2.14</u>. Let $E \overset{\iota}{\rightarrowtail} A' \overset{\pi}{\twoheadrightarrow} A$ be an operator extension of A by E. Let $\tilde{E}$ be a second vector bundle on B and $\tilde{\rho}: A \to CDO(\tilde{E})$ a representation of A on $\tilde{E}$. Then if $\phi: E \to \tilde{E}$ is an A-equivariant morphism of vector bundles over B, there is a unique extension $\tilde{E} \rightarrowtail \tilde{A} \twoheadrightarrow A$ of A by $\tilde{E}$ which induces $\tilde{\rho}$ and which is such that there is a morphism of Lie algebroids $\tilde{\phi}: A \to \tilde{A}$ making

$$\begin{array}{ccccc} E & \rightarrowtail & A' & \twoheadrightarrow & A \\ \phi \downarrow & & \tilde{\phi} \downarrow & & \| \\ \tilde{E} & \rightarrowtail & \tilde{A} & \twoheadrightarrow & A \end{array}$$

commute.

<u>Proof</u>: Choose a transversal $\chi: A \to A'$ and define a bracket on $\Gamma(A \oplus \tilde{E})$ by

$$[X \oplus \tilde{\mu}, Y \oplus \tilde{\nu}] = [X,Y] \oplus \{\tilde{\rho}(X)(\tilde{\nu}) - \tilde{\rho}(Y)(\tilde{\mu}) - \phi\bar{R}_{\chi}(X,Y)\}.$$

With this bracket and with anchor $\tilde{q}(X \oplus \tilde{\mu}) = q(X)$, $A \oplus \tilde{E}$ becomes a Lie algebroid on B; denote it by $\tilde{A}$. It is easily checked that $\tilde{A}$ is independent of the choice of χ, up to isomorphism, that $E \overset{\iota_2}{\rightarrowtail} \tilde{A} \overset{\pi_1}{\twoheadrightarrow} A$ is a $\tilde{\rho}$-operator extension, and that $X' \mapsto \pi(X') \oplus \phi\lambda(X')$ has the properties required of $\tilde{\phi}$. //

$\tilde{E} \rightarrowtail \tilde{A} \twoheadrightarrow A$ is the <u>pushout extension</u> of $E \rightarrowtail A' \twoheadrightarrow A$ along ϕ. It is difficult, and unrewarding, to give a proof of 2.14 without using 2.13.

Now, for an arbitrary Lie algebroid A, and an extension $E \rightarrowtail A' \twoheadrightarrow A$, the map $E \to E$, $\mu \mapsto k\mu$, where $k \in \mathbb{R}$ is a constant, is equivariant and the corresponding pushout is the scalar multiple of $[E \rightarrowtail A' \twoheadrightarrow A] \in \mathcal{O}$pext(A,E) by k. If A = L is totally intransitive, and $u \in C(B)$ is a function, then $E \to E$, $\mu \mapsto u(p(\mu))\mu$ is equivariant, and the pushout extension similarly defines the C(B)-module structure.

To define the addition on $\mathcal{O}$pext(A,E) the construction of pullback extensions is required. This construction can be given in greater generality than the construction of pushouts.

Proposition 2.15. Let $\phi^1 : A^1 \to A$ and $\phi^2 : A^2 \to A$ be morphisms of Lie algebroids over B such that

$$\text{im } \phi^1_x + \text{im } \phi^2_x = A_x \qquad \text{for all} \quad x \in B.$$

Then the pullback vector bundle

$$\tilde{A} = \{X_1 \oplus X_2 \in A^1 \oplus A^2 \mid \phi^1(X_1) = \phi^2(X_2)\}$$

(see C.5) is a Lie algebroid with respect to the anchor $\tilde{q}(X_1 \oplus X_2) = q^1(X_1) = q^2(X_2)$, and the bracket

$$[X_1 \oplus X_2, Y_1 \oplus Y_2] = [X_1,X_2] \oplus [X_2,Y_2].$$

Further, the restrictions to $\tilde{A}$ of the projections $\pi_1 : A^1 \oplus A^2 \to A^1$, $\pi_2 : A^1 \oplus A^2 \to A^2$ are Lie algebroid morphisms over B, and if $\bar{A}$ is any other Lie algebroid on B and $\psi^1 : \bar{A} \to A^1$, $\psi^2 : \bar{A} \to A^2$ are morphisms of Lie algebroids over B such that $\phi^1 \circ \psi^1 = \phi^2 \circ \psi^2$, then there is a unique Lie algebroid morphism $\psi : \bar{A} \to \tilde{A}$ such that $\pi_1 \circ \psi = \psi_1$, $\pi_2 \circ \psi = \psi_2$.

Proof: The pullback vector bundle is defined in C.5. The remainder is straightforward. //

$\tilde{A}$ is called the pullback Lie algebroid, and may be denoted $A^1 \underset{A}{\oplus} A^2$. The direct sum Lie algebroid (III 2.18) is a particular instance. The second paragraph of 2.15 embodies the pullback property. We will normally denote the maps $\tilde{A} \to A^1$, $\tilde{A} \to A^2$ by $\tilde{\phi}_2$, $\tilde{\phi}_1$ respectively.

Proposition 2.16. Let $E \overset{\iota}{\rightarrowtail} A' \overset{\pi}{\twoheadrightarrow} A$ be an extension of A by E and let $\phi : A'' \to A$ be a morphism of Lie algebroids over B. Then

$$E \overset{\tilde{\iota}}{\rightarrowtail} \tilde{A} \overset{\tilde{\pi}}{\twoheadrightarrow} A''$$

is an extension of A" by E, where $\tilde{\iota}$ is $\mu \mapsto \iota(\mu) \oplus 0$, and, further

$$\begin{array}{ccccc} E & \overset{\tilde{\iota}}{\rightarrowtail} & \tilde{A} & \overset{\tilde{\pi}}{\twoheadrightarrow} & A'' \\ \| & & \downarrow{\scriptstyle \tilde{\phi}} & & \downarrow{\scriptstyle \phi} \\ E & \overset{\iota}{\rightarrowtail} & A' & \overset{\pi}{\twoheadrightarrow} & A \end{array}$$

commutes.

Proof: Routine verification. //

$E \overset{\tilde{\iota}}{\rightarrowtail} \tilde{A} \overset{\tilde{\pi}}{\twoheadrightarrow} A''$ is called the pullback extension of $E \rightarrowtail A' \twoheadrightarrow A$ over ϕ.

The addition on $\mathcal{O}\mathrm{pext}(A,E)$, known as the Baer sum, can now be defined in the usual way (see MacLane (1975, pp. 113-114)): given operator extensions $E \rightarrowtail A^s \twoheadrightarrow A$, $s = 1,2$, form firstly $E \oplus E \rightarrowtail A^1 \oplus A^2 \twoheadrightarrow A \oplus A$. If A is transitive, the direct sums $A \oplus A$ and $A^1 \oplus A^2$ are to be read as $\oplus_{TB}$; if A is totally intransitive, the $\oplus$'s are merely vector bundle sums over B; we do not need any other case. Next take the pullback of this over the diagonal map $A \to A \oplus A$, and lastly take the pushout of the result over the sum map $E \oplus E \to E$. The details are left to the reader.

The bijection $\mathcal{H}^2(A,E) \leftrightarrow \mathcal{O}\mathrm{pext}(A,E)$ is now an isomorphism of vector spaces for any Lie algebroid A, and of C(B)-modules when A is totally intransitive.

The zero element of $\mathcal{H}^2(A,E)$ corresponds, of course, to the flat extensions of A by E. We refer to the extension A^0 constructed in 2.13(i) from the zero cocycle as the semidirect product of A by E and denote it by $A \ltimes_\rho E$, or by $A \ltimes E$ if ρ is understood.

Any transversal χ in $A \ltimes E$ has the form $\chi(X) = X \oplus g(X)$ for some $g \in \Gamma C^1(A,E)$, and χ is flat iff $dg = 0$. Thus $\mathcal{H}^1(A,E)$ can be interpreted as the space of flat transversals in $A \ltimes E$ modulo those flat transversals of the form $\chi(X) = X \oplus \rho(X)(\mu)$ for some $\mu \in \Gamma E$.

More generally, recall from 2.13(iv) that in any operator extension $E \overset{\iota}{\rightarrowtail} A' \overset{\pi}{\twoheadrightarrow} A$ of A by E a cochain $g \in \Gamma C^1(A,E)$ induces a permutation of the transversals of A', namely $\chi \mapsto \chi^g = \chi + \iota \circ g$. From the formula $\bar{R}_{\chi^g} = \bar{R}_\chi - dg$ it follows that the cocycles $g \in \mathcal{Z}^1(A,E)$ are precisely those cochains whose permutations preserve the curvature of transversals. If g is a coboundary

$d\mu$, $\mu \in \Gamma E$, then

$$(\chi + \iota\circ d\mu)(X) = \chi(X) + \iota(\rho(X)(\mu))$$

$$= \chi(X) + [\chi(X), \iota(\mu)].$$

Thus we have

Proposition 2.17. Let $E \xrightarrow{\iota} A' \xrightarrow{\pi} A$ be an operator extension of A by E. Then $\mathcal{H}^1(A,\rho,E)$ may be realized as the space of those automorphisms $A' \to A'$ of the form $X' \to X' + \iota g(\pi X')$, where g is a vector bundle map $A \to E$, which preserve the curvature of transversals of A', modulo the space of automorphisms $A' \to A'$ of the form $X' \to X' + [X', \iota(\mu)]$, where $\mu \in \Gamma E$. //

It is interesting to interpret this result for connections in a principal bundle with abelian structure group.

Lastly we have the calculation of $\mathcal{H}^0(A,E)$.

Proposition 2.18:

$$\mathcal{H}^0(A,E) = \{\mu \in \Gamma E \mid \rho(X)(\mu) = 0 \quad \forall X \in \Gamma A\}$$

$$= (\Gamma E)^A. \qquad //$$

If A is totally intransitive then $(\Gamma E)^A$ is a sub $C(B)$-module of ΓE. It need not correspond to a sub vector bundle of E, since the rank of ρ need not be locally constant. This is so even if A is an LAB.

For A not totally intransitive, $(\Gamma E)^A$ is merely an $\mathbb{R}$-vector space. For A transitive, $(\Gamma E)^A$ was calculated in 1.19.

We close this section by establishing that the cohomology defined in 2.1 coincides with that of Rinehart (1963). The importance of this is that Rinehart gives a construction which yields a "universal enveloping algebroid" for a given Lie algebroid and - proving a Poincaré-Birkhoff-Witt theorem for it - obtains the cohomology 2.1 as the derived functors of the appropriate Hom functor (op. cit., Theorem 4.2). Although the results presented here and in the next three sections are best formulated in terms of the standard cochain complex 2.1, it may eventually be necessary to relate it to a Lie groupoid cohomology defined as the derived functors of a fixed-point functor (as in, for example, Mackenzie (1978)).

The point which needs to be established is the equivalence of the two definitions of module.

Let A be a Lie algebroid on B. Denote by $A^{\#}$ the vector bundle $A \oplus (B \times \mathbb{R})$. Define a Lie bracket on $\Gamma(A^{\#})$ by

$$[X \oplus f, Y \oplus g] = [X,Y] \oplus (q(X)(g) - q(Y)(f)).$$

($A^{\#}$ is now itself a Lie algebroid, with anchor $q^{\#}(X \oplus f) = q(X)$.)

<u>Definition 2.19</u> (Rinehart (1963)). A C(B)-<u>regular</u> $A^{\#}$-<u>module</u> is a vector bundle E on B together with an $\mathbb{R}$-linear map $\tilde{\rho}\colon \Gamma(A^{\#}) \oplus \Gamma E \to \Gamma E$ which is a representation of the $\mathbb{R}$-Lie algebra $\Gamma(A^{\#})$ on the $\mathbb{R}$-vector space ΓE, such that

(i) $\tilde{\rho}(f\mathcal{X})(\mu) = f\tilde{\rho}(\mathcal{X})(\mu) \quad \forall f \in C(B),\ \mathcal{X} \in \Gamma(A^{\#}),\ \mu \in \Gamma E$,

(ii) $\tilde{\rho}(0 \oplus 1) = \mathrm{id}_E$. //

<u>Proposition 2.20</u>. Given a representation ρ of A on a vector bundle E, there is a C(B)-regular $A^{\#}$-module $\tilde{\rho}\colon \Gamma(A^{\#}) \oplus \Gamma E \to \Gamma E$ defined by

$$\tilde{\rho}(X \oplus f)(\mu) = \rho(X)(\mu) + f\mu.$$

Conversely, given a C(B)-regular $A^{\#}$-module $\tilde{\rho}$, the map $\rho\colon \Gamma A \oplus \Gamma E \to \Gamma E$, $(X,\mu) \mapsto \tilde{\rho}(X \oplus 0)(\mu)$ defines a Lie algebroid representation of A on E.

The correspondences $\rho \leftrightarrow \tilde{\rho}$ are mutually inverse.

<u>Proof</u>: Note that $\tilde{\rho}(0 \oplus f)(\mu) = \tilde{\rho}(f(0 \oplus 1))(\mu) = f\mu$, by (ii) in 2.19. Given this, the verifications are easy. For example,

$$\begin{aligned}\rho(X)(f\mu) &= \tilde{\rho}(X \oplus 0)(\tilde{\rho}(0 \oplus f)(\mu))\\ &= \tilde{\rho}([X \oplus 0, 0 \oplus f])(\mu) + \tilde{\rho}(0 \oplus f)(\tilde{\rho}(X \oplus 0)(\mu))\\ &= \tilde{\rho}(0 \oplus q(X)(f))(\mu) + f\rho(X)(\mu)\\ &= q(X)(f)\mu + f\rho(X)(\mu). \qquad //\end{aligned}$$

I am grateful to Rui Almeida for pointing out in 1982 the omission of (ii) in 2.19 from the account given in Mackenzie (1979).

§3. Non-abelian extensions of Lie algebroids and the existence of transitive Lie algebroids with prescribed curvature.

We now present the classification theory of non-abelian extensions of Lie algebroids, and its application to the problem of constructing a connection with prescribed curvature form.

Given a non-abelian extension of Lie algebroids $K \overset{\iota}{\rightarrowtail} A' \overset{\pi}{\twoheadrightarrow} A$, each $X' \in \Gamma A'$ induces a Lie covariant differential operator $V \mapsto \iota^{-1}[X', \iota(V)]$ of K. In the abelian case these operators depend only on $\pi(X')$; in the non-abelian case, however, $X'' \in \Gamma A'$ with $\pi(X'') = \pi(X')$ will induce an operator which differs from that induced by X' by an element of $\Gamma ad(K) < \Gamma CDO[K]$. The extension therefore induces, not a representation of A on K, but a morphism $\Xi\colon A \to \frac{CDO[K]}{ad(K)}$ which we call, following Robinson (1982), a coupling of A with K. This section begins with the construction of $\frac{CDO[K]}{ad(K)}$, which we call the Lie algebroid of outer covariant differential operators, and denote by $OutDO[K]$.

Given a coupling $\Xi\colon A \to OutDO[K]$, there is a natural representation, ρ^{Ξ}, of A on ZK, and from 3.2 to 3.12 we are concerned to show that Ξ defines an element of $\mathcal{H}^3(A, \rho^{\Xi}, ZK)$, called the obstruction class of Ξ and denoted $Obs(\Xi)$. From 3.14 through to 3.19 we give the detailed construction of the coupling arising from a (nonabelian) extension and prove (3.18) that for such a coupling the obstruction class is zero. From 3.20 through to 3.31 we are concerned with establishing the converse of 3.18. The main construction result is 3.20 which shows (3.22) that a coupling Ξ with obstruction class zero arises from an extension. From 3.23 through to 3.31 we classify the extensions which induce a given coupling Ξ; the analysis shows that $\mathcal{H}^2(A, \rho^{\Xi}, ZK)$ acts freely and transitively on the set of equivalence classes of Ξ-operator extensions of A by K. In 3.32 to 3.34 we address the question of semidirect (or flat) extensions, and note that not every coupling with zero obstruction class arises from a semidirect extension; on the other hand there may be several inequivalent semidirect extensions inducing the one coupling. The section closes with a brief application of 3.20 to the construction of produced Lie algebroids.

The cohomological formalism developed in this section is of a standard type, and closely follows the theory of nonabelian Lie algebra extensions, as developed by Hochschild (1954a,b), Mori (1953) and Shukla (1966). Some aspects of this formalism have been noted before (Palais (1961b), Hermann (1967), Teleman (1972)), but worked on the level of the modules of sections rather than the Lie algebroids themselves.

What is remarkable is that this cohomological apparatus yields results of considerable geometric significance. From 3.20 we obtain a solution to the algebraic half of a long-standing problem (Weil (1958), Kostant (1970)): When is a 2-form the curvature of a connection? From 3.20 it follows that an LAB-valued 2-form $R \in A^2(B,L)$ on a manifold B is the curvature of a connection in a Lie algebroid iff there is a Lie connection ∇ in L such that $\overline{R}_\nabla = \mathrm{ad}\, R$ and $\nabla(R) = 0$. In Chapter V we will answer the question of when the resulting Lie algebroid can be integrated to a Lie groupoid, in the case where the base is simply-connected.

The second major application of the results of this section is to Theorem 4.1 (see also Theorem 5.1) where the classification of nonabelian extensions is the key to the proof that a transitive Lie algebroid on a contractible base admits a flat connection.

In addition, the reader will probably be pleased to see how the standard identities of infinitesimal connection theory arise naturally in cohomological terms.

The results of this section first appeared in Mackenzie (1979). In the present account, some of the proofs have been reformulated for clarity.

* * * * * * * *

Definiton 3.1. Let K be an LAB on B. Then the quotient Lie algebroid

$$\mathrm{Der}(K)/\mathrm{ad}(K) \xrightarrow{\overline{j}} \mathrm{CDO}[K]/\mathrm{ad}(K) \xrightarrow{\overline{q}} TB$$

is denoted by

$$\mathrm{Out}(K) \xrightarrow{\overline{j}} \mathrm{OutDO}[K] \xrightarrow{\overline{q}} TB$$

and elements of $\Gamma\mathrm{OutDO}[K]$ are called outer covariant differential operators on K. //

Quotient Lie algebroids are defined in 1.11. That $\mathrm{ad}(K) = \mathrm{im}(\mathrm{ad}\colon K \to \mathrm{Der}(K))$ is a sub LAB of Der(K) follows from 1.2. That ad(K) is an ideal of OutDO[K] from the formula

(1) $$[D,(j\circ \mathrm{ad})(V)] = (j\circ \mathrm{ad})(D(V))$$

for $D \in \Gamma CDO[K]$, $V \in \Gamma K$, which is proved by an easy manipulation of the Jacobi identity. We now have

$$\begin{array}{ccccc}
ZK & = & ZK & & \\
{\scriptstyle i}\downarrow & & \downarrow{\scriptstyle i} & & \\
K & = & K & & \\
{\scriptstyle ad}\downarrow & & \downarrow{\scriptstyle ad} & & \\
Der(K) & \xrightarrow{j} & CDO[K] & \xrightarrow{q} & TB \\
{\scriptstyle \natural^{+}}\downarrow & & \downarrow{\scriptstyle \natural} & & \| \\
Out(K) & \xrightarrow{\bar{j}} & OutDO[K] & \xrightarrow{\bar{q}} & TB
\end{array}$$

with both rows and columns exact.

From now until 3.31 we consider a single Lie algebroid A on B. The anchor of A is denoted q^A. It is not assumed that A is transitive.

<u>Definition 3.2</u>. A <u>coupling</u> of A is an LAB K together with a morphism of Lie algebroids Ξ: $A \to OutDO[K]$; we also say that A and K are <u>coupled</u> by Ξ. //

This is what would once have been called an "abstract kernel for A"; the present terminology comes from Robinson (1982). Every transitive Lie algebroid $L \rightarrowtail A \twoheadrightarrow TB$ induces a coupling of TB to L, namely $\natural \circ \nabla^{\gamma}$ for any connection γ: $TB \to A$. See 3.17 below.

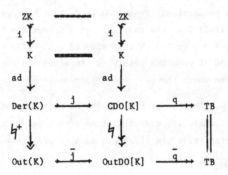

Now fix a coupling Ξ of an LAB K to A until 3.13. Since $\natural$: $CDO[K] \twoheadrightarrow OutDO[K]$ is a surjective submersion, as a map of vector bundles over B, there are vector bundle morphisms ∇: $A \to CDO[K]$, $X \mapsto \nabla_X$, such that $\natural\nabla = \Xi$. We call ∇ a <u>Lie derivation law</u> covering Ξ. (See 3.8 for the formal definition.) Since $\bar{q} \circ \natural = q$ it follows that $q \circ \nabla = q^A$; that is, ∇ is an anchor-preserving map. Therefore the curvature of ∇ is a well-defined map R_{∇}: $A \oplus A \to CDO[K]$. Since $\natural \circ \nabla = \Xi$ is a morphism, it follows that $\natural \circ R_{\nabla} = 0$ and so R_{∇} takes its values in $ad(K) \subseteq Der(K)$; we denote this map $A \oplus A \to ad(K)$ by $\bar{R}_{\nabla}$. Now, as with Ξ above, there are alternating vector bundle morphisms Λ: $A \oplus A \to K$ such that $ad \circ \Lambda = \bar{R}_{\nabla}$. This follows from momentarily considering $\bar{R}_{\nabla}$ to be defined on $\Lambda^2 A$ and lifting it from $\Lambda^2 A \to ad(K)$ to $\Lambda^2 A \to K$ across the surjective vector bundle morphism ad: $K \to ad(K)$. We call any alternating map $\Lambda: A \oplus A \to K$ with $ad \circ \Lambda = \bar{R}_{\nabla}$ a <u>lift</u> of R_{∇}.

$\Xi: A \to OutDO[K]$ induces a representation of A on ZK, the centre of K. To see this, let ∇ be any Lie derivation law covering Ξ. Then for $X \in \Gamma A$, the operator $\nabla_X: \Gamma K \to \Gamma K$ restricts to $\Gamma ZK \to \Gamma ZK$, for if $Z \in \Gamma ZK$ and $V \in \Gamma K$ then

$$[V,\nabla_X(Z)] = \nabla_X([V,Z]) - [\nabla_X(V),Z]$$

$$= \nabla_X(0) - 0 = 0,$$

since Z is central. Write $\rho(X)$ for the restriction of ∇_X to $\Gamma ZK \to \Gamma ZK$. Then ρ defines a vector bundle map $A \to CDO[ZK] = CDO(ZK)$ which is easily seen to be a Lie algebroid morphism. If ∇' is a second Lie derivation law for Ξ then $\nabla'_X - \nabla_X$ is in $\Gamma(ad(K))$, for all $X \in \Gamma A$, and therefore vanishes on ΓZK. Hence ρ is independent of the choice of ∇.

<u>Definition 3.3.</u> The representation $\rho: A \to CDO(ZK)$ just constructed is called the <u>central representation</u> of Ξ and is denoted ρ^{Ξ}. //

Our concern now is to show that every coupling Ξ of A to K defines an element of $\mathcal{H}^3(A,\rho^{\Xi},ZK)$. This will take us until 3.13.

<u>Lemma 3.4.</u> Let ∇ be a Lie derivation law covering Ξ and let Λ be a lift of R_{∇}. Then for all $X,Y,Z \in \Gamma A$ the element

$$\mathfrak{S}\{\nabla_X(\Lambda(Y,Z)) - \Lambda([X,Y],Z)\}$$

of ΓK lies in ΓZK.

<u>Proof:</u> Apply $j \circ ad$. We obtain, firstly,

$$j \circ ad(\nabla_X(\Lambda(Y,Z))) = [\nabla_X,(j \circ ad)(\Lambda(Y,Z))],$$

by (1), and we have $j\circ ad\circ\Lambda = j\circ\bar{R}_\nabla = R_\nabla$ by definition. So $j\circ ad$ of the cyclic sum is

$$\mathfrak{S}\{[\nabla_X, R_\nabla(Y,Z)] - R_\nabla([X,Y],Z)\}$$

and this is zero by the general Bianchi identity 2.9. //

Write $f(X,Y,Z)$ for the element in the lemma. It is easily checked that f is an alternating and $C(B)$-trilinear function of $X,Y,Z \in \Gamma A$, and it therefore defines an element of $\Gamma C^3(A,ZK)$, also denoted f.

Lemma 3.5. $df = 0$ with respect to the coboundary induced by ρ.

Proof: This is a long but straightforward calculation, and requires no ingenuity. //

So $f \in \mathcal{Z}^3(A,\rho,ZK)$. f is called the obstruction cocycle defined by ∇ and Λ for the coupling Ξ. We may write $f = f(\nabla,\Lambda)$.

Lemma 3.6. Fix a Lie derivation law ∇ covering Ξ and let Λ and Λ' be two lifts of R_∇, with corresponding obstructions $f = f(\nabla,\Lambda)$, $f' = f(\nabla,\Lambda')$. Then $\Lambda' - \Lambda = i\circ g$ for some $g \in \Gamma C^2(A,ZK)$ and $dg = f' - f$.

Proof: Since $j\circ ad\circ(\Lambda' - \Lambda) = R_\nabla - R_\nabla = 0$ there is a unique $g\colon A \oplus A \to ZK$ with $i\circ g = \Lambda' - \Lambda$. Since Λ' and Λ are alternating, it follows that g is so. Thus $g \in \Gamma C^2(A,ZK)$. Now

$$\begin{aligned} i(f'(X,Y,Z) - f(X,Y,Z)) &= \mathfrak{S}\{\nabla_X(i\circ g(Y,Z)) - i\circ g([X,Y],Z)\} \\ &= i(\mathfrak{S}\{\rho(X)(g(Y,Z)) - g([X,Y],Z)\}) \\ &= i(dg(X,Y,Z)) \end{aligned}$$

for $X,Y,Z \in \Gamma A$. //

We now need to show that the cohomology class of f is independent of the choice of ∇.

Proposition 3.7. Let ∇ and ∇' be two Lie derivation laws covering Ξ. Then $\nabla' = \nabla + j\circ ad\circ\ell$ for various maps $\ell\colon A \to K$, and

$$\bar{R}_{\nabla'} - \bar{R}_{\nabla} = -\mathrm{ad}(\nabla_X(\ell(Y)) - \nabla_Y(\ell(X)) - \ell[X,Y] + [\ell(X),\ell(Y)]).$$

Proof: The existence of ℓ follows as before. For $X,Y \in \Gamma A$ we have

$$(R_{\nabla'} - R_{\nabla})(X,Y) = \nabla'_{[X,Y]} - \nabla_{[X,Y]} - [\nabla'_X,\nabla'_Y] + [\nabla_X,\nabla_Y]$$

$$= (j\circ \mathrm{ad}\circ\ell)[X,Y] - [(j\circ \mathrm{ad}\circ\ell)(X),\nabla_Y]$$

$$- [\nabla_X,(j\circ \mathrm{ad}\circ\ell)(Y)] - [(j\circ \mathrm{ad}\circ\ell)(X),(j\circ \mathrm{ad}\circ\ell)(Y)].$$

Using (1), this becomes

$$(j\circ \mathrm{ad}\circ\ell)[X,Y] + (j\circ \mathrm{ad})(\nabla_Y(\ell(X))) - (j\circ \mathrm{ad})(\nabla_X(\ell(Y))) - (j\circ \mathrm{ad})[\ell(X),\ell(Y)],$$

whence the result. //

This can be expressed more succinctly by extending the definition of the exterior covariant derivative given in III§5. While we do this, briefly disregard the coupling Ξ.

Definition 3.8. Let A be any Lie algebroid and let K be any LAB on the same base. A Lie derivation law for A with coefficients in K is an anchor-preserving map $\nabla\colon A \to CDO[K]$. //

Definition 3.9. Let ∇ be a Lie derivation law for A with coefficients in K. The (exterior) covariant derivative induced by ∇ is the sequence of operators $\nabla\colon \Gamma C^n(A,K) \to \Gamma C^{n+1}(A,K)$ defined by

$$\nabla(f)(X_1,\ldots,X_{n+1}) = \sum_{r=1}^{n+1} (-1)^{r+1}\nabla_{X_r}(f(X_1,\ldots\hat{}\ldots,X_{n+1}))$$

$$+ \sum_{r<s} (-1)^{r+s} f([X_r,X_s],X_1,\ldots\hat{}\ldots\hat{}\ldots,X_{n+1}). \qquad //$$

The requirement that ∇ be anchor-preserving ensures that the RHS actually is $C(B)$-multilinear. There is a formula for $\nabla\circ\nabla$ in terms of the curvature of the Lie derivation law.

The equation in 3.7 can now be written

$$\bar{R}_{\nabla'} - \bar{R}_{\nabla} = -\mathrm{ad}(\nabla(\ell) + [\ell,\ell]).$$

This clearly resembles III 5.12 and III 5.13. The general result is the following.

Proposition 3.10. Let A be a Lie algebroid on B and let $L' \rightarrowtail A' \twoheadrightarrow TB$ be a transitive Lie algebroid on B. Let $\phi_1, \phi_2 \colon A \to A'$ be two anchor-preserving maps and write $\phi_2 = \phi_1 + j' \circ \ell$, where $\ell \colon A \to L'$. Then

$$\bar{R}_{\phi_2} - \bar{R}_{\phi_1} = -(\nabla(\ell) + [\ell,\ell])$$

where $\nabla \colon A \to CDO[L']$ is the Lie derivation law defined by $j'(\nabla_X(V')) = [\phi_1(X), j'(V')]$.

Proof: Identical to that of 3.7 or III 5.12. //

III 5.13 now follows by applying 3.10 to $\phi_2 = \mathrm{id}_A$ and $\phi_1 = \gamma \circ q \colon A \to A$.

We return to the coupling Ξ. Notice that the definition of f can now be written $f = \nabla(\Lambda)$, and f thus measures the extent to which Λ satisfies the Bianchi identity with respect to ∇.

Proposition 3.11. Let ∇ be a Lie derivation law covering Ξ and let Λ be a lift of R_∇. Let ∇' be a second Lie derivation law covering Ξ and write $\nabla' = \nabla + j \circ \mathrm{ad} \circ \ell$, where ℓ is a map $A \to K$. Then

$$\Lambda' = \Lambda - (\nabla(\ell) + [\ell,\ell])$$

is a lift of $R_{\nabla'}$, and $f(\nabla', \Lambda') = f(\nabla, \Lambda)$.

Proof: Certainly Λ' is alternating, and $\mathrm{ad} \circ \Lambda' = \bar{R}_\nabla - \mathrm{ad}(\nabla(\ell) + [\ell,\ell]) = \bar{R}_{\nabla'}$, by 3.7. It remains to show that $\nabla'(\Lambda') = \nabla(\Lambda)$.

Now ∇' is a linear operator, and so

$$\nabla'(\Lambda') - \nabla(\Lambda) = \nabla'(\Lambda) - \nabla'(\nabla(\ell) + [\ell,\ell]) - \nabla(\Lambda).$$

For $X,Y,Z \in \Gamma A$,

$$(\nabla'(\Lambda) - \nabla(\Lambda))(X,Y,Z) = \mathfrak{S}\,\{(j \circ \mathrm{ad} \circ \ell)(X)(\Lambda(Y,Z))\}$$

$$= \mathfrak{S}\{[\ell(X),\Lambda(Y,Z)]\}.$$

Similarly

$$\nabla'(\nabla(\ell) + [\ell,\ell])(X,Y,Z) = \nabla(\nabla(\ell) + [\ell,\ell])(X,Y,Z)$$

$$+ \mathfrak{S}\{(j\circ ad\circ\ell)(X)((\nabla(\ell) + [\ell,\ell])(Y,Z))\}.$$

First calculate $\nabla(\nabla(\ell))$. By regrouping terms and using the Jacobi identity one obtains

$$\nabla(\nabla(\ell))(X,Y,Z) = -\mathfrak{S}\{R_\nabla(X,Y)(\ell(Z))\},$$

and since $j\circ ad\circ\Lambda = R_\nabla$, this cancels with the term $(\nabla'(\Lambda) - \nabla(\Lambda))(X,Y,Z)$.

Next, by expanding the cyclic sum and using $\nabla_X([V,W]) = [\nabla_X(V),W] + [V,\nabla_X(W)]$ repeatedly, one obtains

$$\nabla([\ell,\ell])(X,Y,Z) = \mathfrak{S}[\nabla(\ell)(X,Y),\ell(Z)].$$

Lastly,

$$\mathfrak{S}\{(j\circ ad\circ\ell)(X)((\nabla(\ell) + [\ell,\ell])(Y,Z))\}$$

$$= \mathfrak{S}\{[\ell(X),\nabla(\ell)(Y,Z)]\} + \mathfrak{S}\{[\ell(X),[\ell(Y),\ell(Z)]]\}.$$

The second term of this vanishes, by the Jacobi identity, and the first cancels with $\nabla([\ell,\ell])(X,Y,Z)$.

Thus we obtain $\nabla'(\Lambda') - \nabla(\Lambda) = 0$, as desired. //

3.11 is a solid calculation, no matter how one approaches it. However in the setting provided here, it is at least the case that $\nabla'(\Lambda') - \nabla(\Lambda)$ decomposes into geometrically significant groups of terms. This insight is not available in Lie algebra cohomology, from where the calculation comes.

Putting 3.6 and 3.11 together, we obtain

Theorem 3.12. Let A be a Lie algebroid on B and let K be an LAB on B. Let Ξ be a coupling of A with K. Then the cohomology class in $\mathcal{H}^3(A,\rho^{\Xi},ZK)$ of $f(\nabla,\Lambda)$, where ∇ is a Lie derivation law covering Ξ, and Λ is a lift of R_{∇}, depends only on Ξ and is independent of the choice of ∇ and Λ. //

This class is called the obstruction class of the coupling Ξ, and will be denoted $\mathrm{Obs}(\Xi)$.

The following observation will be important later.

Proposition 3.13. Let A, K and Ξ be as in 3.12. Let ∇ be any Lie derivation law covering Ξ, and let f' be any cocycle in $\mathrm{Obs}(\Xi)$. Then there is a lift Λ' of R_{∇} such that $f(\nabla,\Lambda') = f'$.

Proof: Let Λ be any lift of R_{∇}, and write $f = f(\nabla,\Lambda)$. Then f and f' are cohomologous; choose any $g \in \Gamma C^2(A,ZK)$ such that $f' = f + dg$. Define $\Lambda' = \Lambda + \iota\circ g$; then $ad\circ\Lambda' = ad\circ\Lambda = \bar{R}_{\nabla}$ and 3.6 shows that $f' = f(\nabla,\Lambda')$. //

* * * * * * * *

We now describe the coupling associated to a general (nonabelian) extension of Lie algebroids, and its obstruction class. Until 3.19, let A be a fixed Lie algebroid on B and let K be an LAB on B.

Definition 3.14. An extension of A by K is an exact sequence of Lie algebroids over B

(2) $$K \overset{\iota}{\rightarrowtail} A' \overset{\pi}{\twoheadrightarrow} A,$$

as defined in III 2.14.

Two extensions of A by K, $K \overset{\iota_r}{\rightarrowtail} A^r \overset{\pi_r}{\twoheadrightarrow} A$, $r = 1,2$, are equivalent if there is a morphism of Lie algebroids over B, $\phi\colon A^1 \to A^2$, necessarily an isomorphism, such that $\phi\circ\iota_1 = \iota_2$ and $\pi_2\circ\phi = \pi_1$. //

We define the concepts of transversal, back-transversal, flat transversal and flat extension exactly as for extensions by vector bundles (see 2.11). The discussion following 2.11 also applies, except that an extension (2) by a general LAB need not induce a representation of A on K. Instead the representation formula defines a Lie derivation law.

The example of an extension of Lie algebroids which is of most importance to us is the extension $L \rightarrowtail A \twoheadrightarrow TB$ associated with a transitive Lie algebroid. All the calculations of this subsection are formally equivalent to results for infinitesimal connections. (See III§5.)

Fix an extension $K \overset{\iota}{\rightarrowtail} A' \overset{\pi}{\twoheadrightarrow} A$ of A by K until 3.19.

__Proposition 3.15.__ Let $\chi\colon A \to A'$ be a transversal in $K \overset{\iota}{\rightarrowtail} A' \overset{\pi}{\twoheadrightarrow} A$. Then

$$\iota(\nabla^{\chi}_X(V)) = [\chi(X), \iota(V)] \tag{3}$$

for $X \in \Gamma A$, $V \in \Gamma K$, defines a Lie derivation law for A with coefficients in K.

__Proof:__ It need only be checked that $\nabla^{\chi}\colon A \to CDO[K]$ is anchor-preserving, and this follows from $q^{A'} \circ \chi = q^{A}$. //

__Lemma 3.16.__ With the notation of 3.15,

$$\bar{R}_{\nabla^{\chi}} = \mathrm{ad} \circ \bar{R}_{\chi}.$$

__Proof:__ For $X, Y \in \Gamma A$ and $V \in \Gamma K$,

$$\begin{aligned}\iota(R_{\nabla^{\chi}}(X,Y)(V)) &= [\chi[X,Y], \iota(V)] - [\chi(X),[\chi(Y),\iota(V)]] + [\chi(Y),[\chi(X),\iota(V)]] \\ &= [\chi[X,Y], \iota(V)] + [\iota(V),[\chi(X),\chi(Y)]] \\ &\qquad \text{(by the Jacobi identity in } \Gamma A'\text{)} \\ &= [R_{\chi}(X,Y), \iota(V)]\end{aligned}$$

and the result follows. //

Hence the composition $A \xrightarrow{\nabla^{\chi}} CDO[K] \twoheadrightarrow OutDO[K]$ is a morphism of Lie algebroids. If $\chi'\colon A \to A'$ is a second transversal then $\chi' = \chi + \iota \circ \ell$ for some map $\ell\colon A \to K$ and $\nabla^{\chi'} = \nabla^{\chi} + \mathrm{ad}\ \ell$. Hence $\natural \circ \nabla^{\chi'} = \natural \circ \nabla^{\chi}$.

Definition 3.17. The coupling $\natural\circ\nabla^{\chi}$: $A \to CDO[K]$, where χ is any transversal in the extension $K \overset{\iota}{\rightarrowtail} A' \overset{\pi}{\twoheadrightarrow} A$, is the coupling of A with K induced by the extension. //

Note that the choice of a transversal χ determines both a Lie derivation law ∇^{χ} covering the coupling induced by the extension, and (by 3.16) a lift $\bar{R}_{\chi}$ of $R_{\nabla^{\chi}}$. Thus each χ determines an obstruction cocycle.

Proposition 3.18. For every transversal χ,

$$f(\nabla^{\chi},\bar{R}_{\chi}) = 0.$$

Proof: $f(X,Y,Z) = \mathfrak{S}\{\nabla^{\chi}_{X}(\bar{R}_{\chi}(Y,Z)) - \bar{R}_{\chi}([X,Y],Z))\}$ and, applying ι this becomes

$$\mathfrak{S}\{[\chi(X),R_{\chi}(Y,Z)] - R_{\chi}([X,Y],Z)\}$$

which is zero by the general Bianchi identity of 2.9. //

This equation, $f(\nabla^{\chi},\bar{R}_{\chi}) = 0$, is called the Bianchi identity for χ.

In particular, the obstruction cohomology class for the coupling detemined by $K \rightarrowtail A' \twoheadrightarrow A$ is zero.

The following result shows that every Lie derivation law covering the coupling determined by an extension, arises from a transversal.

Proposition 3.19. Let ∇ be any Lie derivation law covering the coupling determined by $K \overset{\iota}{\rightarrowtail} A' \overset{\pi}{\twoheadrightarrow} A$. Then there is a transversal χ: $A \to A'$ such that $\nabla^{\chi} = \nabla$.

Proof: Let χ' by any transversal. Since $\natural\circ\nabla = \natural\circ\nabla^{\chi'}$, there is a map $\bar{\ell}$: $A \to ad(K)$ such that $\nabla = \nabla^{\chi'} + j\circ\bar{\ell}$. Since ad: $K \to ad(K)$ is a surjective submersion of vector bundles, $\bar{\ell}$ can be lifted to ℓ: $A \to K$. Define $\chi = \chi' + \iota\circ\ell$. Certainly χ is a transversal, and for all $X \in \Gamma A$ and $V \in \Gamma K$,

$$\iota(\nabla^{\chi}_{X}(V)) = [\chi'(X),\iota(V)] + \iota[\ell(X),V]$$

$$= \iota((\nabla^{\chi'}_{X}(V)) + (j\circ ad\circ\ell)(X)(V))$$

$$= \iota(\nabla_{X}(V)),$$

as required. //

Thus the nonzero elements of the obstruction class arise from having chosen a Lie derivation law ∇^X and having then failed to choose the natural lift, namely $\bar{R}_X$, of R_{∇^X}; see 3.6. On the other hand, it is not true that if Λ is a lift of R_{∇^X}, and if $f(\nabla^X,\Lambda) = 0$, then $\Lambda = \bar{R}_X$; there are usually, for example, many closed non-zero two-forms on a manifold.

* * * * * * * *

We arrive now at the construction and enumeration of extensions corresponding to a coupling which has obstruction class zero. The first result is the basic construction principle.

<u>Theorem 3.20.</u> Let A be a Lie algebroid on base B and let K be an LAB on B. Let ∇: A → CDO[K] be a Lie derivation law such that $\natural\circ\nabla$: A → OutDO[K] has zero curvature, and let R: A ⊕ A → K be an alternating 2-form on A with values in K. Then, if

(i) $\bar{R}_\nabla = \mathrm{ad}\circ R$, and

(ii) the Bianchi identity $\nabla(R) = 0$,

hold, then the formula

$$[X \oplus V, Y \oplus W] = [X,Y] \oplus \{\nabla_X(W) - \nabla_Y(V) + [V,W] - R(X,Y)\}$$

defines a bracket on $\Gamma(A \oplus K)$ which makes $A \oplus K$ a Lie algebroid on B with respect to the arrow $q' = q\circ\pi_1$, and an extension

$$K \xrightarrow{\iota_2} A' \xrightarrow{\pi_1} A$$

of A by K such that ι_1: A → A' is a transversal with $\nabla^{\iota_1} = \nabla$ and $\bar{R}_{\iota_1} = R$.

<u>Proof:</u> This is a straightforward calculation. We verify the Jacobi identity as an example.

Given $X_r \oplus V_r \in \Gamma A'$, r = 1,2,3, we have for the K-component of $\mathfrak{S}[[X_1 \oplus V_1, X_2 \oplus V_2], X_3 \oplus V_3]$ the formula

$$\mathfrak{S}\{\nabla_{[X_1,X_2]}(V_3) - \nabla_{X_3}(\nabla_{X_1}(V_2)) + \nabla_{X_3}(\nabla_{X_2}(V_1))$$
$$- \nabla_{X_3}([V_1,V_2]) + \nabla_{X_3}(R(X_1,X_2)) + [\nabla_{X_1}(V_2),V_3]$$
$$- [\nabla_{X_2}(V_1),V_3] + [[V_1,V_2],V_3] - [R(X_1,X_2),V_3]$$
$$- R([X_1,X_2],X_3)\}.$$

The term $\mathfrak{S}[[V_1,V_2],V_3]$ vanishes by the Jacobi identity in K. The fifth term can be rewritten as $\mathfrak{S}\nabla_{X_1}(R(X_2,X_3))$, by cyclic permutation, and then cancels with the last, by the Bianchi identity (ii). Rewriting the first three terms via cyclic permutations, we have $\mathfrak{S}\, R_\nabla(X_1,X_2)(V_3)$ and this is, by (i), equal to $\mathfrak{S}[R(X_1,X_2),V_3]$, and so cancels with the second-last term. The remaining terms cancel by the equation $\nabla_X([V,W]) = [\nabla_X(V),W] + [V,\nabla_X(W)]$ for $X \in \Gamma A$, $V,W \in \Gamma K$, which characterizes the elements of CDO[K]. //

We will treat the uniqueness of this construction in 3.23 below.

3.20 gives, in particular, the following construction principle for transitive Lie algebroids with a prescribed curvature form.

<u>Corollary 3.21.</u> Let L be an LAB on B and let $R \in \Gamma C^2(TB,L)$ be an alternating 2-form on B with values in L. Then there is a transitive Lie algebroid $L \rightarrowtail A \twoheadrightarrow TB$ and a connection γ in A with $\bar{R}_\gamma = R$ iff there is a Lie connection ∇ in L such that (i) $\bar{R}_\nabla = \mathrm{ad}\circ R$, and (ii) $\nabla(R) = 0$. //

3.21 provides a solution to the algebraic part of a long-standing problem: Given an alternating 2-form on a manifold B, when is it the curvature form of a connection in a bundle P(B,G) over B? In the case of real-valued 2-forms this problem is solved by the integrality lemma of Weil (1958); see also Kostant (1970). Weil's result corresponds to the case of 3.21 in which $L = B \times \mathbb{R}$; note that 3.21 is slightly stronger in that we require only some flat connection ∇ with $\nabla(R) = 0$, whereas Weil requires R to be closed with respect to the standard flat connection (see III 5.17 for the equivalence). In the case of non-abelian coefficients it is obviously necessary for the 2-form to take values in an LAB, rather than in a single Lie algebra, and this necessitates the introduction of the Lie connection ∇.

The presence in 3.20 and 3.21 of the Lie derivation law ∇ certainly makes their application difficult. However ∇, and condition (i), are essential links between the algebraic properties and the curvature properties of the desired Lie algebroid. For example, if in 3.21 L is abelian, then it must be flat as a vector bundle, in order for a transitive Lie algebroid $L \rightarrowtail A \twoheadrightarrow TB$ to exist (by III 5.10(ii)).

This phenomenon should be compared to the situation in III 5.15 where a transitive Lie algebroid is constructed from a family of Maurer-Cartan forms, subject to a cocycle condition and a compatibility condition which involve an Aut($\mathfrak{g}$)-cocycle, which turns out to be a cocycle for the adjoint LAB. The Aut($\mathfrak{g}$)-cocycle in III 5.15 corresponds exactly to the Lie derivation law in 3.21, and conditions (ii) and (iii) of III 5.15 correspond to conditions (ii) and (i) of 3.21.

The remaining part of this question on the realizability as curvature forms of given 2-forms, concerns the integrability of the transitive Lie algebroid found by the method of 3.21. This problem is solved, under the assumption that B is simply-connected, in Chapter V.

Corollary 3.22. Let A be a Lie algebroid on B, let K be an LAB on B, and let Ξ be a coupling of A with K. Then, if $\mathrm{Obs}(\Xi) = 0 \in \mathcal{H}^3(A, ZK)$, there is a Lie algebroid extension

$$K \rightarrowtail A' \twoheadrightarrow A$$

of A by K, inducing the coupling Ξ, namely that constructed in 3.20, using any Lie derivation law ∇ which covers Ξ, and any lift $\Lambda = R$ of R_∇.

Proof: This follows from 3.20 by applying 3.13 to the cocycle 0 in $\mathrm{Obs}(\Xi)$. //

3.22 and 3.18 together show that a coupling $\Xi\colon A \to \mathrm{OutDO}[K]$ arises from an extension of A by K iff $\mathrm{Obs}(\Xi) = 0 \in \mathcal{H}^3(A, \rho^\Xi, ZK)$.

There are usually many distinct such extensions. We come now to the problem of their description.

Proposition 3.23. Let $K \overset{\iota}{\rightarrowtail} A' \overset{\pi}{\twoheadrightarrow} A$ be an extension of Lie algebroids, with K an LAB. Then for any transversal $\chi\colon A \to A'$, there is an equivalence of this extension with the extension $K \overset{\iota_2}{\rightarrowtail} A \oplus K \overset{\pi_1}{\twoheadrightarrow} A$ constructed via 3.20 from ∇^χ and $\bar{R}_\chi$, namely

$$\mathcal{E}_\chi : A \oplus K \to A', \quad X \oplus V \mapsto \chi(X) + \iota(V).$$

Proof: For example, that $\mathcal{E}_\chi$ preserves the brackets merely asserts that

$$[\chi(X) + \iota(V), \chi(Y) + \iota(W)] = \chi[X,Y] + \iota(\nabla^\chi_X(W) - \nabla^\chi_Y(V) + [V,W] - \bar{R}_\chi(X,Y))$$

for all $X,Y \in \Gamma A$, $V,W \in \Gamma K$, and this is easily verified. //

In the terminology of MacLane (1963), 3.23 shows that every extension can be given a "crossed-product" representation. In particular, the extension constructed in 3.20 is determined up to equivalence by ∇ and R.

3.23 is the correct uniqueness result for the problem of constructing transitive Lie algebroids with a preassigned curvature form and adjoint connection. However we will also need in what follows an enumeration of the extensions of A by K in terms of $\mathcal{H}^2(A,ZK)$.

From now until 3.31, let A be a fixed Lie algebroid on B, let K be an LAB on B, and let $\Xi: A \to OutDO[K]$ be a coupling of A with K such that $Obs(\Xi) = 0 \in \mathcal{H}^3(A,\rho^\Xi,ZK)$.

Definition 3.24. An operator extension of A by K is an extension $K \rightarrowtail A' \twoheadrightarrow A$ which induces, via 3.17, the coupling Ξ. //

The set of equivalence classes of operator extensions of A by K is denoted by $\mathcal{O}pext(A,\Xi,K)$, or by $\mathcal{O}pext(A,K)$ if Ξ is understood. We will show that $\mathcal{H}^2(A,\rho^\Xi,ZK)$ acts freely and transitively on $\mathcal{O}pext(A,\Xi,K)$. It will then follow that $\mathcal{O}pext(A,\Xi,K)$ can be put in bijective correspondence with $\mathcal{H}^2(A,\rho^\Xi,ZK)$, by the choice of any extension as reference point.

The first step is to define an action of $\mathcal{Z}^2(A,ZK)$ on the class of all operator extensions.

Definition 3.25. Let $K \overset{\iota}{\rightarrowtail} A' \overset{\pi}{\twoheadrightarrow} A$ be an operator extension, and let g be in $\mathcal{Z}^2(A,ZK)$. Then the action of g on the extension yields the extension

$$K \overset{\iota}{\rightarrowtail} A'_g \overset{\pi}{\twoheadrightarrow} A$$

where $A'_g = A'$ as vector bundles, the maps ι and π are the same in both extensions, the anchors $q': A' \to TB$ and $q'_g: A'_g \to TB$ are equal, and the bracket $[\ ,\ ]_g$ on $\Gamma(A'_g)$

is given by

$$[X,Y]_g = [X,Y] + \iota i g(\pi X, \pi Y). \qquad //$$

The cocycle condition for g ensures that $[\ ,\ ]_g$ obeys the Jacobi identity. Since the values of g are in $ZK \leq K$, the maps ι and π remain morphisms with respect to the new structure, and the coupling is unchanged.

Proposition 3.26. Continue the notation of 3.25. If $\chi: A \to A'$ is a transversal for A', then it is also a transversal for A'_g, and the two Lie derivation laws $A \to CDO[K]$ are equal. The curvatures of χ are related by

$$\bar{R}^g_\chi = \bar{R}_\chi - i\circ g.$$

Proof: Easy calculation. //

In particular, $K \rightarrowtail^{\iota} A'_g \xrightarrow{\pi}\!\!\!\!\to A$ is an operator extension. Clearly $(A'_g)_h = A'_{g+h}$.

Proposition 3.27. Let χ be a transversal in an operator extension $K \rightarrowtail^{\iota} A' \xrightarrow{\pi}\!\!\!\!\to A$, and let $\ell: A \to K$ be a map. Then

$$\bar{R}_{\chi+\iota\circ\ell} = \bar{R}_\chi - (\nabla^\chi(\ell) + [\ell,\ell]).$$

Proof: This is formally identical to III 5.12. //

In particular, if ℓ takes values in ZK then

$$\bar{R}_{\chi+\iota\circ i\circ\ell} = \bar{R}_\chi - i\circ d\ell.$$

From this we deduce that the action of $\mathcal{Z}^2(A,ZK)$ factors to an action of $\mathcal{H}^2(A,ZK)$.

Proposition 3.28. Let $K \rightarrowtail^{\iota} A' \xrightarrow{\pi}\!\!\!\!\to A$ be an operator extension, and let $h \in \Gamma C^1(A,ZK)$. Then $K \rightarrowtail A'_{dh} \twoheadrightarrow A$ is equivalent to $K \rightarrowtail A' \twoheadrightarrow A$.

Proof: Let χ be a transversal of A'. Regarded as a transversal of A'_{dh}, it has curvature $\bar{R}_\chi - i\circ dh$ (by 3.26). So, by 3.27, the transversal $\chi + \iota\circ i\circ h$ in A'_{dh} has curvature $\bar{R}_\chi$. Since h takes values in ZK, the Lie derivation law determined by $\chi + \iota\circ i\circ h$ in A'_{dh} is the same as that determined by χ in A'. So we have two extensions, A' and A'_{dh}, with two transversals, χ and $\chi + \iota\circ i\circ dh$, respectively, which have the same curvature and Lie derivation law. It now follows from 3.23 that the extensions are equivalent. //

The equivalence $A' \to A'_{dh}$ is ε_h: $X \mapsto X - (\iota\circ i\circ h)(\pi X)$. Briefly, $\varepsilon_h = id - h\circ\pi$.

We therefore have an action of $\mathcal{H}^2(A,ZK)$ on the class of all operator extensions. It is easily checked that the action sends equivalent extensions to equivalent extensions, so we in fact have an action of $\mathcal{H}^2(A,ZK)$ on $\mathcal{O}$pext(A,K).

We now prove that this action is free, and transitive.

<u>Theorem 3.29</u>. Let $K \overset{\iota}{\rightarrowtail} A' \overset{\pi}{\twoheadrightarrow} A$ be an operator extension, and let $g \in \mathcal{Z}^2(A,ZK)$ be a cocycle, and suppose that there is an equivalence ϕ: $A' \to A'_g$. Then g is cohomologous to zero, g = dh, and $\phi = \varepsilon_h$, where $h \in \Gamma C^1(A,ZK)$ is the cochain determined by $h = -\lambda\circ\phi\circ\chi$ for any transversal χ and associated back-transversal λ.

<u>Proof</u>: From 3.26 it follows that

$$i\circ g = \bar{R}_\chi - \bar{R}^g_\chi$$

for any transversal χ. Now by exactly the same calculation as in III 5.13, we obtain

$$\bar{R}_\chi(\pi X,\pi Y) = \nabla^\chi_{\pi X}(\lambda Y) - \nabla^\chi_{\pi Y}(\lambda X) - \lambda[X,Y] + [\lambda X,\lambda Y]$$

for $X,Y \in \Gamma A'$.

Similarly, working in A'_g with the same transversal χ, and recalling from 3.26 that the two Lie derivation laws for χ, with respect to A' and A'_g, are equal, we get

$$\bar{R}^g_\chi(\pi X,\pi Y) = \bar{R}^g_\chi(\pi\phi X,\pi\phi Y)$$
$$= \nabla^\chi_{\pi X}(\lambda\phi Y) - \nabla^\chi_{\pi Y}(\lambda\phi X) - \lambda[\phi X,\phi Y] + [\lambda\phi X,\lambda\phi Y].$$

Substituting these in the equation for $i\circ g$, we get

$$(i\circ g)(\pi X,\pi Y) = \nabla^\chi_{\pi X}(\lambda Y - \lambda\phi Y) - \nabla^\chi_{\pi Y}(\lambda X - \lambda\phi X)$$

$$- (\lambda - \lambda\phi)[X,Y] + \{[\lambda X, \lambda Y] - [\lambda\phi X, \lambda\phi Y]\}.$$

Is $\lambda - \lambda\phi$ equal to $\theta\pi$ for any map $\theta: A \to K$? If it is, then $(\lambda - \lambda\phi)\chi = \theta\pi\chi = \theta$ and since $\lambda\chi = 0$, it follows that $\theta = -\lambda\phi\chi$ and $\lambda - \lambda\phi = -\lambda\phi\chi\pi$. Writing now $h = -\lambda\phi\chi$ we have

$$(\iota\circ g)(\pi X,\pi Y) = \nabla^{\chi}_{\pi X}(h\pi Y) - \nabla^{\chi}_{\pi Y}(h\pi X) - h\pi[X,Y]$$

$$+ \{[\lambda\phi X + h\pi X,\ \lambda\phi Y + h\pi Y] - [\lambda\phi X, \lambda\phi Y]\}$$

$$= \{\nabla^{\chi}_{\pi X}(h\pi X) - \nabla^{\chi}_{\pi Y}(h\pi X) - h\pi[X,Y] + [h\pi X, h\pi Y]\}$$

$$+ \{[\lambda\phi X, h\pi Y] + [h\pi X, \lambda\phi Y]\}.$$

Set $\pi Y = 0$. Then the LHS and all terms on the RHS except the last, vanish. So $[h\pi X, \lambda\phi Y] = 0$ for all $X \in \Gamma A'$ and $Y \in \mathrm{im}(\iota)$, and since π is onto A and $\lambda\circ\phi$ is onto K (for λ is a surjective submersion), this proves that h takes values in ZK.

The equation for $\iota\circ g$ now reduces to $(\iota\circ g)(\pi X,\pi Y) = dh(\pi X,\pi Y)$, which proves that $g = dh$, since π is onto A.

After 3.28 it was remarked that $\varepsilon_h = id - \iota\circ i\circ h\circ\pi$. Neglecting the i, we have

$$\varepsilon_h = id + \iota\circ\lambda\circ\phi\circ\chi\circ\pi$$

$$= id + \iota\circ\lambda\circ\phi\circ(id - \iota\circ\lambda)$$

$$= id + \iota\circ\lambda\circ\phi - \iota\circ\lambda \quad (\text{since } \phi\circ\iota = \iota \text{ and } \lambda\circ\iota = id)$$

$$= \chi\circ\pi + \iota\circ\lambda\circ\phi$$

$$= \phi \quad (\text{since } \pi = \pi\circ\phi \text{ and } \chi\circ\pi + \iota\circ\lambda = id). \qquad //$$

<u>Theorem 3.30</u>. Let $K \xrightarrow{\iota_r} A^r \xrightarrow{\pi_r} A$, $r = 1,2$, be two operator extensions. Let χ_1 and χ_2 be transversals of A^1 and A^2, respectively, which induce the same Lie derivation law (see 3.19), and define $g \in \Gamma C^2(A,ZK)$ by $\iota\circ g = \bar{R}_{\chi_1} - \bar{R}_{\chi_2}$. Then g is a cocycle, $g \in \mathcal{Z}^2(A,ZK)$, and $\phi = \iota_2\circ\lambda_1 + \chi_2\circ\pi_1$ is an equivalence $(A^1)_g \to A^2$.

Proof: That $\bar{R}_{\chi_1} - \bar{R}_{\chi_2}$ takes values in ZK follows from the assumption that $\nabla^{\chi_1} = \nabla^{\chi_2}$, since $ad\circ\bar{R}_{\chi_1} = \bar{R}_{\nabla^{\chi_1}}$ and $ad\circ\bar{R}_{\chi_2} = \bar{R}_{\nabla^{\chi_2}}$ (by 3.16).

Now

$$(i\circ dg)(X,Y,Z) = i(\mathfrak{S}\{\rho(X)(g(Y,Z)) - g([X,Y],Z)\}$$

$$= \mathfrak{S}\{\nabla_X^{\chi_1}(\bar{R}_{\chi_1}(Y,Z)) - \bar{R}_{\chi_1}([X,Y],Z)\}$$

$$- \mathfrak{S}\{\nabla_X^{\chi_2}(\bar{R}_{\chi_2}(Y,Z)) - \bar{R}_{\chi_2}([X,Y],Z)\},$$

which is zero by the Bianchi identities for χ_1 and χ_2 (compare 3.18).

To prove that ϕ preserves the brackets, requires manipulations of a type that must now be familiar. //

Putting together 3.25 to 3.30, we have

Theorem 3.31. Let A be a Lie algebroid on B, let K be an LAB on B, and let Ξ be a coupling of A with K such that $\mathrm{Obs}(\Xi) = 0 \in \mathcal{H}^3(A,ZK)$. Then the additive group of $\mathcal{H}^2(A,\rho^{\Xi},ZK)$ acts freely and transitively on $\mathcal{O}\mathrm{pext}(A,\Xi,K)$. //

One may say that $\mathcal{O}\mathrm{pext}(A,\Xi,K)$ is an affine space overlying the vector space $\mathcal{H}^2(A,\rho^{\Xi},ZK)$.

3.31 yields in particular a classification of transitive Lie algebroids up to equivalence. This should be contrasted with the Chern-Weil theory. Classically one tries to distinguish principal bundles by studying their connections. Infinitesimal connections actually belong in the Lie algebroids and one shows that two Lie algebroids are non-isomorphic by exhibiting a connection in one that cannot exist in the other. This is usually done by means of cohomological invariants derived from the curvature of all connections via the Weil morphism and in this context Lie algebroids cannot be distinguished from their reductions: such studies are topological, not geometric.

In view of the application of 3.31 in §4, some comments about semi-direct extensions are in order.

Definition 3.32. Let A be a Lie algebroid on B, let K be an LAB on B, and let $\nabla: A \to CDO[K]$ be a flat Lie derivation law; that is, let ∇ be a representation of A on K. Then the ∇-semidirect extension of A by K is the extension

$$K \overset{\iota_2}{\rightarrowtail} A \underset{\nabla}{\ltimes} K \overset{\pi_1}{\twoheadrightarrow} A$$

where $A \underset{\nabla}{\ltimes} K$ is the vector bundle $A \oplus K$ with anchor $q'(X \oplus V) = q^A(X)$ and bracket

$$[X \oplus V,\ Y \oplus W] = [X,Y] \oplus \{\nabla_X(W) - \nabla_Y(V) + [V,W]\}. \qquad //$$

It follows immediately from 3.23 that every flat extension is equivalent to a semidirect extension.

Not every coupling whose obstruction class is zero has a flat Lie derivation law covering it. A characteristic example follows.

Example 3.33. Let Ω be the Lie groupoid associated to the Hopf bundle $S^7(S^4,SU(2))$. Let Ξ denote the coupling of TS^4 with $L\Omega = \dfrac{S^7 \times \mathfrak{su}(2)}{SU(2)}$ induced by $L\Omega \rightarrowtail A\Omega \twoheadrightarrow TS^4$. Then there is no flat Lie derivation law covering Ξ; in fact, $L\Omega$ admits no flat Lie connection.

To see this, note first that ad: $A\Omega \to CDO[L\Omega]$ is an isomorphism, since $ad^+: L\Omega \to Der(L\Omega)$ is fibrewise the adjoint representation of $\mathfrak{su}(2)$, and $\mathfrak{su}(2)$ is semisimple. Now it is sufficient to observe that $S^7(S^4,SU(2))$ itself admits no flat connection, and this is elementary (since S^4 is simply-connected, a flat connection would trivialize the bundle). //

On the other hand, there may be nonequivalent semidirect extensions associated with the one coupling.

Example 3.34. Let B be a manifold with $H^2_{deRh}(B) \neq 0$, and let $\mathfrak{g}$ be a nonabelian Lie algebra with centre $\mathfrak{z} \neq 0$. Define $K = B \times \mathfrak{g}$. Let ∇^o be the standard flat Lie connection in K and let $\ell: TB \to K$ be a 1-form such that $\Lambda = (\nabla^o(\ell) + [\ell,\ell]) = (\delta\ell + [\ell,\ell])$ is closed, but not exact.

Let A^o denote the semidirect extension $TB \underset{\nabla^o}{\ltimes} K$ and let A' denote $(A^o)_{-\Lambda}$, in the notation of 3.25. Then A' is flat, for it is easily seen that $\chi(X) = X \oplus \ell(X)$ defines a flat transversal. Thus, by 3.23 A' is equivalent to the semidirect extension corresponding to $\nabla^\chi = \nabla^o + j\circ ad\circ\ell$. But, by 3.29, A' is not equivalent to A^o, for $-\Lambda$ is not exact. //

A complete interpretation of $\mathcal{H}^3$ in Lie algebroid cohomology would require a notion of similarity for couplings, according to which couplings would define the same obstruction class iff they are similar, and a notion of effaceability for elements of $\mathcal{H}^3$, which would characterize those realizable as obstructions. Compare Hochschild (1954b). These matters may have a geometric significance of their own, but we have not considered them here since they are not necessary for the results of §4.

We close this section with a concept of produced Lie algebroid. This construction should be compared with II 2.21 (see also the comments in III 7.30).

Theorem 3.35. Let $L \rightarrowtail A \twoheadrightarrow TB$ be a transitive Lie algebroid on B and let $\phi: L \to L'$ be a morphism of LAB's over B. Suppose that there exists a representation $\rho : A \to CDO[L']$ such that

(i) $\rho^+ = ad'\circ\phi: L \to Der(L')$, where ad' is the LAB adjoint $L' \to Der(L')$; and

(ii) ϕ is A-equivariant with respect to the actions ad and ρ of A on L and L'.

Then there is a transitive Lie algebroid $L' \rightarrowtail A' \twoheadrightarrow TB$ and a morphism of Lie algebroids $\tilde{\phi}: A \to A'$ such that $(\tilde{\phi})^+: L \to L'$ is equal to ϕ, and such that $\rho = ad'\circ\tilde{\phi}$, where ad' is now the adjoint representation $A' \to CDO[L']$. Further, A' is uniquely determined up to equivalence by these conditions.

Proof: Let γ be a connection in A and define on the vector bundle $TB \oplus L'$ a bracket structure by

$$[X \oplus V', Y \oplus W'] = [X,Y] \oplus \{\rho(\gamma X)(W') - \rho(\gamma Y)(V') + [V',W'] - \phi\bar{R}_\gamma(X,Y)\}$$

where $X,Y \in \Gamma TB$, $V',W' \in \Gamma L'$. It is easily checked that this makes $TB \oplus L'$ a transitive Lie algebroid on B; denote it by A^γ. Further, $X \mapsto qX \oplus \phi\omega X$ is a Lie algebroid morphism $A \to A^\gamma$, where ω is the back-connection in A corresponding to γ. The required properties are easily verified.

If $L' \rightarrowtail A' \twoheadrightarrow TB$ is another transitive Lie algebroid and $\tilde{\phi}: A \to A'$ a morphism with the required properties, then $\mathcal{E} : A^\gamma \to A'$, $X \oplus V' \mapsto \tilde{\phi}\gamma X + V'$ may be checked to be an equivalence. (Note that $\rho = ad'\circ\tilde{\phi}$ ensures that $\rho(\gamma X)(W') = [\tilde{\phi}\gamma X, j'W']$.) //

The remark following 1.14 shows that every base-preserving morphism of transitive Lie algebroids is of the form constructed in 3.35. The remarks following II 2.23 apply equally in the case of Lie algebroids.

Note that (ii) is, in part, a requirement that $\phi: L \to L'$, regarded as a map $B \to Hom(L,L')$, be constant in the same sense in which one may say that the morphism $\phi: G\Omega \to G\Omega'$ arising from a morphism of Lie groupoids is constant because ϕ commutes with inner automorphisms.

Using this as a model, the reader may like to give a construction of a produced Lie groupoid, given

$$\begin{array}{ccccc} G\Omega & \rightarrowtail & \Omega & \twoheadrightarrow & B \times B \\ {\scriptstyle\phi}\downarrow & & & & \\ M & & & & \end{array}$$

and suitable compatibility conditions on ϕ.

§4. The existence of local flat connections and families of transition forms.

This section is concerned with a single result, and its immediate consequences.

<u>Theorem 4.1.</u> Let B be a contractible manifold and let $L \rightarrowtail A \twoheadrightarrow TB$ be a transitive Lie algebroid on B. Then A admits a flat connection. //

This result first appeared in Mackenzie (1979); the proof given now is a revision. 4.1 is an infinitesimal analogue of the well-known result that a principal bundle - or Lie groupoid - on a contractible base admits a global section. The principal bundle result is achieved by contracting the base, and using the homotopy classification of bundles; the proof of 4.1 achieves a similar end by using the cohomology theory of §3. From 4.1 it will follow that the construction of a transitive Lie algebroid from a family of transition forms (III 5.15) gives a classification of all abstract transitive Lie algebroids.

We begin with some observations which apply to any transitive Lie algebroid $L \rightarrowtail A \twoheadrightarrow TB$ on an arbitrary manifold B.

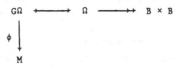

Definition 4.2. Let $L \rightarrowtail A \twoheadrightarrow TB$ be a transitive Lie algebroid. Then the α-connected Lie subgroupoid of $\Pi[L]$ which corresponds to $ad(A) \leq CDO[L]$ is denoted $Int(A)$, and called the Lie groupoid of inner automorphisms of A, or the adjoint groupoid of A. //

For any connection γ in A, the adjoint connection ∇^{γ} takes values in ad(A), and so, by III 7.6, may be said to reduce to Int(A).

Now assume that B is contractible. The idea of the proof is roughly as follows: Since B is contractible, it is presumably true that $\mathcal{H}^2(TB,ZL) = (0)$. Hence there is only one equivalence class of operator extensions of TB by L. So if we can show that there is an operator extension of TB by L which is flat, then the given Lie algebroid, being equivalent to it, must also be flat.

To carry out this idea, two matters must be arranged rather carefully. Firstly, in order to obtain $\mathcal{H}^2(TB,ZL) = (0)$, it is necessary to realize $\mathcal{H}^2(TB,ZL)$ as the de Rham cohomology of B. Since B is contractible, it is certainly true that ZL is isomorphic to the trivial bundle $B \times \mathfrak{z}$; it must also be shown that the representation of TB on ZL transports to the trivial representation. Since B is simply-connected, this can be achieved by III 5.20.

Secondly, in order to obtain a flat Lie algebroid equivalent to the given one, it must be shown that the coupling $TB \to OutDO[L]$ of the given Lie algebroid is covered by a flat Lie connection $TB \to CDO[L]$ (see 3.32 to 3.34). Although L is isomorphic as an LAB to a trivial bundle $B \times \mathfrak{g}$, one cannot apply III 5.20, for it is not known that the given Lie algebroid has an adjoint connection which is flat. Neither can one apply III 5.20 to the coupling itself, for there is no known general construction of a Lie groupoid whose Lie algebroid is OutDO[L]. (This is so even in the case of Lie algebras.) However, all these difficulties can be circumvented simultaneously.

Consider, then, a transitive Lie algebroid $L \overset{j}{\rightarrowtail} A \overset{q}{\twoheadrightarrow} TB$ on a contractible base B. We may as well assume from the outset that L is a trivial LAB $B \times \mathfrak{g}$. Since Int(A) is a Lie groupoid on the contractible base B, it is trivializable; equivalently, it admits a global decomposing section $\sigma\colon B \to Int(A)_b$, for some $b \in B$. From σ we obtain, as in III§5, a global morphism $\theta\colon B \times B \to Int(A)$ and a flat connection θ_* in ad(A).

Since $ad(A) \leq CDO[L]$, we may also consider θ_* to be a flat Lie connection in $L = B \times \mathfrak{g}$. Therefore, by III 5.20, there is an automorphism $\psi\colon B \times \mathfrak{g} \to B \times \mathfrak{g}$ which maps θ_* to the standard flat connection ∇^0. That is,

(1) $$\psi(\theta_*(X)(V)) = X(\psi(V))$$

for $X \in \Gamma TB$, $V: B \to \mathfrak{g}$.

Let $\gamma: TB \to A$ be a connection in A. Since ∇^{γ} and θ_* are both connections in ad(A), there is a map $\ell: TB \to B \times \mathfrak{g}$ such that

(2) $$\nabla^{\gamma} = \theta_* + \mathrm{ad}\circ\ell.$$

Our intention is to transform (2) into $\nabla^{\gamma} = \nabla^{0} + \mathrm{ad}\circ\ell'$ by using (1).

Define a new embedding of $L = B \times \mathfrak{g}$ into A by $j' = j\circ\psi^{-1}$.

<u>Lemma 4.3.</u> Let B be an arbitrary manifold and $L \xrightarrow{j} A \xrightarrow{q} TB$ a transitive Lie algebroid on B. Let $\psi: L \to L$ be an LAB automorphism of L, and let $j' = j\circ\psi^{-1}$.

Then for any connection $\gamma: TB \to A$, the adjoint connections ∇^{γ} and $\nabla^{',\gamma}$ induced in L by the two Lie algebroids $L \xrightarrow{j} A \xrightarrow{q} TB$ and $L \xrightarrow{j'} A \xrightarrow{q} TB$, are related by

$$\nabla_X^{',\gamma}(\psi(V)) = \psi(\nabla_X^{\gamma}(V)).$$

<u>Proof:</u> For $X \in \Gamma TB$ and $V: B \to \mathfrak{g}$,

$$\begin{aligned} j'\left(\nabla_X^{',\gamma}(\psi(V))\right) &= [\gamma(X), j'(\psi(V))] \\ &= [\gamma(X), j(V)] \\ &= j\left(\nabla_X^{\gamma}(V)\right) \\ &= j'\left(\psi(\nabla_X^{\gamma}(V))\right). \qquad // \end{aligned}$$

Returning to the proof of 4.1, we have from (2) that

$$\nabla_X^{\gamma}(V) = \theta_*(X)(V) + [\ell(X), V].$$

Therefore

$$\psi\left(\nabla_X^{\gamma}(V)\right) = (\psi\circ\theta_*)(X)(V) + \psi([\ell(X), V])$$

and therefore by the lemma, and equation (1),

$$\nabla^{\prime,\gamma}_X(\psi(V)) = X(\psi(V)) + [(\psi\circ\ell)(X),\psi(V)].$$

Since ψ is surjective, this establishes that

$$\nabla^{\prime,\gamma} = \nabla^{0} + \mathrm{ad}(\ell') \tag{3}$$

where $\ell' = \psi\circ\ell$.

We now work exclusively with $B\times\mathfrak{g} \xrightarrow{j'} A \xrightarrow{q} TB$. Equation (3) implies immediately that, firstly, the representation of TB on $B\times\mathfrak{z}$ induced by A is the trivial one, and, secondly, that the coupling Ξ of TB with $B\times\mathfrak{g}$ induced by A admits ∇^0 as a Lie connection covering it.

We therefore have $\mathcal{H}^2(TB,\nabla^0,B\times\mathfrak{z}) \cong H^2_{deRh}(B,\mathfrak{z}) = (0)$ and so, by 3.31, there is a unique equivalence class of Ξ-operator extensions of TB by $B\times\mathfrak{g}$. Since Ξ admits ∇^0 as a covering connection we can, by 3.32, construct the semidirect Ξ operator extension

$$B\times\mathfrak{g} \to TB \ltimes_{\nabla^0} (B\times\mathfrak{g}) \to TB,$$

and it must be in the same equivalence class as $B\times\mathfrak{g} \xrightarrow{j'} A \xrightarrow{q} TB$. So there is an isomorphism of Lie algebroids $TB \ltimes_{\nabla^0} (B\times\mathfrak{g}) \to A$, and A therefore admits a flat connection. This completes the proof of 4.1. //

The proof has actually established that A is isomorphic to a trivial Lie algebroid $TB \oplus (B\times\mathfrak{g})$. This stronger formulation can in any case be deduced from 4.1 by an application of III 5.20.

Consider now a transitive Lie algebroid $L \xrightarrow{j} A \xrightarrow{q} TB$ on an arbitrary base B. Given any cover $\{U_i\}$ of B by contractible open sets, there is an isomorphism of Lie algebroids over U_i

$$S_i\colon TU_i \oplus (U_i\times\mathfrak{g}) \to A_{U_i}.$$

In this sense, 4.1 has established that transitive Lie algebroids are locally trivial.

$S_i^+\colon U_i\times\mathfrak{g} \to L_{U_i}$ is an LAB chart; denote it by ψ_i. Denote by θ^i the flat connection $X \mapsto S_i(X\oplus 0)$ induced in A_{U_i} by S_i. We refer to the θ^i as local flat connections in A. Exactly as in III§5, we define transition forms χ_{ij} by

$$\theta^j = \theta^i + j\circ \ell_{ij}, \quad \chi_{ij} = \psi_i^{-1}\circ \ell_{ij}$$

for every nonvoid U_{ij}. (Note that the χ_{ij} are not, strictly speaking, defined in terms of the θ^i alone.)

Proposition 4.4. $\chi_{ij} \in A^1(U_{ij},\mathfrak{g})$, defined above, is a Maurer-Cartan form.

Proof: Since θ^i and θ^j are two flat connections in $A_{U_{ij}}$, it follows from III 5.12 that

$$\nabla^{\theta^i}(\ell_{ij}) + [\ell_{ij},\ell_{ij}] = 0.$$

(Here ∇^{θ^i} is the covariant differentiation induced by the flat Lie connection ∇^{θ^i}.) Now, by expanding out the bracket-preservation equation for S^i, it easily follows that

(4) $$\nabla_X^{\theta^i}(\psi_i(V)) = \psi_i(X(V)),$$

for $X \in \Gamma TU_i$, $V: U_i \to \mathfrak{g}$.

Since $\ell_{ij} = \psi_i \circ \chi_{ij}$, the Maurer-Cartan equation for χ_{ij} follows. //

Exactly as in III§5, we obtain the compatibility equation

$$X(a_{ij}(V)) - a_{ij}(X(V)) + [\chi_{ij}(X),a_{ij}(V)] = 0,$$

where $a_{ij}: U_{ij} \to \mathrm{Aut}(\mathfrak{g})$ are the transition functions for the atlas $\{\psi_i\}$ of L.

We have thus established a converse to III 5.15 - that every transitive Lie algebroid generates a family of transition forms χ_{ij} and an LAB cocycle $\{a_{ij}\}$ such that the three conditions of III 5.15 are satisfied, namely

(i) each χ_{ij} is a Maurer-Cartan form,

(ii) $\chi_{ik} = \chi_{ij} + a_{ij}(\chi_{jk})$ whenever $U_{ijk} \neq \emptyset$,

(iii) $\Delta(a_{ij}) = \mathrm{ad}\circ\chi_{ij}$ for all $U_{ij} \neq \emptyset$.

If $\{\theta_i': TU_i \to A_{U_i}\}$ is a second system of local flat connections with respect to the same open cover, and $\{\psi_i': U_i \times \mathfrak{g} \to L_{U_i}\}$ is an LAB atlas with

$[\theta'_i(X),\psi'_i(V)] = \psi'_i(X(V))$ for $X \in \Gamma TU_i$, $V: U_i \to \mathfrak{g}$, then, writing

$$\theta'_i = \theta_i + \psi_i \circ m_i, \qquad m_i \in A^1(U_i,\mathfrak{g})$$

and

$$\psi'_i = \psi_i \circ n_i, \qquad n_i: U_i \to \mathrm{Aut}(\mathfrak{g}),$$

we obtain

(5a) $$\delta m_i + [m_i,m_i] = 0,$$

(5b) $$\Delta(n_i) = \mathrm{ad}\circ m_i,$$

(5c) $$\chi'_{ij} = n_i^{-1}\{-m_i + \chi_{ij} + a_{ij}m_j\},$$

(5d) $$a'_{ij} = n_i^{-1}a_{ij}n_j,$$

(compare equations (10) of III§5).

Definition 4.5. (i) Let $L \rightarrowtail A \twoheadrightarrow TB$ be a transitive Lie algebroid on B. A system of local flat connections $\{\theta_i: TU_i \to A_{U_i}\}$ and an LAB atlas $\{\psi_i: U_i \times \mathfrak{g} \to L_{U_i}\}$ are compatible if

$$[\theta_i(X),\psi_i(V)] = \psi_i(X(V))$$

holds identically; we then refer to the θ_i and ψ_i collectively as a compatible system of local data.

(ii) Let $\mathfrak{g}$ be a Lie algebra and let B be a manifold. A system of transition forms $\{\chi_{ij} \in A^1(U_{ij},\mathfrak{g})\}$ and an LAB cocycle $\{a_{ij}: U_{ij} \to \mathrm{Aut}(\mathfrak{g})\}$ for some open cover $\{U_i\}$ of B, which satisfy equations (i)-(iii) above are called a system of transition data on B with values in $\mathfrak{g}$.

(iii) Let $\mathfrak{g}$ be a Lie algebra and let B be a manifold. Let $\{\chi_{ij},a_{ij}\}$ and $\{\chi'_{ij},a'_{ij}\}$ be two systems of transition data on B with values in $\mathfrak{g}$ and with respect to the one open cover $\{U_i\}$ of B. Then $\{\chi_{ij},a_{ij}\}$ and $\{\chi'_{ij},a'_{ij}\}$ are equivalent if there exists a system of Maurer-Cartan forms $m_i \in A^1(U_i,\mathfrak{g})$ and a system of functions $n_i: U_i \to \mathrm{Aut}(\mathfrak{g})$ such that (5b), (5c), (5d) are satisfied. //

Evidently systems of transition data which arise from the one Lie algebroid, or from equivalent Lie algebroids, are equivalent.

Proposition 4.6. Let $\{\chi_{ij}, a_{ij}\}$ and $\{\chi'_{ij}, a'_{ij}\}$ be two equivalent systems of transition data on a manifold B with values in a Lie algebra $\mathfrak{g}$ and with respect to the same open cover. Then the Lie algebroids constructed from $\{\chi_{ij}, a_{ij}\}$ and $\{\chi'_{ij}, a'_{ij}\}$ are equivalent.

Proof: Let $m_i \in A^1(U_i, \mathfrak{g})$, $n_i: U_i \to \mathrm{Aut}(\mathfrak{g})$ be the data establishing the equivalence. Let A and A' denote the Lie algebroids constructed via III 5.15 from $\{\chi_{ij}, a_{ij}\}$ and $\{\chi'_{ij}, a'_{ij}\}$.

Define $\phi: A' \to A$ locally by

$$(i,\ X \oplus V)' \mapsto (i,\ X \oplus (m_i(X) + n_i(V)));$$

it is easily verified that this is well-defined, and gives the desired equivalence. //

Clearly one may modify 4.5(iii) and 4.6 to take account of systems of transition data defined with respect to different open covers, and one may take an inductive limit; we leave the working-out of this to the reader.

Examples of transition forms may be obtained easily, by taking the right-derivative of transition functions of known examples of Lie groupoids. For example, with Ω the Lie groupoid of $SU(2)(\mathbf{S}^2, U(1))$ and charts defined by stereographic projection in the usual way, the transition form for $\mathbf{S}^2 \times R \rightarrowtail \frac{TSU(2)}{U(1)} \twoheadrightarrow TS^2$ is essentially the Maurer-Cartan form for U(1). Transition forms for CDO(E), CDO[L], CDO<E>, etc., can be constructed directly from transition functions for the bundles E, L, etc. This method was actually used by Teleman (1972, §6) to construct CDO(E).

Using local flat connections one can prove a stronger version of 1.6 which is of independent interest.

Theorem 4.7. Let $\phi: A \to A'$ be a morphism of transitive Lie algebroids over B. Then there is an open cover $\{U_i\}$ of B and isomorphisms

$$S_i: TU_i \oplus (U_i \times \mathfrak{g}) \to A_{U_i}, \quad S'_i: TU_i \oplus (U_i \times \mathfrak{g}') \to A'_{U_i}$$

such that $(S'_i)^{-1} \circ \phi \circ S_i$ is characterized (as in III 2.4) by the zero Maurer-Cartan form and a map $U_i \to \mathrm{Hom}(\mathfrak{g}, \mathfrak{g}')$ which is constant.

Proof: Let $\{U_i\}$ by any cover by contractible open sets, and let $\{S_i\}$ be given by 4.1. Denote by θ^i and ψ_i the flat connections and LAB charts associated with S_i. Define $\theta'^i = \phi\circ\theta^i$; then θ'^i is a flat connection in A'_{U_i}. Now the adjoint connection $\nabla^{\theta'^i}$ is a flat Lie connection in a trivializable LAB L'_{U_i} on a simply-connected base U_i; by an obvious modification of III 5.20 there is an LAB chart $\psi'_i: U_i \times \mathfrak{g}' \to L'_{U_i}$ which maps the standard flat connection ∇^o to $\nabla^{\theta'^i}$. Denote by S'_i the isomorphism $TU_i \oplus (U_i \times \mathfrak{g}') \to A'_{U_i}$ defined by ψ'_i and θ'^i.

We now have a morphism of trivial Lie algebroids

$$(S'_i)^{-1}\circ\phi\circ S_i: TU_i \oplus (U_i \times \mathfrak{g}) \to TU_i \oplus (U_i \times \mathfrak{g}');$$

denote by $f_i: U_i \times \mathfrak{g} \to U_i \times \mathfrak{g}'$ its restriction $(\psi'_i)^{-1}\circ\phi^+\circ\psi_i$. To determine the Maurer-Cartan form for $(S'_i)^{-1}\circ\phi\circ S_i$, calculate

$$\begin{aligned}((S'_i)^{-1}\circ\phi\circ S_i)(X \oplus V) &= (S'_i)^{-1}(\phi(\theta^i(X) + j\psi_i(V))\\ &= (S'_i)^{-1}(\theta'^i(X) + j'\circ\phi^+\circ\psi_i(V))\\ &= X \oplus f_i(V).\end{aligned}$$

Thus the Maurer-Cartan form is zero.

Now the compatibility equation for $\omega_i = 0$ and f_i is

$$X(f_i(W)) - f_i(X(W)) = 0$$

and, by B§2 equation (6), this implies that the Lie derivatives $X(f_i)$ of f_i as a $\text{Hom}(\mathfrak{g},\mathfrak{g}')$-valued map on U_i, are zero. Since U_i is connected, it follows that f_i is constant. //

This is the proof given in Mackenzie (1979). A similar proof can be given for 1.15.

The following is an abstract reduction theorem.

Theorem 4.8. Let $L \rightarrowtail A \twoheadrightarrow TB$ be a transitive Lie algebroid. Then if A has a connection γ whose curvature $\bar{R}_\gamma$ takes values in a sub LAB K which is stable under ∇^γ, then A has a system of transition forms taking values in the fibre type of K.

Proof: Immediate from III 7.25 and 4.1. //

§5. The spectral sequence of a transitive Lie algebroid.

For a transitive Lie algebroid $L \rightarrowtail A \twoheadrightarrow TB$ and an arbitrary representation of A on a vector bundle E we construct a natural spectral sequence $\mathcal{H}^s(TB,H^t(L,E)) \Rightarrow \mathcal{H}^n(A,E)$ which gives the cohomology of A in terms of those of TB and L. The construction follows closely the construction of the spectral sequence of an extension of Lie algebras, due to Hochschild and Serre (1953), and is at the same time a generalization of the Leray-Serre spectral sequence of a principal bundle in de Rham cohomology as constructed by, for example, Greub et al (1976).

The construction of this spectral sequence was originally motivated by a problem in groupoid cohomology. In Mackenzie (1978) a cohomology for locally trivial groupoids is constructed. This cohomology has the characteristic property that it classifies only those extensions $E \rightarrowtail \Omega' \twoheadrightarrow \Omega$ which admit a global transversal $\Omega \to \Omega'$, and on this account it is called rigid; it is nonetheless what is generally known as a continuous cohomology theory. Such extensions are determined by their restriction to the vertex groups and it is shown that the rigid cohomology is in fact naturally isomorphic in all degrees to the continuous cohomology of any vertex group.

For the case of coefficients in vector bundles, this construction actually suffices (op.cit., Theorem 4). For coefficients in general group bundles, however, more general extensions exist and one desires a cohomology which will classify all extensions which are, in a suitable sense, locally trivial. (We will detail elsewhere why this is an interesting question.)

In approaching a general cohomology theory for locally trivial groupoids, one expects that a vertex group extension will correspond to a plurality of groupoid extensions; one must also consider the possibility that not all vertex group extensions will lift to a groupoid extension. One also expects that certain groupoid extensions will arise from the cohomological structure of the base manifold. This section addresses the corresponding questions in the theory of transitive Lie algebroids, as a first-order approximation and guide to the situation for locally trivial groupoids. Thus we study the images and kernels of the restriction and inflation maps: given an extension of transitive Lie algebroids $E \rightarrowtail A' \twoheadrightarrow A$, its restriction is the induced extension of the adjoint bundles, $E \rightarrowtail L' \twoheadrightarrow L$; the inflation map constructs certain extensions $E \rightarrowtail A' \twoheadrightarrow A$ from transitive Lie algebroids on B with adjoint bundle E^L. The images and kernels of these maps have natural expressions in terms of the spectral sequence.

The question of the image of the restriction map is an infinitesimal version of a previously studied question (for example, Greub and Petry (1978), Haefliger (1956)): when can a principal bundle P(B,G) be lifted to a group H given as the domain of a surjective morphism $H \twoheadrightarrow G$? In groupoid terms this is the problem of lifting a vertex group extension $K \rightarrowtail H \twoheadrightarrow G$, to a groupoid extension. In 5.15 we obtain one simple criterion for the lifting to be always possible on the Lie algebroid level. This technique will be developed elsewhere.

A second major reason for the study of this spectral sequence is that it provides an abstraction and algebraization of the Leray-Serre spectral sequence of a principal bundle in de Rham cohomology. Because the construction is algebraic, and because coefficients in general vector bundles are permitted, it is possible to apply techniques developed for the Lyndon-Hochschild-Serre spectral sequence for an extension of discrete groups, or of Lie algebras, to the de Rham spectral sequence of a principal bundle. In this section we give only a single and elementary instance of this, 5.10. In a future paper we will take this process further.

The results of this section are from Mackenzie (1979); in a few cases the statement of results has been sharpened.

<u>Definition 5.1</u>. Let A and A' be Lie algebroids on the same base B and let $\rho: A \to CDO(E)$ and $\rho': A' \to CDO(E')$ be representations of A and A'. Then a <u>change of Lie algebroids</u> from (A,ρ,E) to (A',ρ',E') is a pair (ϕ,ψ) where $\phi: A \to A'$ is a morphism of Lie algebroids over B and $\psi: E' \to E$ is a morphism of vector bundles over B, such that

$$\psi(\rho'(\phi(X))(\mu')) = \rho(X)(\psi(\mu'))$$

for $\mu' \in E'$, $X \in A$. //

A change of Lie algebroids induces a morphism of cochain complexes

$$(\phi,\psi)^{\#}: C^n(A',E') \to C^n(A,E) \qquad f' \mapsto \psi \circ f' \circ \phi^n$$

and hence morphisms $(\phi,\psi)^*: \mathcal{H}^n(A',E') \to \mathcal{H}^n(A,E)$. If A and A' are transitive, then (ϕ^+,ψ) is also a change of Lie algebroids and induces morphisms $(\phi^+,\psi)^*: H^n(L',E') \to H^n(L,E)$. If $E = E'$ and $\psi = id$ we write $\phi^\#$, ϕ^* for $(\phi,id)^\#$, $(\phi,id)^*$.

Associated with a given transitive Lie algebroid $L \overset{j}{\rightarrowtail} A \overset{q}{\twoheadrightarrow} TB$ and representation $\rho: A \to CDO(E)$ there are two natural changes, $(j,id_E): (L,\rho^+,E) \to (A,\rho,E)$ and $(q,\natural): (A,\rho,E) \to (TB,\bar{\rho},E^L)$. Here $\bar{\rho}$ is the representation of TB on E^L induced by ρ (see 1.17). These induce maps

$$j^*: \mathcal{H}^n(A,\rho,E) \to \Gamma H^n(L,\rho^+,E) \quad \text{and} \quad q^* = (q,\natural)^*: \mathcal{H}^n(TB,\bar{\rho},E^L) \to \mathcal{H}^n(A,\rho,E)$$

which, following MacLane (1963), we call the *restriction* and *inflation* maps of (A,ρ,E). Note that $j^* \circ q^* = 0$. Our chief concern is with the kernels and cokernels of these maps.

For $n = 2$ the restriction and inflation maps have natural definitions in terms of extensions. Given a ρ-operator extension $E \overset{\iota}{\rightarrowtail} A' \overset{\pi}{\twoheadrightarrow} A$, the extension $E \overset{\iota^+}{\rightarrowtail} L' \overset{\pi^+}{\twoheadrightarrow} L$ is a ρ^+-operator extension of L by E and for any transversal $\chi: A \to A'$ with cocycle $\bar{R}_\chi$, the map $\chi^+: L \to L'$ is a transversal of π^+ and has cocycle $j^\#(\bar{R}_\chi)$. (Note that χ^+ is defined by virtue of $q' \circ \chi = q$.) It is easy to see that this defines a map $\mathcal{O}pext(A,\rho,E) \to \Gamma\mathcal{O}pext(L,\rho^+,E)$ which represents j^* modulo 2.13. We call $E \overset{\iota^+}{\rightarrowtail} L' \overset{\pi^+}{\twoheadrightarrow} L$ the *restriction* of $E \overset{\iota}{\rightarrowtail} A' \overset{\pi}{\twoheadrightarrow} A$. There is a commutative diagram

$$\begin{array}{ccccc}
E & \overset{\iota^+}{\rightarrowtail} & L' & \overset{\pi^+}{\twoheadrightarrow} & L \\
\| & & \downarrow{\scriptstyle j'} & & \downarrow{\scriptstyle j} \\
E & \overset{\iota}{\rightarrowtail} & A' & \overset{\pi}{\twoheadrightarrow} & A \\
 & & \downarrow{\scriptstyle q'} & & \downarrow{\scriptstyle q} \\
 & & TB & = & TB
\end{array}$$

Similarly the inflation map q^* can be realized in terms of $\mathcal{O}pext(TB,\bar{\rho},E^L) \to \mathcal{O}pext(A,\rho,E)$. Given an extension $E^L \rightarrowtail J \twoheadrightarrow TB$, construct the pullback extension (2.16)

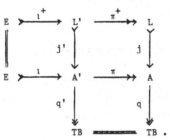

$$\begin{array}{ccccc} E^L & \rightarrowtail & J & \twoheadrightarrow & TB \\ \| & & \uparrow & & \uparrow q \\ E^L & \rightarrowtail & \bar{J} & \twoheadrightarrow & A \end{array}$$

and then the pushout (2.14)

$$\begin{array}{ccccc} E^L & \rightarrowtail & \bar{J} & \twoheadrightarrow & A \\ \downarrow & & \downarrow & & \| \\ E & \rightarrowtail & \tilde{J} & \twoheadrightarrow & A \; ; \end{array}$$

or the steps may be interchanged. We call $E \rightarrowtail \tilde{J} \twoheadrightarrow A$ the <u>inflation</u> of the transitive Lie algebroid $E^L \rightarrowtail J \twoheadrightarrow TB$. It is easy to see directly that the restriction of the inflation $E \rightarrowtail \tilde{J} \twoheadrightarrow A$ is semi-direct; in this sense the extension $E \rightarrowtail \tilde{J} \twoheadrightarrow A$ has no algebraic component to its curvature and we will therefore also call it a <u>geometric</u> extension of A by E. An extension $E \rightarrowtail A' \twoheadrightarrow A$ whose restriction is semi-direct will be called a <u>restriction semi-direct</u>, or <u>RSD</u>, extension. Not all RSD extensions are geometric; see 5.14 below. The quotient space $\frac{\ker j^*}{\operatorname{im} q^*}$ is given by the term $E_3^{1,1}$ of the spectral sequence.

For the image of j* there is first of all the following result.

<u>Proposition 5.2</u>. $j^*: \mathscr{H}^n(A,\rho,E) \to \Gamma H^n(L,\rho^+,E)$ takes values in $(\Gamma H^n(L,\rho^+,E))^A$, where A acts on $H^n(L,E)$ via the Lie derivative θ of 2.5.

<u>Proof</u>: Observe that $j^\# \circ \theta_X = \theta_X \circ j^\#: \Gamma C^n(A,E) \to \Gamma C^n(L,E)$ for $X \in \Gamma A$. Here the θ_X on the left is defined in 2.2 and that on the right is the action of A on $C^n(L,E)$ defined in 2.4. It follows that for $f \in \mathscr{Z}^n(A,E)$,

$$\begin{aligned} \theta_X(j^\#(f)) &= j^\#(\theta_X(f)) \\ &= j^\#(d \circ \iota_X(f)) + 0 \qquad \text{by 2.3(v)} \\ &= d(j^\# \circ \iota_X(f)) \end{aligned}$$

and so $\theta_X(j^*[f]) = [0]$. //

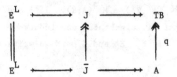

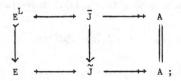

For $n = 2$ the action θ can be transported to an action of A on the vector bundle $\mathrm{Opext}(L,\rho^{+},E)$ via 2.13. Since $\theta^{+} = 0$ (by 2.6) the equation $\theta_X(E \rightarrowtail L' \twoheadrightarrow L) = 0$ may be interpreted as signifying that the Lie algebra extension $E_x \rightarrowtail L'_x \twoheadrightarrow L_x$, as a function of $x \in B$, has zero derivatives in all directions $q(X) \in \Gamma TB$. Thus 5.2 implies that a necessary condition for an extension $E \rightarrowtail L' \twoheadrightarrow L$ of the adjoint bundle L to be the restriction of an extension of A is that the Lie algebra extensions $E_x \rightarrowtail L'_x \twoheadrightarrow L_x$ be constant with respect to x, where constancy is taken to refer to θ. As in 1.13, this constancy is an abstraction of equivariance with respect to actions of adjoint type. See also III 7.30.

There are two further conditions necessary on an element of $(\Gamma H^2(L,E))^A$ before it can be guaranteed to lie in the image of j^*. These are most naturally formulated in terms of the spectral sequence, to which we now turn.

* * * * * * *

Our references for spectral sequences are Cartan and Eilenberg (1956), MacLane (1975) and Greub et al (1976). We deal with the spectral sequence of a canonically bounded descending filtration; thus $F^kC^n \supseteq F^{k+1}C^n$, $F^oC^n = C^n$ and $F^{n+1}C^n = (0)$. We use an explicit approach; thus $E_r^{s,t} = Z_r^{s,t}/B_r^{s,t}$ where $Z_r^{s,t} = \{f \in C^{s+t} \mid df \in F^{s+r}C^{s+t+1}\}$, $B_r^{s,t} = dZ_{r-1}^{s-r+1,t+r-2} + Z_{r-1}^{s+1,t-1}$ for $r \geq 1$, and $B_0^{s,t} = F^{s+1}C^{s+t}$. The isomorphism $E_{r+1}^{s,t} \to H^{s,t}(E_r^{*,*})$ induced by the inclusion $Z_{r+1}^{s,t} \subseteq Z_r^{s,t}$ is denoted $\sigma_r^{s,t}$. Such a spectral sequence is strongly convergent (Cartan and Eilenberg (1956, XV.4)); the filtration $F^sH^n(C^*) = \mathrm{im}(H^n(F^sC^*) \to H^n(C^*))$ is also canonically bounded and the isomorphism $E_\infty^{s,t} \cong \dfrac{F^sH^{s+t}(C^*)}{F^{s+1}H^{s+t}(C^*)}$ is, on the cochain level, the identity map. Here $E_\infty^{s,t} = Z_\infty^{s,t}/B_\infty^{s,t}$ where $Z_\infty^{s,t} = \{f \in F^sC^{s+t} \mid df = 0\}$ and $B_\infty^{s,t} = (F^sC^{s+t} \cap dC^{s+t-1}) + Z_\infty^{s+1,t-1}$.

Note that $E_r^{s,t} = E_{r+1}^{s,t} = E_\infty^{s,t}$ for $r > \max\{s,t+1\}$. The edge morphisms are denoted by $e_B\colon E_2^{s,0} \twoheadrightarrow E_\infty^{s,0} \subseteq H^s(C^*)$ and $e_F\colon H^t(C^*) \twoheadrightarrow E_\infty^{0,t} \rightarrowtail E_2^{0,t}$. The transgression relation $E_2^{0,n} \rightsquigarrow E_2^{n+1,0}$ is denoted tg^n. For $n = 1$ it is a well- and fully-defined map, namely $d_2^{0,1}$. If $E_{n+2}^{0,n} = (0) = E_{n+2}^{n+1,0}$ then $d_{n+1}^{0,n}$ is an isomorphism and the composite $E_2^{n+1,0} \twoheadrightarrow E_{n+1}^{n+1,0} \rightarrow E_{n+1}^{0,n} \rightarrowtail E_2^{0,n}$ may be considered a Cartan map κ^n with tg^n as inverse relation.

Let $L \overset{j}{\rightarrowtail} A \overset{q}{\twoheadrightarrow} TB$ be a transitive Lie algebroid and let $\rho: A \to CDO(E)$ be a representation of A.

We will filter the cochain complex $\Gamma C^*(A,E)$; to cut down on notation, denote $\Gamma C^n(A,E)$ by $\Gamma^n(A,E)$. Define $F^s\Gamma^n(A,E) = \{f \in \Gamma^n(A,E) \mid f(X_1,\ldots,X_n) = 0$ whenever $(n - s + 1)$ or more of the X_i are in ker q$\}$. Then $F^s\Gamma^n(A,E) \supseteq F^{s+1}\Gamma^n(A,E)$, $F^0\Gamma^n(A,E) = \Gamma^n(A,E)$, $F^{n+1}\Gamma^n(A,E) = (0)$ and, by convention, $F^s\Gamma^n(A,E) = (0)$ for $s > n + 1$. This is the standard filtration on an exact sequence of Lie algebras (see Hochschild and Serre (1953)) and is also of the same type as the filtration associated with a $\mathfrak{g}$-DGA (Greub et al (1976, 9.1)).

Define $a^{s,t}: F^s\Gamma^{s+t}(A,E) \to \Gamma^s(TB,C^t(L,E))$ by

$$a^{s,t}(f)(X_1,\ldots,X_s)(V_1,\ldots,V_t) = f(jV_1,\ldots,jV_t,\gamma X_1,\ldots,\gamma X_s)$$

for any connection $\gamma: TB \to A$. It is easy to see that $a^{s,t}$ is independent of the choice of γ and has kernel $F^{s+1}\Gamma^{s+t}(A,E)$. It is also surjective; to see this, define

$$e^{s,t}: \Gamma^s(TB,C^t(L,E)) \to F^s\Gamma^{s+t}(A,E)$$

by

$$e^{s,t}(f)(X_1,\ldots,X_{s+t}) = \frac{(-1)^{st}}{s!t!} \sum_{\sigma} \varepsilon_\sigma f(qX_{\sigma(t+1)},\ldots,qX_{\sigma(t+s)})(\omega X_{\sigma(1)},\ldots,\omega X_{\sigma(t)})$$

where $\omega: A \to L$ is any back-connection and the summation is over all permutations of $\{1,\ldots,s + t\}$. It is straightforward to verify that $a^{s,t}(e^{s,t}(f)) = f$.

Therefore, $a^{s,t}$ quotients to an isomorphism $\alpha_0^{s,t}: E_0^{s,t} \to \Gamma^s(TB,C^t(L,E))$.

<u>Proposition 5.3.</u>

$$\begin{array}{ccc} E_0^{s,t} & \xrightarrow{\;d_0^{s,t}\;} & E_0^{s,t+1} \\ {\scriptstyle \alpha_0^{s,t}}\downarrow & & \downarrow{\scriptstyle \alpha_0^{s,t+1}} \\ \Gamma^s(TB,C^t(L,E)) & \xrightarrow{\;\Gamma^s(TB,d)\;} & \Gamma^s(TB,C^{t+1}(L,E)) \end{array}$$

commutes.

Proof: Take $f \in Z_0^{s,t} = F^s\Gamma^{s+t}(A,E)$. We have to show that $a^{s,t+1}(df)(X_1,\ldots,X_s)$ is equal to $d(a^{s,t}(f)(X_1,\ldots,X_s))$, for all $X_1,\ldots,X_s \in \Gamma TB$.

Take $V_1,\ldots,V_{t+1} \in \Gamma L$. Expanding out $a^{s,t+1}(df)(X_1,\ldots,X_s)(V_1,\ldots,V_{t+1})$, we get

$$df(V_1,\ldots,V_{t+1},\gamma X_1,\ldots,\gamma X_s)$$

$$= \sum_{i=1}^{t+1} (-1)^{i+1}\rho^+(V_i)f(V_1,\ldots\hat{\ldots},V_{t+1},\gamma X_1,\ldots,\gamma X_s)$$

$$+ \sum_{1\leq i<j\leq t+1} (-1)^{i+j}f([V_i,V_j],V_1,\ldots\hat{\ }\hat{\ }\ldots,V_{t+1},\gamma X_1,\ldots,\gamma X_s)$$

$$+ \sum_{i=1}^{s} (-1)^{i+t}\rho(\gamma X_i)f(V_1,\ldots,V_{t+1},\gamma X_1,\ldots\hat{\ldots},\gamma X_s)$$

$$+ \sum_{1\leq i<j\leq s} (-1)^{i+j}f([\gamma X_i,\gamma X_j],V_1,\ldots,V_{t+1},\gamma X_1,\ldots\hat{\ }\hat{\ }\ldots,\gamma X_s)$$

$$+ \sum_{i=1}^{t+1}\sum_{j=1}^{s} (-1)^{i+j+t+1}f([V_i,\gamma X_j],V_1,\ldots\hat{\ldots},V_{t+1},\gamma X_1,\ldots\hat{\ldots},\gamma X_s).$$

Here each term in each of the last three summations vanishes, for each term has $(t+1)$ arguments in ΓL and $f \in F^s\Gamma^{s+t}(A,E)$. The first two terms can be rewritten as

$$\sum_{i=1}^{t+1} (-1)^{i+1}\rho^+(V_i)\big(a^{s,t}(f)(X_1,\ldots,X_s)(V_1,\ldots\hat{\ldots},V_{t+1})\big)$$

$$+ \sum_{1\leq i<j\leq t+1} (-1)^{i+j}a^{s,t}(f)(X_1,\ldots,X_s)([V_i,V_j],V_1,\ldots\hat{\ }\hat{\ }\ldots,V_{t+1})$$

and are therefore equal to $d(a^{s,t}(f)(X_1,\ldots,X_s))(V_1,\ldots,V_{t+1})$, as required. //

In the case of Lie algebras, Hochschild and Serre (1953) have an elegant device by which to simplify this proof, but it cannot be properly formulated in the case of transitive Lie algebroids.

Identifying $E_1^{s,t}$ with $H^{s,t}(E_0^{*,*})$, we now have isomorphisms

$$\alpha_1^{s,t}: E_1^{s,t} \to \Gamma^s(TB,H^t(L,E)).$$

Following through the identifcations, if $f \in Z_1^{s,t}$ represents a chosen element of $E_1^{s,t}$, then f is in $F^s\Gamma^{s+t}(A,E)$ and $a^{s,t}(f) \in \Gamma^s(TB,C^t(L,E))$ takes values in $\Gamma Z^t(L,E)$; the cohomology class $[a^{s,t}(f)(X_1,\ldots,X_s)] \in \Gamma H^t(L,E)$ is $\alpha_1^{s,t}([f])(X_1,\ldots,X_s)$.

Proposition 5.4.

$$\begin{array}{ccc} E_1^{s,t} & \xrightarrow{d_1^{s,t}} & E_1^{s+1,t} \\ {\scriptstyle \alpha_1^{s,t}}\downarrow & & \downarrow{\scriptstyle \alpha_1^{s+1,t}} \\ \Gamma^s(TB,H^t(L,E)) & \xrightarrow{(-1)^t d} & \Gamma^{s+1}(TB,H^t(L,E)) \end{array}$$

commutes.

Proof: Similar to the proof of 5.3. //

Thus $\alpha_1^{*,t}$ induces isomorphisms $E_2^{s,t} \to \mathcal{H}^s(TB,H^t(L,E))$. Given $f \in Z_2^{s,t} \subseteq F^s\Gamma^{s+t}(A,E)$ representing $[f] \in E_2^{s,t}$, the class $\alpha_2^{s,t}([f])$ is represented by the cocycle which to $(X_1,\ldots,X_s)$ assigns the cohomology class in $\Gamma H^t(L,E)$ represented by $a^{s,t}(f)(X_1,\ldots,X_s)$.

Theorem 5.5. For a transitive Lie algebroid $L \rightarrowtail A \twoheadrightarrow TB$ and representation $\rho: A \to CDO(E)$, there is a natural convergent spectral sequence

$$\mathcal{H}^s(TB,H^t(L,E)) \Rightarrow \mathcal{H}^n(A,E).$$

Proof: Only the naturality remains to be described. If $(\phi,\psi): (A,\rho,E) \to (A',\rho',E')$ is a change of transitive Lie algebroids, then $(\phi,\psi)^{\#}: \Gamma^*(A',E') \to \Gamma^*(A,E)$ preserves the filtrations and so induces a morphism of spectral sequences $E_*^{*,*}(A',E') \to E_*^{*,*}(A,E)$. Also, there is a change of Lie algebroids $(id_{TB},(\phi^+,\psi)^*)$ from $TB \to CDO(H^*(L,E))$ to $TB \to CDO(H^*(L',E'))$. It is now straightforward to show that the induced morphisms $\mathcal{H}^s(TB,H^t(L',E')) \to \mathcal{H}^s(TB,H^t(L,E))$ commute with the morphisms of the spectral sequences, and similarly on the E_∞ level. //

<u>Remark.</u> If $K \rightarrowtail A' \twoheadrightarrow A$ is an exact sequence of arbitrary Lie algebroids, then there is a similar spectral sequence $\mathcal{H}^s(A,H^t(K,E)) \Rightarrow \mathcal{H}^n(A',E)$ for any representation of A' on a vector bundle E.

<u>Proposition 5.6.</u>

$$\begin{array}{ccccc}
\mathcal{H}^n(TB,E^L) & & \xrightarrow{\quad q^* \quad} & & \mathcal{H}^n(A,E) \\
\alpha_2^{n,0}\uparrow\cong & & \overset{e_B}{\nearrow} & & \uparrow \\
E_2^{n,0} & \twoheadrightarrow & E_\infty^{n,0} & \xrightarrow{\cong} & F^n\mathcal{H}^n(A,E)
\end{array}$$

and

$$\begin{array}{ccccc}
\mathcal{H}^n(A,E) & & \xrightarrow{\quad j^* \quad} & & (\Gamma H^n(L,E))^A \\
\uparrow & & \overset{e_F}{\searrow} & & \cong\uparrow\alpha_2^{0,n} \\
\dfrac{\mathcal{H}^n(A,E)}{F^1\mathcal{H}^n(A,E)} & \xrightarrow{\cong} & E_\infty^{0,n} & \rightarrowtail & E_2^{0,n}
\end{array}$$

commute.

<u>Proof</u>: Take $[f] \in E_2^{n,0}$ with $f \in Z_2^{n,0}$. Then $f \in F^n\Gamma^n(A,E)$ and $df \in F^{n+2}\Gamma^{n+1}(A,E) = (0)$. So $f \in \mathcal{Z}^n(A,E)$ and $[f] \in \mathcal{H}^n(A,E)$ is $e_B([f])$. On the other hand, $\alpha_2^{n,0}([f])$ is represented by $a^{n,0}(f)$, which lies in $\mathcal{Z}^n(TB,E^L)$. Now $q^{\#}(a^{n,0}(f))(X_1,\ldots,X_n) = f(\gamma q X_1,\ldots,\gamma q X_n)$ for $X_i \in \Gamma A$, and since $\gamma q X = X - j\omega X$ and $f \in F^n\Gamma^n(A,E)$ vanishes whenever $n - n + 1$ or more arguments are in $j(L)$, it follows that $q^{\#}(a^{n,0}(f)) = f$.

The second half is proved similarly. Note that $\mathcal{H}^0(TB,H^n(L,E)) = (\Gamma H^n(L,E))^{TB}$ and this is equal to $(\Gamma H^n(L,E))^A$ by 2.6. //

We can now express the images and kernels of q^* and j^* in terms of the spectral sequence. The image of q^* is $E_\infty^{n,0} \cong F^n\mathcal{H}^n(A,E)$ and this, for $n = 2$, characterizes the geometric extensions of A by E. Similarly the kernel of j^* is $F^1\mathcal{H}^n(A,E)$, so the RSD extensions of A by E are precisely those in $F^1\mathcal{H}^2(A,E)$. Of course $F^1\mathcal{H}^2(A,E) \supseteq F^2\mathcal{H}^2(A,E)$ and the quotient, which represents the failure of RSD extensions to be geometric, is isomorphic to $E_\infty^{1,1}$.

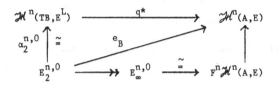

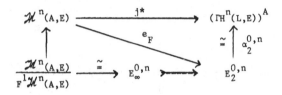

Lastly, the image of j^* is isomorphic to $E_\infty^{0,n}$ and so $E_\infty^{0,2}$ represents those extensions of L which can be lifted to extensions of A.

For $n = 2$ these E_∞ spaces reduce: $E_\infty^{2,0} \cong E_3^{2,0}$, $E_\infty^{1,1} \cong E_3^{1,1}$, and $E_\infty^{0,2} \cong E_4^{0,2}$. In turn, $E_3^{2,0}$ is the cokernel of $d_2^{0,1}: E_2^{0,1} \to E_2^{2,0}$, while $E_3^{1,1}$ is the kernel of $d_2^{1,1}: E_2^{1,1} \to E_2^{3,0}$ and $E_4^{0,2}$ is the kernel of $d_3^{0,2}: E_3^{0,2} \to E_3^{3,0}$. Thus these spaces are all accessible in terms of d_2 and d_3.

The situation is summarized in the following diagram.

$$\begin{array}{ccccc}
\text{Geometric extensions} & & \text{RSD extensions} & & \\
E_3^{2,0} \cong E_\infty^{2,0} \cong F^2\mathcal{H}^2(A,E) & \rightarrowtail & F^1\mathcal{H}^2(A,E) & \twoheadrightarrow & \dfrac{F^1\mathcal{H}^2(A,E)}{F^2\mathcal{H}^2(A,E)} \cong E_\infty^{1,1} \cong E_3^{1,1} \\
\uparrow\!\!\uparrow & & \downarrow & & \\
E_2^{2,0} \cong \mathcal{H}^2(TB,E^L) & \xrightarrow{q^\#} & \mathcal{H}^2(A,E) & \xrightarrow{j^*} & (\Gamma H^2(L,E))^A \cong E_2^{0,2} \\
 & & \downarrow & & \uparrow \\
 & & \dfrac{\mathcal{H}^2(A,E)}{F^1\mathcal{H}^2(A,E)} & \xrightarrow{\cong} & E_\infty^{0,2} \cong E_4^{0,2} \\
 & & & & \text{Liftable extensions of L}
\end{array}$$

We now calculate $d_2^{r,1}$, $r \geq 0$, when the action of L on E is trivial. For this we need the concept of pairing of spectral sequences. We summarize briefly the details.

Let A be an arbitrary Lie algebroid and let ρ^M, ρ^N, ρ^P be representations of A on vector bundles M, N, P. Then M and N are paired to P if there is a bilinear vector bundle map $M \oplus N \to P$, denoted $\wedge$, such that

$$\rho^P(X)(\mu \wedge \nu) = \rho^M(X)(\mu) \wedge \nu + \mu \wedge \rho^N(X)(\nu) \tag{1}$$

for all $X \in \Gamma A$, $\mu \in \Gamma M$, $\nu \in \Gamma N$.

If M and N are paired to P and K is an ideal of A, then there is an induced pairing

$$C^m(K,M) \oplus C^n(K,N) \to C^{m+n}(K,P)$$

defined by

$$(2)\quad (f_\wedge g)(X_1,\ldots,X_{m+n}) = \frac{1}{m!n!} \sum \varepsilon_\sigma f(X_{\sigma(1)},\ldots,X_{\sigma(m)})_\wedge g(X_{\sigma(m+1)},\ldots,X_{\sigma(m+n)})$$

where the sum is over all permutations on $\{1,\ldots,m+n\}$. Here the representation of A on $C^m(K,M)$ is by the Lie derivative 2.2 - if A is transitive then K, being an ideal, must be a sub bundle of the adjoint bundle L of A and so, by 2.3(iii), $\theta\colon A \to CDO(C^m(K,M))$ is indeed a representation of A.

As for differential forms, we have

$$(3)\qquad d(f_\wedge g) = df_\wedge g + (-1)^m f_\wedge dg,$$

where $f \in \Gamma C^m(K,M)$, $g \in \Gamma C^n(K,N)$. There is therefore an induced pairing

$$H^m(K,M) \oplus H^n(K,N) \to H^{m+n}(K,P),$$

still denoted by $\wedge$.

Now consider a transitive Lie algebroid $L \rightarrowtail A \twoheadrightarrow TB$. Applying the above, there are pairings $C^m(L,M) \oplus C^n(L,N) \to C^{m+n}(L,P)$ and $H^m(L,M) \oplus H^n(L,N) \to H^{m+n}(L,P)$. There is also a map

$$C^m(A,M) \oplus C^n(A,N) \to C^{m+n}(A,P),$$

defined as in (2), which is bilinear and satisfies (3). Here the vector bundles do not (generally) admit representations of A and (1) has no meaning. We will call this map the <u>formal pairing</u> induced by the pairing $M \oplus N \to P$.

In particular there is a formal pairing

$$C^s(TB,H^m(L,M)) \oplus C^{s'}(TB,H^n(L,N)) \to C^{s+s'}(TB,H^{m+n}(L,E))$$

and, applying a form of (3), there is a bilinear map

$$\mathcal{H}^s(TB,H^m(L,M)) \times \mathcal{H}^{s'}(TB,H^n(L,E)) \to \mathcal{H}^{s+s'}(TB,H^{m+n}(L,E)).$$

<u>Proposition 5.7</u>. Let $L \rightarrowtail A \twoheadrightarrow TB$ be a transitive Lie algebroid, and let ρ^M, ρ^N, ρ^P be representations of A on vector bundles M, N, P. Let E(M), E(N),

E(P) denote the corresponding spectral sequences. Then

$$F^s\Gamma^m(A,M)\wedge F^{s'}\Gamma^n(A,N) \subseteq F^{s+s'}\Gamma^{m+n}(A,P)$$

and

$$\begin{array}{ccc} E_2^{s,t}(M) \times E_2^{s',t'}(N) & \longrightarrow & E_2^{s+s',t+t'}(P) \\ \downarrow & & \downarrow \\ \mathcal{H}^s(TB,H^t(L,M)) \times \mathcal{H}^{s'}(TB,H^{t'}(L,N)) & \xrightarrow{(-1)^{st'}} & \mathcal{H}^{s+s'}(TB,H^{t+t'}(L,P)) \end{array}$$

commutes, where the bottom row is the map described above multiplied by $(-1)^{st'}$.

<u>Proof:</u> Exactly follows Hochschild and Serre (1953); one proves the corresponding result at the E_0 level and then follows through the formation of the homologies. //

We also need two elementary observations. Fix a transitive Lie algebroid $L \rightarrowtail A \twoheadrightarrow TB$ and a representation $\rho: A \to CDO(E)$.

<u>Lemma 5.8.</u> Let $L_{ab} \rightarrowtail A_{ab} \twoheadrightarrow TB$ be the quotient Lie algebroid $A_{ab} = A/[L,L]$ (see 1.1 and 1.11), and let $\natural: A \to A_{ab}$ denote the natural projection. Then each connection $\gamma: TB \to A$ maps to a connection $\natural\circ\gamma$ in A_{ab} whose curvature $\bar{R}_{\natural\circ\gamma} = \natural^{+}\circ\bar{R}_{\gamma}$ belongs to $\mathcal{Z}^2(TB,L_{ab})$ and which represents the cohomology class of $L_{ab} \rightarrowtail A_{ab} \twoheadrightarrow TB$.

<u>Proof:</u> Immediate from 2.13. //

Denote the class of $L_{ab} \rightarrowtail A_{ab} \twoheadrightarrow TB$ in $\mathcal{H}^2(TB,L_{ab})$ by R_{ab}.

<u>Lemma 5.9.</u> Let L be any totally intransitive Lie algebroid and let E be a vector bundle on the same base. Then, with respect to the zero representation of L on E,

$$H^1(L,E) \cong C^1(L_{ab},E).$$

<u>Proof:</u> The coboundary $d: C^0(L,E) \to C^1(L,E)$ is zero and the next coboundary, $d: C^1(L,E) \to C^2(L,E)$ is $df(X,Y) = -f([X,Y])$. From the first formula it follows that $H^1(L,E) = Z^1(L,E)$ and from the second it follows that $\Gamma Z^1(L,E)$ consists of

those maps $\Gamma L \to \Gamma E$ which vanish on $\Gamma[L,L] = [\Gamma L,\Gamma L]$. Hence the result. //

Theorem 5.10. Let $L \rightarrowtail A \twoheadrightarrow TB$ be a transitive Lie algebroid and ρ a representation of A on a vector bundle E for which $E^L = E$. Then the map

$$\mathcal{H}^n(TB,H^1(L,E)) \to \mathcal{H}^{n+2}(TB,E)$$

corresponding to $d_2^{n,1}$ is given by

$$F \mapsto (-1)^n F_\wedge R_{ab}$$

where $R_{ab} \in \mathcal{H}^2(TB,L_{ab})$ is the class characterizing $L_{ab} \rightarrowtail A_{ab} \twoheadrightarrow TB$ and the pairing is that induced by $H^1(L,E) \oplus L/[L,L] \to E$ via 5.9.

Remark: The condition $E^L = E$ forces E to be flat.

Proof: Let $f \in \mathcal{Z}^n(TB,Z^1(L,E))$ represent F. Then $(\alpha_2^{n,1})^{-1}(F)$ is represented by $e^{n,1}(f) \in \Gamma^{n+1}(A,E)$, where

$$e^{n,1}(f)(X_1,\ldots,X_{n+1}) = (-1)^n \sum_{i=1}^{n+1} (-1)^i f(qX_1,\ldots\hat{},\ldots,qX_{n+1})(\omega X_i).$$

Therefore $(\alpha_2^{n+2,0} \circ d_2^{n,1} \circ (\alpha_2^{n,1})^{-1})(F)$ is represented by $a^{n+2,0}(d(e^{n,1}(f)))$, and this reduces to

$$\sum_{i<j} (-1)^{i+j}((-1)^n f(X_1,\ldots\hat{}\ldots\hat{}\ldots,X_{n+2})(\omega[\gamma X_i,\gamma X_j]))$$

since all other terms in $d(e^{n,1}(f))(\gamma X_1,\ldots,\gamma X_{n+2})$ involve $\omega\circ\gamma$, and $\omega\circ\gamma = 0$. Now $\omega[\gamma X_i,\gamma X_j] = -\bar{R}_\gamma(X_i,X_j)$ so, recalling the definition of the pairing $H^1(L,E) \oplus L_{ab} \to E$, the sum

$$(-1)^n \sum_{i<j} (-1)^{i+j+1} f(X_1,\ldots\hat{}\ldots\hat{}\ldots,X_{n+2})(\bar{R}_\gamma(X_i,X_j))$$

is seen to be the (value at $(X_1,\ldots,X_{n+2})$ of the) cocycle representing $(-1)^n F_\wedge R_{ab}$. //

This result is a direct analogue of Theorem 8 of Hochschild and Serre (1953) for extensions of Lie algebras. It decomposes $d_2^{n,1}$ into the pairing - which concerns only the adjoint bundle L and the coefficient bundle E and may be regarded as a purely algebraic matter - and the class R_{ab}, which is a topological invariant. Thus we have the following corollaries.

Corollary 5.11. (i) If $L = [L,L]$, for example if L is semisimple, then $d_2^{n,1} = 0$ for all $n > 0$.

(ii) If the abelianized Lie algebroid $L_{ab} \rightarrowtail A_{ab} \twoheadrightarrow TB$ is flat, then $d_2^{n,1} = 0$ for all $n > 0$. //

The sequence of terms of low degree is (without any assumption on the coefficients)

$$\mathcal{H}^1(TB,E^L) \xrightarrow{q^*} \mathcal{H}^1(A,E) \xrightarrow{j^*} (\Gamma H^1(L,E))^A \xrightarrow{tg^1} \mathcal{H}^2(TB,E^L) \xrightarrow{q^*} \mathcal{H}^2(A,E).$$

The transgression here is simply $d_2^{0,1}$ and so if either of the conditions of 5.11 is satisfied, it follows that $\mathcal{H}^2(TB,E)$ is injected into $\mathcal{H}^2(A,E)$ and the space of geometric extensions of A by E may be identified with $\mathcal{H}^2(TB,E)$.

In the general case, the map $(\Gamma H^1(L,E))^A \to \mathcal{O}pext(TB,E)$ can be interpreted as assigning to suitable $f \in \Gamma Z^1(L,E)$ the pushout of $L \rightarrowtail A \twoheadrightarrow TB$ over $f: L \to E^L$. These pushouts are those extensions $E^L \rightarrowtail J \twoheadrightarrow TB$ which, when inflated, give the semidirect extension of A by E.

Corollary 5.12. If $E^L = E$ and either condition of 5.11 is satisfied, then the space of geometric extensions of A by E is isomorphic to $\mathcal{H}^2(TB,E)$. //

Corollary 5.13. If $E^L = E$ and either condition of 5.11 is satisfied, then the space of RSD extensions, modulo the space of geometric extensions, is isomorphic to $\mathcal{H}^1(TB,H^1(L,E)) \cong \mathcal{H}^1(TB,C^1(L_{ab},E))$. //

In general $d_2^{1,1}$ maps $\mathcal{H}^1(TB,H^1(L,E))$ into $\mathcal{H}^3(TB,E^L)$; if the element of $\mathcal{H}^3(TB,E^L)$ is zero, then there is an extension of TB by E whose inflation is the given RSD extension. This phenomenon may thus be considered a species of obstruction theory.

The following example illustrates the circumstances in which $E_3^{1,1} \neq 0$.

Example 5.14. Let $\mathfrak{g}$ be a reductive Lie algebra with a one-dimensional centre, for example $\mathfrak{gl}(n,\mathbb{R})$. Let B be a manifold with $H^3_{deRh}(B) = 0$ and $H^1_{deRh}(B) \neq 0$. Let A be a transitive Lie algebroid on B with adjoint bundle $L = B \times \mathfrak{g}$ and let $\rho: A \to CDO(B \times \mathbb{R})$ be the trivial representation q. Then $E_2^{1,1} \cong$

$\mathcal{H}^1(TB, H^1(L, B \times \mathbb{R}) = \mathcal{H}^1(TB, B \times \mathbb{R})$ (since $\mathfrak{g} \cong [\mathfrak{g},\mathfrak{g}] \oplus \mathfrak{z}$) $\cong H^1_{deRh}(B) \neq 0$ and $E_2^{3,0} \cong H^3_{deRh}(B) = 0$. So $E_3^{1,1} \cong \ker d_2^{1,1} = E_2^{1,1} \neq 0$.

It is interesting that the Lie algebroids of the examples of Milnor (1958) with $g > 0$ are instances. //

Concerning general criteria for an extension of the adjoint bundle L to lift to an extension of A, we can at present only say that in addition to the condition that $E \rightarrowtail L' \twoheadrightarrow L$ lie in $(\Gamma H^2(L,E))^A$, there are the two consecutive conditions that $d_2^{0,2}$ and $d_3^{0,2}$ map the extension to the zero extension.

<u>Proposition 5.15</u>. Let $L \rightarrowtail A \twoheadrightarrow TB$ be an arbitrary transitive Lie algebroid and let ρ be any representation of A on E. Then if $\mathcal{H}^2(TB, H^1(L,E)) = 0$ and $\mathcal{H}^3(TB, E^L) = 0$, every extension of L by E which lies in $(\Gamma H^2(L,E))^A$ lifts to an extension of A by E.

<u>Proof</u>: The space of such extensions has been identified with $E_4^{0,2}$ and this is isomorphic to the kernel of $d_3^{0,2}\colon E_3^{0,2} \to E_3^{3,0}$. In turn, $E_3^{3,0}$ is a quotient of $E_2^{3,0}$ and so is zero, and $E_3^{0,2}$ is the kernel of $d_2^{0,2}\colon E_2^{0,2} \to E_2^{2,1}$. Now $E_2^{2,1} = 0$ also, by hypothesis, so we finally have $E_4^{0,2} \cong E_2^{0,2} \cong (\Gamma H^2(L,E))^A$.
//

The conditions of 5.15 are fulfilled if B is simply-connected and $H^2_{deRh}(B) = H^3_{deRh}(B) = 0$, or if B is simply-connected, L is semisimple and $H^3_{deRh}(B) = 0$. In particular, we have the following result.

<u>Corollary 5.16</u>. Let $L \rightarrowtail A \twoheadrightarrow TB$ be a transitive Lie algebroid on a simply-connected base B for which $H^3_{deRh}(B) = 0$ and for which either $H^2_{deRh}(B) = 0$ or L is semisimple. Let ρ be a representation of A on a vector bundle E. Then if $V \rightarrowtail \mathfrak{g}' \twoheadrightarrow \mathfrak{g}$ is an operator extension of the fibre type of L by the fibre type of E, there is an operator extension $E \rightarrowtail A' \twoheadrightarrow A$ whose restriction $E \rightarrowtail L' \twoheadrightarrow L$ has fibre type $V \rightarrowtail \mathfrak{g}' \twoheadrightarrow \mathfrak{g}$.

<u>Proof</u>: Notice that, because B is simply-connected, we have $(\Gamma H^2(L,E))^A \cong H^2(\mathfrak{g},V)^{\mathfrak{g}} = H^2(\mathfrak{g},V)$ by 1.19 and 2.6. //

For the case where $\pi_1 B \neq 0$, the comments in III 7.30 apply.

Concerning the multiplicity of lifts possible for a given $E \rightarrowtail L' \twoheadrightarrow L$, this is measured by ker j^*, which can in principle be constructed from $E_3^{2,0}$ and $E_3^{1,1}$; when $E^L = E$ and one of the condtions of 5.11 is satisfied, we have seen that these spaces reduce to $E_2^{2,0}$ and $E_2^{1,1}$.

These techniques will be developed further elsewhere.

* * * * * * *

For the Lie algebroid of a Lie groupoid, this spectral sequence is closely related to the Leray-Serre spectral sequence in de Rham cohomology of the associated principal bundle.

Let Ω be a Lie groupoid on B, and let $\rho: \Omega \to \Pi(E)$ be a representation of Ω on a vector bundle E. Choose $b \in B$ and write $P = \Omega_b$, $G = \Omega_b^b$, $V = E_b$.

There is a natural action of Ω on $H^*(L\Omega,E)$. Each $\xi \in \Omega$ induces a change of Lie algebras (as in 5.1) $(\mathrm{Ad}\xi^{-1},\rho(\xi))$ from $(\rho_{\beta\xi}^{\beta\xi})_*: L\Omega|_{\beta\xi} \to \mathrm{End}(E_{\beta\xi})$ to $(\rho_{\alpha\xi}^{\alpha\xi})_*: L\Omega|_{\alpha\xi} \to \mathrm{End}(E_{\alpha\xi})$. (See III 4.15 for the necessary formula.) Hence ξ induces an isomorphism $(\mathrm{Ad}\xi^{-1},\rho(\xi))^*: H^*(L\Omega,E)_{\alpha\xi} \to H^*(L\Omega,E)_{\beta\xi}$, and it is routine to verify that this defines a smooth action. In particular, G acts on $H^*(\mathfrak{g},V)$ and it follows from II 4.9 that $H^*(L\Omega,E)$ is equivariantly isomorphic to $\frac{P \times H^*(\mathfrak{g},V)}{G}$. Note that this bundle is flat.

In A 4.13 it is shown that the cochain complex $\Gamma C^*(A\Omega,E)$ is naturally isomorphic to the G-equivariant de Rham complex $A^*(P,V)^G$ and it follows (2.7) that $\mathcal{H}^*(A\Omega,\rho_*,E) \cong H^n_{\mathrm{deRh}}(P,V)^G$. The following result is now immediate.

<u>Theorem 5.17.</u> Let P(B,G) be a principal bundle and let G act on a vector space V. Then there is a natural convergent spectral sequence

$$\mathcal{H}^s\left(TB, \frac{P \times H^t(\mathfrak{g},V)}{G}\right) \Rightarrow H^n_{\mathrm{deRh}}(P,V)^G. \qquad //$$

If B is simply-connected then the E_2 term simplifies to $H^s_{\mathrm{deRh}}(B,H^t(\mathfrak{g},V)) = H^s_{\mathrm{deRh}}(B) \otimes H^t(\mathfrak{g},V)$. If in addition G is compact then $H^*_{\mathrm{deRh}}(P,V)^G = H^*_{\mathrm{deRh}}(P,V)$ and $H^t(\mathfrak{g},V) \cong H^t_{\mathrm{deRh}}(G) \otimes V$ and we obtain the standard Leray-Serre spectral sequence in de Rham cohomology.

For the Leray-Serre spectral sequence of a principal bundle with compact group, there is an extensive and deep theory (see Greub et al (1976)). It seems reasonable to expect that for general groups, the equivariant spectral sequence 5.17 will admit generalizations of the structure theorems already established in the case of compact groups.

The most fundamental result one wishes to prove is that the spectral sequence collapses from E_2 on when A admits a flat connection. For the Leray-Serre spectral sequence of a principal bundle with compact group, or of a $\mathfrak{g}$-DGA with $\mathfrak{g}$ reductive, this is a deep result (Greub et al (1976, 3.17)). It depends strongly on the fact that when $\mathfrak{g}$ is reductive the primitive elements in $H^*(\mathfrak{g})$ are precisely the universally transgressive ones and that, using the Weil homomorphism, the problem can be reduced to showing that the transgressions are zero. The Weil homomorphism for a transitive Lie algebroid has been set up by N. Teleman (1972) but it remains to be seen if it is worthwhile to rewrite more of the theory of $\mathfrak{g}$-DGA's in terms of transitive Lie algebroids.

We close this section with some brief comments on cases in which the Lie algebroid spectral sequence collapses to a Gysin sequence. Let $L \rightarrowtail A \twoheadrightarrow TB$ be an arbitrary transitive Lie algebroid and let ρ be any representation of A on a vector bundle E.

Assume firstly that $H^n(L,E) = 0$ for $n \geq 2$. Then the sequence of terms of low degree can be continued

$$\cdots \longrightarrow \mathscr{H}^1(TB,H^1(L,E)) \xrightarrow{d_2^{1,1}} \mathscr{H}^3(TB,E^L) \xrightarrow{q^*} \mathscr{H}^3(A,E) \longrightarrow \cdots$$

$$\cdots \longrightarrow \mathscr{H}^n(TB,E^L) \xrightarrow{q^*} \mathscr{H}^n(A,E) \longrightarrow \mathscr{H}^{n-1}(TB,H^1(L,E))$$

$$\xrightarrow{d_2^{n-1,1}} \mathscr{H}^{n+1}(TB,E^L) \longrightarrow \cdots .$$

This is essentially Theorem 7 of Hochschild and Serre (1953). When $E^L = E$, 5.10 applies and the identification of $d_2^{n,1}$ with $(-1)^n\text{-}\wedge R_{ab}$ shows that R_{ab} may - in this case - be regarded as a generalization of the Euler class of a circle bundle: Suppose that $E = E^L = B \times \mathbf{R}$ with ρ the trivial representation, and also assume that $H^1(L, B \times \mathbf{R}) \cong B \times \mathbf{R}$. By 5.9, this last assumption is equivalent to $L/[L,L] \cong B \times \mathbf{R}$. Now $R_{ab} \in \mathscr{H}^2(TB, B \times \mathbf{R}) = H^2_{deRh}(B)$ and the pairing $H^1(L, B \times \mathbf{R}) \oplus L_{ab} \to B \times \mathbf{R}$ is reduced to the multiplication $\mathbf{R} \times \mathbf{R} \to \mathbf{R}$.

Also, $tg^1 : (\Gamma H^1(L, B \times \mathbb{R}))^A \to \mathcal{H}^2(TB, B \times \mathbb{R})$ becomes $\mathbb{R} \to H^2_{deRh}(B)$, $t \mapsto tR_{ab}$ and so R_{ab} is the image under tg^1 of a generator of $(\Gamma H^1(L, B \times \mathbb{R}))^A$. It may thus be considered to be the Euler class of $L \rightarrowtail A \twoheadrightarrow TB$ (compare Greub et al (1973, 6.23)).

Secondly, assume that $H^n(L,E) = 0$ for $n = 1,2$. By the Whitehead lemmas for Lie algebras, this is the case for semisimple L. We then have $\mathcal{H}^2(TB,E^L) \overset{q^*}{\cong} \mathcal{H}^2(A,E)$, so that all extensions are geometric, and an exact sequence

$$\mathcal{H}^3(TB,E^L) \xrightarrow{q^*} \mathcal{H}^3(A,E) \xrightarrow{i^*} (\Gamma H^3(L,E))^A \xrightarrow{tg^3} \mathcal{H}^4(TB,E^L) \xrightarrow{q^*} \mathcal{H}^4(A,E).$$

If, in addition, $H^n(L,E) = 0$ for $n \geq 4$, then this sequence can be continued, and it then includes the Gysin sequence for SU(2)-bundles.

CHAPTER V AN OBSTRUCTION TO THE INTEGRABILITY OF TRANSITIVE LIE ALGEBROIDS

For many years the major outstanding problem in the theory of differentiable groupoids and Lie algebroids was to provide a full proof of a result announced by Pradines (1968b), that every Lie algebroid is (isomorphic to) the Lie algebroid of a differential groupoid. This problem was resolved recently in the most unexpected manner by Almeida and Molino (1985) who announced the existence of transitive Lie algebroids which are not the Lie algebroid of any Lie groupoid. (It is easily seen (III 3.16) that a differential groupoid on a connected base whose Lie algebroid is transitive must be a Lie groupoid.) The examples of Almeida and Molino (1985) arise as infinitesimal invariants attached to transversally complete foliations, and represent an entirely new insight into the subject.

We now construct a single cohomological invariant, attached to a transitive Lie algebroid on a simply-connected base, which gives a necessary and sufficient condition for integrability. The method is from Mackenzie (1980), which gave the construction of the elements here denoted e_{ijk} and the fact that if the e_{ijk} lie in a discrete subgroup of the centre of the Lie group involved, then the Lie algebroid is integrable. (In particular, a semisimple Lie algebroid on a simply-connected base is always integrable.) However in Mackenzie (1980) the author believed that sufficient work would show that the e_{ijk} could always be quotiented out.

With the discovery of counterexamples to the general result by Almeida and Molino (1985), it is easy to see that the e_{ijk} form a cocycle; it should be noted that Almeida and Molino independently made this observation for the corresponding elements in Mackenzie (1980). The method now yields a cohomological obstruction to the problem of realizing a transitive Lie algebroid on a simply-connected base as the Lie algebroid of a Lie groupoid.

A construction related to that given here was announced by Aragnol (1957). This reference was pointed out by Professor Molino, after the completion of the work recorded here.

The results of §1, taken together with those of those of IV§3, give necessary and sufficient conditions for an LAB-valued 2-form to be the curvature of a connection in a principal bundle, providing that the manifold on which the form is defined is simply-connected. These results thus generalize - and reformulate - a classical result of Weil (1958) (see also Kostant (1970)) on the integrality of Chern classes. The case of real-valued forms was also noted in Almeida and Molino (1985).

§1. Results

Throughout this section, all base manifolds are connected. We call a Lie algebroid A <u>integrable</u> if there is a differentiable groupoid Ω such that $A\Omega \cong A$.

In III§5 we showed that the right derivatives $\Delta(s_{ij})$ of a cocycle $\{s_{ij}\}$ for a Lie groupoid Ω are transition forms for the Lie algebroid $A\Omega$ of Ω. Then in IV§4 we showed that an abstract transitive Lie algebroid A on an arbitrary base B admits a system of transition forms χ_{ij}. Our problem now is the following: Given an abstract transitive Lie algebroid A and a system of transition forms χ_{ij}, is it possible to integrate the χ_{ij} to functions s_{ij} which obey the cocycle condition? If this can be accomplished, then the resulting Lie groupoid will have A as its Lie algebroid, by the classification theorem III 5.15.

Consider, therefore, a transitive Lie algebroid $L \rightarrowtail A \twoheadrightarrow TB$ on an arbitrary (connected) base B. Let $\mathfrak{g}$ denote the fibre type of L, and let $\{U_i\}$ be a simple open cover of B.

By IV§4, there are local flat connections $\theta_i : TU_i \to A_{U_i}$ and LAB charts $\psi_i : U_i \times \mathfrak{g} \to L_{U_i}$ such that $\nabla^{\theta_i}_X(\psi_i(V)) = \psi_i(X(V))$ identically. Let $\chi_{ij} \in A^1(U_{ij},\mathfrak{g})$ and $a_{ij} : U_{ij} \to \operatorname{Aut}(\mathfrak{g})$ be the resulting system of transition data. From IV§4 we have

(1) $$\delta\chi_{ij} + [\chi_{ij},\chi_{ij}] = 0, \text{ whenever } U_{ij} \neq \emptyset,$$

(2) $$\chi_{ik} = \chi_{ij} + a_{ij}(\chi_{jk}), \text{ whenever } U_{ijk} \neq \emptyset,$$

(3) $$\Delta(a_{ij}) = \mathrm{ad}\circ\chi_{ij}, \text{ whenever } U_{ij} \neq \emptyset.$$

Let $\tilde{G}$ be the connected and simply-connected Lie group with Lie algebra $\mathfrak{g}$. From (1) and the simple-connectivity of U_{ij}, it follows that there are functions $s_{ij} : U_{ij} \to \tilde{G}$ such that $\Delta(s_{ij}) = \chi_{ij}$; such functions are unique, up to right-translation by constants. From equation (8) of B§2, it follows that

$$\Delta(\mathrm{Ad}\circ s_{ij}) = \mathrm{ad}\circ\Delta(s_{ij}) = \mathrm{ad}\circ\chi_{ij} = \Delta(a_{ij}),$$

where the first and last Δ refer to the group $\mathrm{Aut}(\mathfrak{g})$, and so, by the uniqueness result just referred to, there are elements $\phi_{ij} \in \mathrm{Aut}(\mathfrak{g})$ such that

$$a_{ij} = (\mathrm{Ad}\circ s_{ij})\circ\phi_{ij}. \tag{4}$$

If the a_{ij} take values in $\mathrm{Ad}(\tilde{G}) = \mathrm{Int}(\mathfrak{g}) < \mathrm{Aut}(\mathfrak{g})$, the equation $\Delta(\mathrm{Ad}\circ s_{ij}) = \Delta(a_{ij})$ may be solved with respect to $\mathrm{Ad}(\tilde{G})$, and it will be possible to take $\phi_{ij} = \mathrm{id}$ in (4). We now show that this can be done whenever B is simply-connected. The key is the following general result, which is a refinement of IV 4.1.

Theorem 1.1. Let $L \rightarrowtail A \twoheadrightarrow TB$ be a transitive Lie algebroid on an arbitrary base B. Let $\psi\colon U \to \mathrm{Int}(A)_b$ be a decomposing section of the Lie groupoid $\mathrm{Int}(A)$ over a contractible open set U. Let ψ also denote the chart $U \times \mathfrak{g} \to L_U$ obtained by regarding $\mathrm{Int}(A)$ as a reduction of $\Pi[L]$. (Here $\mathfrak{g} = L_b$.) Then there is a local flat connection $\theta\colon TU \to A_U$ such that $\nabla^\theta = \psi_*(\nabla^0)$.

Proof: The decomposing section $\psi\colon U \to \mathrm{Int}(A)_b$ induces, as in III§5, a local flat connection $\kappa\colon TU \to \mathrm{ad}(A)_U$. By equation (8) of III§5, applied to $\Omega = \mathrm{Int}(A)$, the induced chart

$$\mathrm{Ad}(\psi)\colon U \times \mathrm{ad}(\mathfrak{g}) \to \mathrm{ad}(L)_U$$

maps ∇^0 to ∇^κ. Here Ad refers to the Lie groupoid $\mathrm{Int}(A)$: note that $\mathrm{Ad}(\psi)_x\colon \mathrm{ad}(\mathfrak{g}) \to \mathrm{ad}(L_x)$, for $x \in U$, is $T(I_{\psi(x)})_{id}\colon T(\mathrm{Int}(\mathfrak{g}))_{id} \to T(\mathrm{Int}(L_x))_{id}$ and since $I_{\psi(x)}$ is linear, it is its own tangent and so

$$\mathrm{Ad}(\psi)_x(\phi) = \psi(x)\circ\phi\circ\psi(x)^{-1}$$

for $\phi \in \mathrm{ad}(\mathfrak{g})$.

Choose any connection γ in A; since the adjoint connection ∇^γ and $\psi_*(\nabla^0)$ are both in ad(A) we can write

$$\nabla^\gamma = \psi_*(\nabla^0) + \mathrm{ad}(\psi\circ\ell) \tag{5}$$

for some $\ell \in A^1(U,\mathfrak{g})$. Equation (5) shows that $\psi_*(\nabla^0)$ and ∇^γ cover the same coupling and so, by IV 3.19, there is a connection $\gamma'\colon TU \to A_U$ such that $\nabla^{\gamma'} = \psi_*(\nabla^0)$. It remains to show that there is a flat such connection.

Let A' be the semidirect extension $TU \ltimes_{\nabla} L_U$, where $\nabla = \nabla^{\gamma'} = \psi_*(\nabla^0)$. Now A' and A_U both define elements of $\mathcal{O}\mathrm{pext}(TU,L_U)$ and since $\mathcal{H}^2(TU,ZL_U) \cong H^2_{deRh}(U,\mathfrak{z}) = (0)$ (by the contractibility of U, as in IV§4) it follows that A' and A_U are equivalent. Thus there is an isomorphism $\phi\colon A' \to A_U$ such that

$$\begin{array}{ccccc} L_U & \xrightarrow{\;j\;} & A_U & \xrightarrow{\;q\;} & TU \\ \| & & \uparrow{\scriptstyle\phi} & & \| \\ L_U & \xrightarrow{\;j'\;} & A' & \xrightarrow{\;q'\;} & TU \end{array}$$

commutes. Define $\theta\colon TU \to A_U$ by $\theta(X) = \phi(X \oplus 0)$. Then θ is a flat connection and

$$\begin{aligned} j(\nabla^{\theta}_X(V)) &= [\theta(X), j(V)] \\ &= [\phi(X \oplus 0), \phi j'(V)] \\ &= \phi(0 \oplus \nabla_X(V)) \\ &= \phi j'(\nabla_X(V)) \\ &= j(\nabla_X(V)). \end{aligned}$$

So $\nabla^{\theta} = \nabla = \psi_*(\nabla^0)$, as required. //

It is interesting to note that the full force of the classification of extensions by $\mathcal{H}^2$ and $\mathcal{H}^3$ is used in this proof.

We can now make a fresh start. Assume that B is simply-connected. Given any simple cover $\{U_i\}$ of B we obtain decomposing sections $\psi_i\colon U_i \to \mathrm{Int}(A)_b$ and, from Theorem 1.1, local flat connections θ_i with $(\psi_i)_*(\nabla^0) = \nabla^{\theta_i}$. Proceeding as before, we now have $a_{ij}\colon U_{ij} \to \mathrm{Int}(A)^b_b$. Since Int(A) is α-connected and B is simply-connected, it follows that $\mathrm{Int}(A)^b_b$ is connected. It is therefore equal to $\mathrm{Int}(\mathfrak{g})$, and so to $\mathrm{Ad}(\tilde{G})$. Now the equation $\Delta(\mathrm{Ad}\circ s_{ij}) = \Delta(a_{ij})$ may be solved with respect to the group $\mathrm{Ad}(\tilde{G})$ and we have $\phi_{ij} = \mathrm{Ad}(g_{ij})$ for $g_{ij} \in G$. So, redefining s_{ij} as $s_{ij}g_{ij}$, we have

(6) $$a_{ij} = \mathrm{Ad}\circ s_{ij}.$$

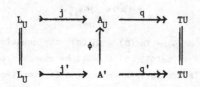

Consider the map $s_{jk}s_{ik}^{-1}s_{ij}: U_{ijk} \to \tilde{G}$ for a nonvoid U_{ijk}. Using formulas (1) and (2) in B§2 and the cocycle equation (2), we have

$$\begin{aligned}\Delta(s_{jk}s_{ik}^{-1}s_{ij}) &= \Delta(s_{jk}) + \mathrm{Ad}(s_{jk})\Delta(s_{ik}^{-1}s_{ij})\\ &= \chi_{jk} + a_{jk}\{\Delta(s_{ik}^{-1}) + a_{ik}^{-1}\Delta(s_{ij})\}\\ &= \chi_{jk} + a_{jk}\{-a_{ik}^{-1}\chi_{ik} + a_{ik}^{-1}\chi_{ij}\}\\ &= \chi_{jk} + a_{ji}(\chi_{ij} - \chi_{ik})\\ &= \chi_{jk} - a_{ji}a_{ij}\chi_{jk} = 0.\end{aligned}$$

Since U_{ijk} is connected, it follows that $s_{jk}s_{ik}^{-1}s_{ij}$ is constant; denote its value by e_{ijk}. Clearly $e_{ijk} \in Z\tilde{G}$, for $\mathrm{Ad}(e_{ijk}) = a_{jk}a_{ik}^{-1}a_{ij} = \mathrm{id}_{\mathfrak{g}}$. In fact, $\{e_{ijk}\}$ is a Čech 2-cocycle. For if $U_{ijk\ell} \neq \emptyset$, then

$$e_{jk\ell}e_{ik\ell}^{-1}e_{ij\ell}e_{ijk}^{-1} = (s_{k\ell}s_{j\ell}^{-1}s_{jk})(s_{ik}^{-1}s_{i\ell}s_{k\ell}^{-1})(s_{j\ell}s_{i\ell}^{-1}s_{ik}s_{jk}^{-1}).$$

Interchanging the first two bracketed terms, this becomes

$$(s_{ik}^{-1}s_{i\ell}s_{j\ell}^{-1}s_{jk})(s_{j\ell}s_{i\ell}^{-1}s_{ik}s_{jk}^{-1}).$$

Now the second bracketed term is $e_{ij\ell}e_{ijk}^{-1}$ and is central, so we can interchange s_{jk} with it, and the expression then collapses to the identity.

Thus we have $e = \{e_{ijk}\} \in \check{H}^2(B, Z\tilde{G})$. It remains to prove that e is well-defined. This requires a little care. Let $\{\psi_i': U_i \to \mathrm{Int}(A)_b\}$ be a second section-atlas for Int(A) with respect to the same open cover. Write $\psi_i' = \psi_i n_i$ where $n_i: U_i \to \mathrm{Int}(A)_b^b = \mathrm{Int}(\mathfrak{g}) = \mathrm{Ad}(\tilde{G})$. Let $\{\theta_i'\}$ be a second family of local flat connections, compatible with $\{\psi_i'\}$. Write $\theta_i' = \theta_i + \psi_i \circ m_i$. Then, repeating equations (5) of IV§4, we have

(7a) $$\delta m_i + [m_i, m_i] = 0,$$

(7b) $$\Delta(n_i) = \mathrm{ad} \circ m_i,$$

(7c) $$\chi_{ij}' = n_i^{-1}\{-m_i + \chi_{ij} + a_{ij} \circ m_j\},$$

(7d) $$a'_{ij} = n_i^{-1} a_{ij} n_j.$$

Since m_i is a Maurer-Cartan form, we can integrate and get $r_i: U_i \to \tilde{G}$ with $\Delta(r_i) = m_i$. Then $\Delta(\mathrm{Ad}\circ r_i) = \mathrm{ad}\circ\Delta(r_i) = \mathrm{ad}\circ m_i = \Delta(n_i)$ so there exists $\phi_i \in \mathrm{Ad}\tilde{G}$ such that $n_i = (\mathrm{Ad}\circ r_i)\phi_i$. Writing $\phi_i = \mathrm{Ad}g_i$ and redefining r_i to be $r_i g_i$, we now have

(8) $$n_i = \mathrm{Ad}\circ r_i.$$

Now, by using equations (1) and (2) of B§2, one easily sees that 7(c) is equivalent to

$$\Delta(s'_{ij}) = \Delta(r_i^{-1} s_{ij} r_j)$$

so there are elements $c_{ij} \in \tilde{G}$ such that

$$s'_{ij} = r_i^{-1} s_{ij} r_j c_{ij}.$$

Applying Ad to this equation, we find that $c_{ij} \in Z\tilde{G}$. It is now straight-forward to verify that

$$e'_{ijk} = e_{ijk}(c_{jk} c_{ik}^{-1} c_{ij})$$

and so $\{e'_{ijk}\}$ and $\{e_{ijk}\}$ represent the same element of $\check{H}^2(B, Z\tilde{G})$. It is also straightforward to show that this element is well-defined with respect to the inductive limit. There is thus a well-defined element $e \in \check{H}^2(B, Z\tilde{G})$, independent of the choice of section-atlas for Int(A). We call e the <u>integrability obstruction</u> of A on account of the following theorem.

<u>Theorem 1.2</u>. Let $L \rightarrowtail A \twoheadrightarrow TB$ be a transitive Lie algebroid on a simply-connected base B. Then there is a Lie groupoid Ω such that $A\Omega \cong A$ iff e lies in $\check{H}^2(B, D)$ for some discrete subgroup D of $Z\tilde{G}$.

<u>Proof</u>: (⇒) This requires some work. First note that Ω may be assumed to be α-connected, and it then follows that $\mathrm{Ad}(\Omega) = \mathrm{Int}(A\Omega)$. Choose $b \in B$ and denote Ω_b^b by G; since B is simply-connected and Ω is α-connected, G is connected. Let $\tilde{G}$ denote the simply-connected covering group.

Choose an atlas $\{\sigma_i: U_i \to \Omega_b\}$ for Ω. Then $\{\psi_i = \mathrm{Ad}\sigma_i\}$ is an atlas for $\mathrm{Ad}(\Omega) = \mathrm{Int}(A\Omega)$. Let θ_i denote the local morphism $U_i \times U_i \to \Omega_{U_i}^{U_i}$ defined in III§5. Then, by (8) of III§5, $(\theta_i)_*$ is compatible with ψ_i, and so we can use $\{\psi_i, (\theta_i)_*\}$ to define e. Write $\chi_{ij} = \Delta(\tilde{s}_{ij})$ where $\tilde{s}_{ij}: U_{ij} \to \tilde{G}$. Note that the $\tilde{s}_{ij}$ are the would-be cocycles found from $A\Omega$, whereas s_{ij} are the actual cocycles for Ω. Now $\Delta(p\circ\tilde{s}_{ij}) = \chi_{ij} = \Delta(s_{ij})$, where $p: \tilde{G} \to G$ is the covering projection. So $p\circ\tilde{s}_{ij} = s_{ij}w_{ij}$ for some element $w_{ij} \in G$. Now, applying Ad to this and noting that $Ad\circ s_{ij} = a_{ij}$ and $Ad\circ p\circ \tilde{s}_{ij} = Ad\circ\tilde{s}_{ij} = a_{ij}$ (equation (6) of the construction), we find that $w_{ij} \in ZG$. Now the covering projection p maps $Z\tilde{G}$ onto ZG, by general Lie group considerations, and so we can write $w_{ij} = p(\tilde{w}_{ij})$ where $\tilde{w}_{ij} \in Z\tilde{G}$. Now redefine $\tilde{s}_{ij}$ to be $\tilde{s}_{ij}\tilde{w}_{ij}^{-1}$ and we have

$$p\circ\tilde{s}_{ij} = s_{ij}.$$

(Note that this redefinition does not affect condition (6), since $\tilde{w}_{ij}$ is central.) Now

$$\begin{aligned} p(e_{ijk}) &= p(\tilde{s}_{jk}\tilde{s}_{ik}^{-1}\tilde{s}_{ij}) \\ &= s_{jk}s_{ik}^{-1}s_{ij} \\ &= 1 \in G \end{aligned}$$

so e takes its values in the discrete subgroup ker(p) of $Z\tilde{G}$.

(<=) Assume that $e \in \check{H}^2(B,D)$ for a discrete subgroup D of $Z\tilde{G}$. Define $G = \tilde{G}/D$ and let p be the covering projection. Let $s_{ij}: U_{ij} \to \tilde{G}$ be the system of maps which define a representative $\{e_{ijk}\}$ for which $e_{ijk} \in D$. Define $\bar{s}_{ij} = p\circ s_{ij}$. Then $\{\bar{s}_{ij}: U_{ij} \to G\}$ satisfies the cocycle condition $\bar{s}_{jk}\bar{s}_{ik}^{-1}\bar{s}_{ij} = 1 \in G$; let Ω be the resulting Lie groupoid. The Lie algebroid $A\Omega$ has transition forms $\Delta(\bar{s}_{ij}) = p_*\circ\Delta(s_{ij}) = \Delta(s_{ij})$ and so, by III 5.15, is isomorphic to A. //

In particular, if the centre of $\mathfrak{g}$ is trivial, then $Z\tilde{G}$ itself is discrete:

Corollary 1.3. Let $L \rightarrowtail A \twoheadrightarrow TB$ be a transitive Lie algebroid on a simply-connected base B, with semisimple adjoint bundle L. Then A is integrable. //

For the construction of nonintegrable transitive Lie algebroids from transversely complete foliations, see Almeida and Molino (1985).

Weil (1958) (see also Kostant (1970)) proves that a closed, real-valued 2-form on an arbitrary manifold is the curvature of a connection in a **C***-bundle (where **C*** is the group of nonzero complex numbers) iff the 2-form is integral. His proof is largely an application of the isomorphism between de Rham and Čech cohomology. Our construction of e and proof of Theorem 1.2 is in certain respects a generalization of Weil's proof; however our proof does not use, even implicitly, a global connection in the Lie algebroid. The differences between the two proofs merit exploration.

Putting Theorem 1.2 together with IV 3.21, we obtain

Theorem 1.4. Let B be a simply-connected manifold, and let L be an LAB on B. Let R be an L-valued 2-form on B. Then R is the curvature of a connection in a principal bundle P(B,G) with $\frac{P \times \mathfrak{g}}{G} \cong L$ iff

(1) there exists a Lie connection ∇ in L such that $\bar{R}_\nabla = \mathrm{ad} \circ R$ and $\nabla(R) = 0$;

and

(2) the integrability obstruction $e \in \check{H}^2(B, Z\tilde{G})$ defined by the transitive Lie algebroid corresponding to ∇ and R, lies in $\check{H}^2(B,D)$ for some discrete subgroup D of $Z\tilde{G}$. //

If L is abelian (and B is simply-connected), then ∇ must be flat, L must be trivializable, and there is a trivialization $B \times V \cong L$ which maps the standard flat connection in $B \times V$ to ∇ (see III 5.20). Thus in this case, the Lie algebroid is determined uniquely by the closed 2-form R (IV 2.13) and the connections in that Lie algebroid with curvature R are determined by $\mathcal{H}^1(TB, B \times V) \cong H^1_{deRh}(B,V) = 0$ (IV 2.17). There is therefore a uniquely determined Lie algebroid with a connection having curvature R and, when the Lie algebroid is integrable, there is a unique principal bundle up to local isomorphism. This includes a result of Kostant (1970).

If L is not abelian, or if B is not simply-connected, then uniqueness in this strong sense fails. The appropriate results can be obtained by following back through the results of IV§3.

There are two differences between Weil's result and the specialization of Theorem 1.4 to $L = B \times \mathbb{R}$. Firstly, we allow R to be closed with respect to any flat connection in $B \times \mathbb{R}$; if B is not simply-connected, this is a genuine generalization. Secondly, we do not insist that $[R] \in H^2_{deRh}(B,\mathbb{R}) \cong \check{H}^2(B,\mathbb{R})$ be integral, but allow it to take values in any discrete subgroup $\alpha\mathbb{Z}$, $\alpha \in \mathbb{R}$, of $\mathbb{R}$.

This is explained by the fact that $e \in \check{H}^2(B, Z\tilde{G})$ is not an isomorphism invariant of a transitive Lie algebroid, but only an invariant up to equivalence. (Compare the classification of principal bundles P(B,G) by $\check{H}^1(B,G)$.) Let

$L \overset{j}{\rightarrowtail} A \overset{q}{\twoheadrightarrow} TB$ be an abelian transitive Lie algebroid and let α be a nonzero real number. Define $\phi\colon L \to L$ by $\phi(V) = \alpha V$. Then ϕ is an isomorphism of LAB's and $L \overset{j\circ\phi}{\rightarrowtail} A \overset{q}{\twoheadrightarrow} TB$ is a second Lie algebroid, isomorphic to $L \overset{j}{\rightarrowtail} A \overset{q}{\twoheadrightarrow} TB$ under $\mathrm{id}\colon A \to A$,

$$\begin{array}{ccccc} L & \overset{j}{\rightarrowtail} & A & \overset{q}{\twoheadrightarrow} & TB \\ {\scriptstyle\phi}\uparrow & & \| & & \| \\ L & \overset{j\circ\phi}{\rightarrowtail} & A & \overset{q}{\twoheadrightarrow} & TB\ . \end{array}$$

If γ is a connection $TB \to A$ and R is its curvature form with respect to the first Lie algebroid, then its curvature form with respect to the second is $\frac{1}{\alpha} R$. Clearly one Lie algebroid is integrable iff the other is so. If e is the integrability obstruction of the first Lie algebroid, then $\frac{1}{\alpha} e$ is the integrability obstruction of the other. The fact is that integrability is a notion invariant under isomorphism whereas the integrability obstruction is an invariant only up to equivalence.

Note that once a transitive Lie algebroid is known to be integrable, it has a natural adjoint bundle, arising from its presentation as $L\Omega \rightarrowtail A\Omega \twoheadrightarrow TB$ or as $\frac{P \times \mathfrak{g}}{G} \rightarrowtail \frac{TP}{G} \twoheadrightarrow TB$, and the curvature of any connection $\gamma\colon TB \to A$ is then well-defined. However, for an abstract transitive Lie algebroid $L \rightarrowtail A \twoheadrightarrow TB$, the curvature of a connection $\gamma\colon TB \to A$ depends not merely on $A \twoheadrightarrow TB$ but on the choice of L.

This nuisance could be avoided by rephrasing the question as follows: Given $L \rightarrowtail A \twoheadrightarrow TB$ and a connected Lie group G with Lie algebra the fibre type of L, is there a principal bundle $P(B,G)$ with Atiyah sequence A? The answer would then be: Iff the class $e_G \in \check{H}^2(B,ZG)$, defined with respect to G rather than $\tilde{G}$, vanishes.

Work on the question of the representation of 2-forms as the curvature of general, non Riemannian connections has also been done by Jacobowitz (1978) and Tsarev (1983). I am grateful to Iain Aitchison for the reference to Tsarev's work.

The integrability obstruction is essentially a feature of the cohomology of Lie groupoids. If M is a Lie group bundle on B, one may ask whether there is a Lie groupoid $M \rightarrowtail \Omega \twoheadrightarrow B \times B$ with respect to a certain notion of coupling of $B \times B$ with M. This question is resolved by an intrinsically defined version of $e \in \check{H}^2(B,ZG)$ (where G is the fibre type of M) in the same way as the obstruction class of IV§3 resolves the corresponding question for Lie algebroid extensions. In this respect, the integrability obstruction is related to the obstruction classes found by Greub and Petry (1978). The general theory underlying these constructions is dealt with in a coming paper.

§2. Epilogue

The construction of the invariant e, and the cohomology theory of Chapter IV on which it depends, grew out of a strategy, developed by the author in the mid-late 1970's, for proving the integrability of transitive Lie algebroids by generalizing the cohomological proof of the integrability of Lie algebras due to van Est (1953,1955b). I believe it will be of interest to describe this process and, by so doing, to set this particular integrability result in the context of the ongoing evolution of the group concept.

The result which asserts the integrability of Lie algebras is commonly referred to as Lie's third theorem. This theorem has a long and continuing history in modern mathematics, corresponding to the evolution of the group concept from its first rigorous formulation to the point where it is again capable of being applied to the study of partial differential equations. Concerning Lie groups in the now standard meaning of the term, prior to the work of van Est (1953,1955b) there were essentially two different methods of proving Lie's third theorem. One method, which I will call the structural proof, first uses the Levi-Mal'cev decomposition to reduce the problem to the two separate cases of solvable and semi-simple Lie algebras. Lie's third theorem for these two cases is straightforward: the solvable case is reduced, by virtue of the chain condition, to the case of 1-dimensional Lie algebras, where the result is trivial; in the semi-simple case, the adjoint representation is faithful and the result follows from the subgroup/subalgebra correspondence for a general-linear group. (For details, see, for example, Varadarajan (1974, 3.15).) This proof is essentially a rigorous reformulation of Cartan's 1930 proof, in which Cartan was chiefly concerned to complete a proof of Lie's, valid only when the adjoint representation is faithful. The other method, for which we know no classical reference, integrates the given general Lie algebra directly to a Lie group germ, and must then show that this Lie group germ can be globalized. For the integration step, see, for example Malliavin (1972, pp. 232-4), or Greub et al (1973, pp. 368-9); the globalization may be accomplished by the method of P. A. Smith (1952) - note that this depends on the fact that $\pi_2(G) = (0)$ for a (semisimple) Lie group G. We will call this second method the geometric proof since it depends on $\pi_2(G) = (0)$, rather than on the structure theory of Lie algebras.

We have described the two proofs so as to stress that Lie's third theorem is an integration result; so it was to Lie (for example, Cohn (1957, Chapter V)), and so it remains. The geometric method divides the proof into an 'integration' step which yields a local group, and a 'globalization' step which depends on a deep topological result; the structural proof uses deep results of Lie algebra structure theory to reduce the integration to that of the subgroup/subalgebra correspondence - that is, to the Frobenius theorem. (Similarly Ado's theorem, an even deeper result from the structure theory of Lie algebras and itself depending on the Levi-Mal'cev decomposition, can be used for the same purpose.) Having noted the element of integration present in both proofs, note also that the other major steps in the two proofs are formally analogous: The existence of Lie subalgebras depends on the semi-directness of the extension resulting from quotienting the Lie algebra over its radical; that the extension is semi-direct follows from the second Whitehead lemma (and the first, via the theorem of Weyl; see, for example, Varadarajan (1974, 3.14.1)), and the second Whitehead lemma, that is, $H^2(\mathfrak{g},V) = (0)$ for semisimple $\mathfrak{g}$, may be regarded as analogous to $\pi_2(G) = (0)$ for semisimple G, the condition which is crucial for the globalization step of the geometric proof.

To describe van Est's proof, it is necessary to first summarize the results from which he deduces Lie's third theorem.

van Est (1955b) in a note reformulating earlier results (1953,1955a) constructed two convergent spectral sequences

$$\text{(I)} \qquad H^s(G,V) \otimes H^t_{deRh}(G) \Rightarrow H^{s+t}(\mathfrak{g},V)$$

$$\text{(II)} \qquad H^s(G,V) \otimes H^t_{deRh}(G/K) \Rightarrow H^{s+t}(\mathfrak{g},\mathfrak{k},V)$$

for a Lie group G and a representation of G on a vector space V. Here $H^*_{deRh}(M)$ denotes the de Rham cohomology of the manifold M, $H^*(G,V)$ denotes the smooth Eilenberg-MacLane cohomology, and K is a compact subgroup of G with Lie algebra $\mathfrak{k} \subseteq \mathfrak{g}$. Both spectral sequences arise from double complexes in the standard manner; the double complex for (II) is a K-variant subcomplex of that for (I).

When G is connected and K is a maximal compact subgroup, G/K is diffeomorphic to a Euclidean space (the Iwasawa decomposition) and the second spectral sequence therefore collapses to isomorphisms $H^*(G,V) \simeq H^*(\mathfrak{g},\mathfrak{k},V)$. This result was reproved in a more modern context by Hochschild and Mostow (1962) and was generalized extensively in the 1970's to model pseudogroups on $\mathbb{R}^n$ (see, for example, the surveys of Lawson (1974, §6), and Stasheff (1978, §7)).

We are exclusively concerned with the first spectral sequence of van Est. In the first paper of the series (1953), van Est noted that for a Lie group G with $H^1_{deRh}(G) = H^2_{deRh}(G) = (0)$ a similar collapse leads to $H^2(G,V) \simeq H^2(\mathfrak{g},V)$. Now in general $\pi_2(G) = (0)$ (Browder (1961)), and so if G is connected and simply-connected the Hurewicz theorem gives $H^1_{deRh}(G) = H^2_{deRh}(G) = (0)$ and so one has $H^2(G,V) \overset{e_B}{\simeq} H^2(\mathfrak{g},V)$ or, equivalently, $\text{Opext}(G,V) \simeq \text{Opext}(\mathfrak{g},V)$ under the map which assigns to $V \overset{\iota}{\rightarrowtail} G \overset{\pi}{\twoheadrightarrow} G$ the differentiated extension $V \overset{\iota_*}{\rightarrowtail} \mathfrak{h} \overset{\pi_*}{\twoheadrightarrow} \mathfrak{g}$. Hochschild (1951) proved that $\text{Opext}(G,V) \simeq \text{Opext}(\mathfrak{g},V)$ for connected and simply-connected G, by use of Lie's third theorem; van Est now shows that the process can be reversed: any Lie algebra $\mathfrak{h}$ is an extension $\mathfrak{z} \rightarrowtail \mathfrak{h} \twoheadrightarrow \text{ad}\,\mathfrak{h}$ and thus defines an element of $H^2(\mathfrak{g},\mathfrak{z})$ where the representation of $\mathfrak{g} = \text{ad}\,\mathfrak{h}$ on $\mathfrak{z}$ is the trivial one. With $\tilde{G}$ the universal covering group of $\text{Int}(\mathfrak{h})$, represented trivially on the vector space $\mathfrak{z}$, he thus obtains an extension $\mathfrak{z} \rightarrowtail H \twoheadrightarrow \tilde{G}$ with $\mathfrak{h}$ the Lie algebra of H. Since H is an extension of a connected and simply-connected Lie group by a vector space, it is itself connected and simply-connected. Thus Lie's third theorem is proved.

Once again the integration step has been reduced to the subgroup/subalgebra correspondence for a general linear group. van Est's procedure thus uses the (deep) topological fact that $\pi_2(G) = (0)$ for any Lie group G but avoids the direct consideration of local groups needed for the geometric proof, and uses no deep result of Lie algebra theory.

A clear analysis of the importance of $\pi_2(G) = (0)$, and other relevant points is given in van Est (1962). For other, more general, forms of the group concept, theorems of "Lie third theorem" type have since been proved (for example, Goldschmidt (1972), Ngô and Rodrigues (1975) (Lie equations and transitive Lie algebras), Pommaret (1977)). These results also represent a combination of cohomology and integration.

The author's original strategy to prove the integrability of transitive Lie algebroids was to construct a cohomology theory for Lie groupoids by means of which a straightforward generalization of van Est's spectral sequence (I) and of the ensuing argument, could be given. Much of this argument does hold for Lie groupoids and Lie algebroids. For example, given a transitive Lie algebroid, the exact sequence $ZL \rightarrowtail A \twoheadrightarrow A/ZL \cong adA$ exists, and, by III 6.1, adA can be integrated to the Lie subgroupoid Int(A) of $\Pi[L]$. The induced representation $ad(A) \to CDO(ZL)$ cannot be said to be trivial, since there is no concept of trivial representation unless ZL is trivializable as a vector bundle, but the representation does integrate to $MInt(A) \to \Pi(ZL)$.

The crucial problem therefore is to construct a satisfactory cohomology theory for Lie groupoids. A cohomology theory for locally trivial topological groupoids with coefficients in vector bundles was presented in Mackenzie (1978), and the corresponding constructions for Lie groupoids can be given and follow the same pattern. It is proved in §7 (op. cit.) that this cohomology, called the rigid cohomology, classifies all extensions (satisfying some natural weak conditions) of locally trivial groupoids by vector bundles.

Nonetheless this theory was not adequate for the application to the Lie third theorem. If $L \rightarrowtail A \twoheadrightarrow TB$ is an abelian Lie algebroid on a simply-connected base then $ZL = L$ and $MInt(A) = B \times B$, and the problem is to find a Lie groupoid Ω on B with $A\Omega \cong A$. The rigid cohomology can only deal with groupoids which are extensions $L \rightarrowtail \Omega \twoheadrightarrow B \times B$ and all such groupoids are trivializable.

The explanation is, of course, that the coefficient bundle ZL must itself be integrated, and a cohomology theory which will classify all extensions of Lie groupoids by Lie group bundles is needed. It is now reasonably clear how to do this, but for the integrability question, a full cohomology theory is not needed.

Out of this strategy the existence of transition forms (Mackenzie (1979); see IV 4.1 and Theorem 1.1) and the construction of the elements e_{ijk} in Mackenzie (1980) emerged, and it is interesting to observe that these results themselves divide into the same two steps. Namely, for adjoint Lie algebroids the problem is (comparatively) easily solved - for every transitive Lie algebroid A the adjoint ad(A) is integrable by III 6.1 - and the problem is to lift this across $ZL \rightarrowtail A \twoheadrightarrow adA$. Further, in Theorem 1.1 and IV 4.1, the existence of the required flat connections is easily established on the adjoint level; the difficulty is in lifting these connections to the given Lie algebroid. It is this lifting process which the cohomological apparatus describes.

APPENDIX A ON PRINCIPAL BUNDLES AND ATIYAH SEQUENCES

This Appendix is an account of the Atiyah sequence of a principal bundle, and its use in the elementary aspects of connection theory. The Appendix is independent of the main text, and requires no knowledge of groupoids; it is assumed that the reader is familiar with the accounts of connections and their curvature forms in Kobayashi and Nomizu (1963) or Greub et al (1973).

What is now known as the Atiyah sequence of a principal bundle was first constructed by Atiyah (1957), and was, from the first, used to construct cohomological obstructions - originally to the existence of complex analytic connections. In the case of real differentiale bundles, it provides a neat encapsulation of the two definitions of a connection, and a conceptually clear and workable definition of the curvature form. These are the only points with which we are concerned here. Beyond this, the concept of Atiyah sequence - and its abstraction, the concept of transitive Lie algebroid - has a multiplicity of virtues; see Chapters III, IV and V.

The main purpose of this Appendix is to provide a lexicon for the correspondence between the infinitesimal connection theory of III§5 and the standard theory of connections in principal bundles. For this reason, most of this Appendix is devoted to establishing the correspondence between the Atiyah sequence formulation of connection theory and the standard theory; the actual definitions of connections and their curvature forms are extremely concise. To the best of my knowledge, this is the first detailed account of this correspondence to appear in print.

Throughout, the Lie algebra of a Lie group is equipped with the right-hand Lie bracket $[\ ,\ ]_R$, which is the negative of the usual bracket. This is necessary for the groupoid theory of the main text, but is in any case more logically consistent: it is a curious anomaly of the standard presentations of principal bundle theory that right-invariant vector fields are used on the bundle space, but left-invariant vector fields are used to define the Lie algebra of the structure group. This is especially curious in the case of homogeneous bundles G(G/H,H). A brief resumé of the right-handed formulation of the elementary formulas of Lie group theory is given in Appendix B.

There are thus some sign-changes in this account and it is partly for this reason that we have given the curvature calculations in §4 in full detail.

§1 is a brief resumé on principal bundles and associated fibre bundles, for reference throughout the text. In §2 we desribe, in general, the process by which the vector bundle $\frac{TP}{G} \to B$ corresponding to a principal bundle P(B,G) is found; this construction includes the construction of associated vector bundles. In §3 the Atiyah sequence of a principal bundle is constructed, and its full structure as a transitive Lie algebroid is delineated. The case of homogeneous bundles G(G/H,H) is treated. In §4 the two definitions of a connection and the definition of the curvature of a connection are given, and their correspondence with the standard account is established. Two examples of working with the Atiyah sequence in specific problems are given.

§1. Principal bundles and fibre bundles.

This section is a brief resumé of standard material (see, for example, Kobayashi and Nomizu (1963)) together with a few definitions (TGB,LGB) which belong to the same circle of ideas.

__Definition 1.1.__ A __Cartan principal bundle__ is a quadruple $P(B,G,\pi)$ where P and B are spaces, G is a topological group acting freely on P to the right through $P \times G \to P$, $(u,g) \mapsto ug$, and $\pi: P \to B$ is a surjective map, subject to the following conditions:

(i) the fibres of π equal the orbits of G, that is, for $u,v \in P$, the statement $\pi(u) = \pi(v)$ is equivalent to $\exists g \in G: v = ug$;

(ii) the division map $\delta: P \underset{\pi}{\times} P \to G$, $(ug,u) \mapsto g$, resulting from (i), is continuous. Here $P \underset{\pi}{\times} P = \{(v,u) \in P \times P \mid \pi(v) = \pi(u)\}$ has the subspace topology;

(iii) $\pi: P \to B$ is an identification map. //

Note that π is automatically open, since $\pi^{-1}(\pi(U)) = \bigcup_{g\in G} Ug$ for any $U \subseteq P$.

For this more general concept, in which local triviality is not assumed, see Palais (1961a). (I am grateful to Iain Raeburn for this reference.) Its main virtue for us is that it includes all homogeneous spaces of topological groups by closed subgroups.

__Example 1.2.__ If G is a topological group and H is a closed subgroup, then $G(G/H,H,\natural)$ is a Cartan principal bundle, where H acts on G by right multiplication, and $\natural: G \to G/H$ is $g \mapsto gH$. We call it a __homogeneous bundle__. //

__Definition 1.3.__ A __morphism__ of Cartan principal bundles, from $P(B,G,\pi)$ to $P'(B',G',\pi')$, is a trio of maps $F: P \to P'$, $f: B \to B'$, $\phi: G \to G'$ where F and f are continuous and ϕ is a continuous morphism, such that $\pi' \circ f = f \circ \pi$ and $F(ug) = F(u)\phi(g)$ for $u \in P$, $g \in G$. If $B = B'$ and $f = id_{B}$, we say that $F: P \to P'$ and $\phi: G \to G'$ constitute a morphism over B. //

Definition 1.4. Let $P(B,G,\pi)$ be a Cartan principal bundle. It is locally trivial if there is an open cover $\{U_i\}$ of B and continuous maps $\sigma_i: U_i \to P$ such that $\pi \circ \sigma_i = id_{U_i}$. //

The σ_i are called local sections of π, or of $P(B,G,\pi)$. Each σ_i induces an isomorphism from $U_i \times G(U_i,G)$ to $\pi^{-1}(U_i)(U_i,G)$, namely $\bar{\sigma}_i(id,id)$ where $\bar{\sigma}_i(x,g) = \sigma_i(x)g$. We will call the collection $\{\sigma_i : U_i \to P\}$ a section-atlas for $P(B,G,\pi)$.

The maps $s_{ij}: U_{ij} \to G$ defined by $\sigma_i(x)s_{ij}(x) = \sigma_j(x)$ or, equivalently, by $s_{ij}(x) = \delta(\sigma_j(x),\sigma_i(x))$, are the transition functions for $P(B,G)$ corresponding to the section-atlas $\{\sigma_i\}$.

Definition 1.5. A principal bundle is a quadruple $P(B,G,\pi)$ where P and B are spaces, G is a topological group acting effectively on P to the right through $P \times G \to P$, $(u,g) \mapsto ug$, and $\pi: P \to B$ is a surjective map, subject to the following conditions:

(i) the fibres of π equal the orbits of G,

(ii) there is an open cover $\{U_i\}$ of B and continuous maps $\sigma_i: U_i \to P$ such that $\pi \circ \sigma_i = id_{U_i}$. //

It is easily verified that a principal bundle is a Cartan principal bundle. A morphism of principal bundles is defined as in 1.3.

Definition 1.6. A fibre bundle is a triple (M,p,B) in which M and B are spaces and $p: M \to B$ is a continuous surjection with the property that there is a space F, called the fibre type of M, and an open cover $\{U_i\}$ of B together with homeomorphisms $\psi_i: U_i \times F \to p^{-1}(U_i)$ such that $\pi(\psi_i(x,a)) = x$, for $x \in U_i$, $a \in F$, and such that the maps $U_{ij} \to \text{Homeom}(F)$ defined by

$$x \dashrightarrow \psi_{i,x}^{-1} \circ \psi_{j,x}$$

are continuous. Here Homeom(F) is the group of homeomorphisms $F \to F$ with the g-topology of Arens (1946), and $\psi_{i,x}$ denotes the restriction of ψ_i to $F \to \{x\} \times F \to M_x = p^{-1}(x)$.

A <u>morphism of fibre bundles</u> from (M,p,B) to (M',p',B') is a pair of continuous maps $F: M \to M'$, $f: B \to B'$ such that $p' \circ F = f \circ p$. If $B = B'$ and $f = id_B$, we say that F is a morphism of fibre bundles <u>over</u> B. //

The continuity condition on the maps $U_{ij} \to \mathrm{Homeom}(F)$ is necessary because in general it is not true that the g-topology of Arens coincides with the compact-open topology. However if F is locally compact, Hausdorff and locally connected, then the two topologies coincide and the continuity condition may be dropped. This suffices for the application in the text (II 1.13).

<u>Proposition 1.7.</u> Let $P(B,G,\pi)$ be a principal bundle and let $G \times F \to F$ be an action of G on a space F which is locally compact, Hausdorff and locally connected. Define a left action on G on the product space $P \times F$ by $g(u,a) = (ug^{-1},ga)$ and let M be the orbit space $\frac{P \times F}{G}$. Denote the orbit of $(u,a) \in P \times F$ by $\langle u,a\rangle$, and define a map $p: M \to B$ by $p(\langle u,a\rangle) = \pi(u)$. Then (M,p,B) is a fibre bundle.

<u>Proof:</u> Charts for (M,p,B) are given by $\psi_i(x,a) = \langle\sigma_i(x),a\rangle$. //

(M,p,B) is called the <u>associated fibre bundle</u> for $P(B,G)$ and the action $G \times F \to F$. The standard notation for $M = \frac{P \times F}{G}$ is $E(B,F,G,P)$.

<u>Proposition 1.8.</u> Let $P(B,G,\pi)$ be a principal bundle, and let $G \times F \to F$ and $G \times F' \to F'$ be actions of G on two locally compact, Hausdorff and locally connected spaces F and F'. Let $f: F \to F'$ be a G-equivariant map; then

$$\tilde{f}: \frac{P \times F}{G} \to \frac{P \times F'}{G}, \qquad \langle u,a\rangle \mapsto \langle u,f(a)\rangle$$

is a well-defined morphism of fibre bundles over B.

<u>Proof:</u> Easy. //

Not all morphisms of associated fibre bundles are of this form. A criterion for them is given in II§4.

<u>Proposition 1.9.</u> Let $P(B,G,\pi)$ be a principal bundle and let $M = \frac{P \times F}{G}$ be an associated fibre bundle. Then if $\phi: P \to F$ is G-equivariant in the sense that $\phi(ug) = g^{-1}\phi(u)$, for $u \in P$, $g \in G$, the formula

$$\mu(x) = \langle u,\phi(u)\rangle \quad \text{where} \quad \pi(u) = x,$$

defines a (global) section of M. Every section of M is of this form.

Proof: The first statement is easy to verify. Conversely, given a section $\mu: B \to M$, each $u \in P$ determines an element $\phi(u)$ of F by the condition

$$\mu(x) = \langle u, \phi(u)\rangle \quad \text{where} \quad x = \pi(u).$$

It is easy to verify that ϕ is equivariant. //

Let $C(P,F)^G$ denote the set of G-equivariant maps $P \to F$. Then 1.9 establishes a bijective correspondence $\Gamma(\frac{P \times F}{G}) \to C(P,F)^G$ which we will usually denote by $\mu \mapsto \tilde{\mu}$.

Definition 1.10. Let $P(B,G,\pi)$ be a principal bundle. A reduction of $P(B,G)$ is a principal bundle $P'(B,G',\pi')$, on the same base B, together with a morphism $F(id_B,\phi): P'(B,G') \to P(B,G)$ for which $\phi: G' \to G$ and $F: P' \to P$ are injections. //

The concept of reduction is the concept of subobject appropriate to the study of principal bundles. One may accordingly define a notion of equivalence for reductions, similar to that for submanifolds or Lie subgroups.

In the situation of 1.10 one also says that $P'(B,G')$ is a reduction of $P(B,G)$ to G'.

Proposition 1.11. Let $P(B,G,\pi)$ be a principal bundle and let $\phi: G \to H$ be a morphism of topological groups. Let $Q = \frac{P \times H}{G}$ be the associated fibre bundle with respect to the action $G \times H \to H$, $(g,h) \mapsto \phi(g)h$. Then $Q(B,H,p)$ is a principal bundle with respect to the action $Q \times H \to Q$, $(\langle u,h\rangle,h') \mapsto \langle u,hh'\rangle$ and projection $p(\langle u,h\rangle) = \pi(u)$.

Proof: Straightforward. //

The principal bundle $Q = \frac{P \times H}{G}(B,H)$ is usually called the prolongation or the extension of $P(B,G)$ along ϕ. Both these words have alternative meanings within bundle and groupoid theory, and we propose to call $Q(B,H)$ the produced principal bundle, or the production of $P(B,G)$ along ϕ. This term is suitably antithetical to "reduced", and may also remind the reader of the process in elementary geometry where one continues a line in an already existing direction, without adding anything which is not already implicit.

If F in 1.7 is a vector space V and the action of G on V is linear, then the associated fibre bundle $\frac{P \times V}{G}$ is a vector bundle, in an obvious way, and is usually denoted E.

Definition 1.12. A topological group bundle, or TGB, is a fibre bundle (M,p,B) in which each fibre $M_x = p^{-1}(x)$, and the fibre type F, has a topological group structure, and for which there is an atlas $\{\psi_i : U_i \times F \to M_{U_i}\}$ such that each $\psi_{i,x}: F \to M_x$, $x \in U_i$, is an isomorphism of topological groups.

A morphism of TGB's from (M,p,B) to (M',p',B') is a morphism (F,f) of fibre bundles such that each $F_x: M_x \to M'_{f(x)}$ is a morphism of topological groups. //

If the space F in 1.7 is a topological group H and G acts on H through topological group automorphisms then the associated fibre bundle is a TGB. For example, given any principal bundle P(B,G), the group G acts on itself by inner-automorphisms; the resulting TGB $\frac{P \times G}{G}$ is sometimes called the gauge bundle in the physics literature. We will call it the inner group bundle or inner TGB.

The preceding results are all valid in the case of C^∞ differentiable manifolds. There are some simplifications: firstly, there is no analogue of "Cartan principal bundle" since the analogue of the topological concept of identification map is the concept of submersion, and a submersion automatically has local right-inverses. Secondly, there is no need in 1.6 for a separate condition on the maps $U_{ij} \to \text{Homeom}(F)$, since all manifolds are locally compact, locally connected and Hausdorff. The analogue of a TGB is of course called a Lie group bundle and the name is abbreviated to LGB.

§2. Quotients of vector bundles by group actions.

Throughout this section $P(B,G,\pi)$ is a given principal bundle.

Proposition 2.1. Let (E,p^E,P) be a vector bundle over P, on which G acts to the right

$$E \times G \to E, \qquad (\xi,g) \mapsto \xi g$$

with the following two properties:

(i) G acts on E by vector bundle isomorphisms; that is, each map $\xi \mapsto \xi g$, $E \to E$ is a vector bundle isomorphism over the right translation $R_g: P \to P$;

(ii) E is covered by the ranges of equivariant charts, that is, around each $u_0 \in P$ there is a π-saturated open set $\mathcal{U} = \pi^{-1}(U)$, where $U \subseteq B$ is open, and a vector bundle chart

$$\psi: \mathcal{U} \times V \to E_{\mathcal{U}}$$

for E, which is <u>equivariant</u> in the sense that

$$\psi(ug,v) = \psi(u,v)g \qquad \forall\, u \in \mathcal{U},\ v \in V,\ g \in G.$$

Then the orbit set E/G has a unique vector bundle structure over B such that the natural projection $\natural: E \to E/G$ is a surjective submersion, and a vector bundle morphism over $\pi: P \to B$. Further,

$$\begin{array}{ccc} E & \xrightarrow{\ \natural\ } & E/G \\ {\scriptstyle p^E}\downarrow & & \downarrow{\scriptstyle p^{E/G}} \\ P & \xrightarrow{\ \pi\ } & B \end{array}$$

is a pullback.

We call $(E/G, p^{E/G}, B)$ the <u>quotient vector bundle</u> of (E, p^E, P) by the action of G.

<u>Proof</u>. Denote the orbit of $\xi \in E$ by $\langle\xi\rangle$. Define $\bar{p} = p^{E/G}: E/G \to B$ by $\bar{p}(\langle\xi\rangle) = \pi(p^E(\xi))$; it is clear from (i) that $\bar{p}$ is well-defined. We will give $(E/G, \bar{p}, B)$ the structure of a vector bundle by constructing local charts for it, which will simultaneously give the manifold structure for E/G (see Greub et al, 1972, §2.5).

Firstly, make each $E/G|_x = \bar{p}^{-1}(x)$, $x \in B$, into a vector space: if $\langle\xi\rangle$, $\langle\eta\rangle \in E/G|_x$ then $p^E(\xi)$, $p^E(\eta)$ lie in the same fibre of P so $\exists!\ g \in G$ such that $p^E(\eta) = p^E(\xi)g$. Define

$$\langle\xi\rangle + \langle\eta\rangle = \langle\xi g + \eta\rangle$$

and, for $t \in \mathbb{R}$,

$$t\langle\xi\rangle = \langle t\xi\rangle.$$

It is easily verified that these operations are well-defined and make $E/G|_x$ a vector space. The restriction $\natural_u$ of $\natural$ to $E_u \to E/G|_x$ $(x = \pi(u))$ is clearly linear; it is in fact an isomorphism. For if $\xi, \eta \in E_u$ and $\langle\xi\rangle = \langle\eta\rangle$ then $\exists\, g \in G$: $\eta = \xi g$ and, by (i), it follows that $u = ug$. So $g = 1$ and $\eta = \xi$.

Given $x_o \in B$, choose $u_o \in \pi^{-1}(x_o)$ and let $\psi_1: \mathcal{U}_1 \times V \to E_{\mathcal{U}_1}$ be an equivariant chart for E defined around u_o. Assume, by shrinking $\mathcal{U}_1$ if necessary, that $\mathcal{U}_1$ is the range of a chart $U_i \times G \to \mathcal{U}_i$ for $P(B,G)$, so that there is a section $\sigma_i : U_i \to \mathcal{U}_i \subsetneq P$.

Define

$$\psi_i^{/G} : U_i \times V \to E/G|_{U_i}, \qquad (x,v) \mapsto \langle \psi_i(\sigma_i(x),v) \rangle .$$

Then $\psi_{i,x}^{/G} : V \to E/G|_x$ is the composite

$$V \xrightarrow{\psi_{i,\sigma_i(x)}} E_{\sigma_i(x)} \xrightarrow{\natural_{\sigma_i(x)}} E/G|_x$$

and is thus an isomorphism. If $\psi_j^{/G} : U_j \times V \to E/G|_{U_j}$ is another chart constructed in the same way from an equivariant chart ψ_j for E and a section σ_j of P, then it is easily seen that

$$(\psi_{i,x}^{/G})^{-1} \circ (\psi_{j,x}^{/G}) = \psi_{i,u}^{-1} \circ \psi_{j,u} \qquad \text{where} \quad u = \sigma_j(x).$$

Thus the charts $\{\psi_i^{/G}\}$ define smooth transition functions $x \mapsto (\psi_{i,x}^{/G})^{-1} \circ (\psi_{j,x}^{/G})$ and so, by the reference quoted above, there is a unique manifold structure on E/G which makes $(E/G, \bar{p}, B)$ a vector bundle with the $\psi_i^{/G}$ as charts.

We next prove that $\natural : E \to E/G$ is a surjective submersion. Let $\psi : \mathcal{U} \times V \to E$ be an equivariant chart for E, let $\sigma : U \to \mathcal{U}$ be a section of P, and $\psi^{/G} : U \times V \to E/G|_U$ the chart for E/G constructed as above. Then

$$\begin{array}{ccc} E_{\mathcal{U}} & \xleftarrow{\ \psi\ } & \mathcal{U} \times V \\ \natural_{\mathcal{U}} \downarrow & & \downarrow \pi \times \mathrm{id}_V \\ E/G|_U & \xleftarrow{\ \psi^{/G}\ } & U \times V \end{array}$$

commutes, for if $(u,v) \in \mathcal{U} \times V$, then $\psi^{/G}((\pi \times \mathrm{id}_V)(u,v)) = \langle \psi(\sigma(\pi(u)),v) \rangle =$ $\langle \psi(ug,v) \rangle$ (for $\exists\, g \in G$: $ug = \sigma(\pi(u))$) $= \langle \psi(u,v)g \rangle = \langle \psi(u,v) \rangle = \natural(\psi(u,v))$. Since $\pi \times \mathrm{id}_V$ is smooth, and a submersion, it follows that $\natural$ is smooth, and a submersion. It is clear that $\bar{p} \circ \natural = \pi \circ p^E$, so $\natural$ is a vector bundle morphism. Since $\natural_u$ is an isomorphism, $u \in P$, it follows that $(\natural, \pi)$ is a pullback (see C.2).

The uniqueness assertion follows from the facts that there is at most one manifold structure on the range of a surjection which makes it a submersion (e.g., Greub et al, 1972, §3.9), and at most one vector space structure on the range of a surjection which makes it linear. //

Proposition 2.2. (i) If (E,p^E,P) is a vector bundle over $P(B,G)$ together with an action of G on E which satisfies the conditions of 2.1, and $(E',p^{E'},M)$ is any vector bundle, then given a vector bundle morphism $\phi\colon E \to E'$ over a map $f\colon P \to M$ such that $\phi(\xi g) = \phi(\xi)$ $\forall\ \xi \in E$, $g \in G$, and $f(ug) = f(u)$, $\forall\ u \in P$, $g \in G$, there is a unique vector bundle morphism

$$\begin{array}{ccc} E/G & \xrightarrow{\ \phi^{/G}\ } & E' \\ {\scriptstyle p^{E/G}}\downarrow & & \downarrow{\scriptstyle p^{E'}} \\ B & \xrightarrow{\ f^{/G}\ } & M \end{array}$$

such that $\phi = \phi^{/G}\circ\natural$ and $f = f^{/G}\circ\pi$.

(ii) Consider vector bundles (E,p^E,P) and $(E',p^{E'},P')$ over principal bundles $P(B,G)$ and $P'(B',G')$, respectively, together with actions of G and G' on E and E', respectively, which satisfy the conditions of 2.1. If $\Phi\colon E \to E'$ is a vector bundle morphism over a principal bundle morphism $F(f,\phi)\colon P(B,G) \to P'(B',G')$ which is equivariant in the sense that $\Phi(\xi g) = \Phi(\xi)\phi(g)$, $\forall\ \xi \in E$, $g \in G$, then there is a unique morphism of vector bundles

$$\begin{array}{ccc} E/G & \xrightarrow{\ \Phi^{/G}\ } & E'/G' \\ {\scriptstyle p^{E/G}}\downarrow & & \downarrow{\scriptstyle p^{E'/G'}} \\ B & \xrightarrow{\ f\ } & B' \end{array}$$

such that $\Phi^{/G}\circ\natural = \natural'\circ\Phi$.

Proof: We prove (ii) only; (i) is a special case of (ii).

Define $\Phi^{/G}\colon E/G \to E'/G'$ by $\langle\xi\rangle \mapsto \langle\Phi(\xi)\rangle$. That $\Phi^{/G}$ is well-defined and fibrewise linear is clear. Clearly $\Phi^{/G}\circ\natural = \natural'\circ\Phi$ so since $\natural'\circ\Phi$ is smooth, and $\natural$ is a surjective submersion, $\Phi^{/G}$ is smooth. That $p^{E'/G'}\circ\Phi^{/G} = f\circ p^{E/G}$ is immediate, and since $\natural$ is onto, the condition $\Phi^{/G}\circ\natural = \natural'\circ\Phi$ determines $\Phi^{/G}$ uniquely. //

Remarks 2.3. (i) This quotienting process includes the construction of an associated vector bundle $\frac{P \times V}{G}$ for a representation $\rho\colon G \to GL(V)$ of G on a vector space V. Namely, let G act on the product bundle $E = P \times V$ over P by $(u,v)g = (ug, \rho(g^{-1})v)$; this action clearly satisfies (i) of 2.1. If $\sigma: U \to P$ is a section of P and $\psi: U \times G \to \mathcal{U} = \pi^{-1}(U)$ is the associated chart $\psi(x,g) = \sigma(x)g$, then it is easy to verify that the chart $\mathcal{U} \times V \to E_{\mathcal{U}}$ defined by

$$\begin{array}{ccc} \mathcal{U} \times V & \longrightarrow & E_{\mathcal{U}} = \mathcal{U} \times V \\ {\scriptstyle \psi \times id_V} \big| & & \big| {\scriptstyle \psi \times id_V} \\ U \times G \times V & \xrightarrow[(x,g,v) \mapsto (x,g,\rho(g^{-1})v)]{} & U \times G \times V \end{array}$$

is equivariant. (Although $E \to P$ is a trivial bundle, it does not admit global equivariant charts in general.)

Clearly the quotient vector bundle $E/G \to B$ coincides with the associated vector bundle $\frac{P \times V}{G} \to B$ of 1.7.

(ii) The construction 2.1 and the universality property 2.2 can easily be extended to equivariant actions of G on general fibre bundles over P. If this is done, then the construction includes all associated fibre bundles $\frac{P \times F}{G} \to B$. //

<u>Proposition 2.4</u>. Let (E,p,P) be a vector bundle over P together with an action of G on E which satisfies the conditions of 2.1. Denote by $\Gamma^G E$ the set of (global) sections X of E which are <u>invariant</u> in the sense that

$$X(ug) = X(u)g \qquad \forall\, u \in P, \quad g \in G.$$

Then $\Gamma^G E$ is a C(B)-module where $fX = (f \circ \pi)X$, $\forall\, f \in C(B)$, $X \in \Gamma^G E$, and the map

$$X \mapsto \bar{X}, \qquad \Gamma(E/G) \to \Gamma^G E$$

where
$$\bar{X}(u) = (\natural_u)^{-1}(X(\pi u))$$

is an isomorphism of C(B)-modules with inverse

$$X \mapsto \underline{X}, \qquad \Gamma^G E \to \Gamma(E/G)$$

where
$$\underline{X}(x) = \natural_u(X(u)) = \langle X(u) \rangle \qquad (\text{any } u \in \pi^{-1}(x)).$$

<u>Proof</u>: $X \mapsto \bar{X}$ is the C(B)-morphism $\natural^{\#}: \Gamma(E/G) \to \Gamma E$ of C.3. It is easily checked that, in fact, $\bar{X} \in \Gamma^G E$. Given $X \in \Gamma^G E$ it is clear that $\underline{X}$ is well-defined, and since $\underline{X} \circ \pi = \natural \circ X$ and π is a surjective submersion, $\underline{X}$ is smooth. It is straightforward to check that $X \mapsto \bar{X}$ and $X \mapsto \underline{X}$ are mutual inverses. //

This result can of course be localized: if $\mathcal{U} = \pi^{-1}(U)$ is a saturated open subset of P, where $U \subseteq B$ is open, then the same formulas for $\bar{X}$, $\underline{X}$ define mutually inverse C(U)-isomorphisms between $\Gamma_U(E/G)$ and $\Gamma^G(E)$.

§3. The Atiyah sequence of a principal bundle.

Throughout this section, P(B,G) is a given principal bundle.

Proposition 3.1. (i) The action of G on the tangent bundle $TP \to P$ induced by the action of G on P, namely

$$Xg = T(R_g)_u(X), \qquad X \in T(P)_u,$$

satisfies the conditions of 2.1.

(ii) The action of G on $TP \to P$ restricts to an action of G on the vertical subbundle $T^{\pi}P \to P$, and this action also satisfies the condtions of 2.1.

Proof: (i) It is clear that G acts on TP by vector bundle isomorphisms. To construct equivariant charts for TP, let $\phi\colon U \times G \to \mathcal{U} = \pi^{-1}(U)$ be a chart for P(B,G) in which U is the range of a chart $\theta\colon \mathbb{R}^n \to U$ for the manifold B. Now $T(U) \cong U \times \mathbb{R}^n$ and $T(G) \cong G \times \mathfrak{g}$, so $T(\phi)\colon T(U) \times T(G) \to T(P)_{\mathcal{U}}$ can be regarded as a map $(U \times G) \times (\mathbb{R}^n \times \mathfrak{g}) \to T(P)_{\mathcal{U}}$ and, identifying $U \times G$ with $\mathcal{U}$ by ϕ, this gives the required equivariant chart.

Precisely, define

$$(U \times G) \times (\mathbb{R}^n \times \mathfrak{g}) \to T(P)_{\mathcal{U}}$$

$$(x,g,\underset{\sim}{t},X) \mapsto T(\phi)_{(x,g)}\big(T(\theta)_{\theta^{-1}(x)}(\underset{\sim}{t}),\ T(R_g)_1(X)\big)$$

and define $\psi\colon \mathcal{U} \times V \to T(P)_{\mathcal{U}}$ (where $V = \mathbb{R}^n \times \mathfrak{g}$) as the composition of $(\phi^{-1} \times \mathrm{id}_V)$ and this map. To show that $\psi(uh,v) = \psi(u,v)h$, it suffices to show that

$$T(\phi)_{(x,gh)}\big(Y,\ T(R_{gh})_1(X)\big) = T(R_h)_{\phi(x,g)}\big(T(\phi)_{(x,g)}(Y,\ T(R_g)_1(X))\big),$$

where $Y \in T(U)_x$, and this is the derivative of the identity $\phi(x,R_h(g)) = R_h(\phi(x,g))$.

(ii) That $T(R_g)\colon TP \to TP$ sends $T^{\pi}P$ to $T^{\pi}P$ follows from $\pi \circ R_g = \pi$. In the notation above, an equivariant chart for $T^{\pi}P$ over $\mathcal{U} \cong U \times G$ is the composite of $\phi^{-1} \times \mathrm{id}_{\mathfrak{g}}$ with

$$(U \times G) \times \mathfrak{g} \to T^{\pi}(P)_{\mathcal{U}}$$

$$(x,g,X) \mapsto T(\phi)_{(x,g)}\big(0_x,\ T(R_g)_1(X)\big). \qquad //$$

Note that in both cases the identification of T(G) with $G \times \mathfrak{g}$ must be made using right translations.

The inclusion map $T^{\pi}P \subseteq TP$ is manifestly equivariant so, by 2.2(ii), it induces a morphism $\frac{T^{\pi}P}{G} \to \frac{TP}{G}$ of vector bundles over B, which is clearly an injection and which we also regard as an inclusion.

On the other hand, from $\pi \circ R_g = \pi$ it follows that $T(\pi)_{ug} \circ T(R_g)_u = T(\pi)_u$ where $u \in P$, $g \in G$, so by 2.2(i) it follows that the vector bundle morphism

$$\begin{array}{ccc} TP & \xrightarrow{T(\pi)} & TB \\ \downarrow & & \downarrow \\ P & \xrightarrow{\pi} & B \end{array}$$

quotients to a map $\pi_* = T(\pi)^{/G}\colon \frac{TP}{G} \to TB$ which is a vector bundle morphism over B. It is clear that π_*, like $T(\pi)$, is fibrewise surjective and so general vector bundle theory (e.g., Greub et al, 1972, 2.23) shows that π_* is a surjective submersion. Alternatively, it is easy to see that π_* is given locally by

$$\begin{array}{ccc} \frac{TP}{G}\Big|_U & \xrightarrow{\pi_*} & T(B)_U \\ {\scriptstyle \psi^{/G}}\uparrow & & \uparrow{\scriptstyle T(\theta)} \\ U \times (\mathbb{R}^n \times \mathfrak{g}) & \xrightarrow[(x,\underset{\sim}{t},X) \mapsto (x,\underset{\sim}{t})]{} & U \times \mathbb{R}^n \end{array}$$

where the notation is that of 3.1.

The kernel of $\pi_*\colon \frac{TP}{G} \twoheadrightarrow TB$ is clearly $\frac{T^{\pi}P}{G}$, since $T^{\pi}P \to P$ is the kernel of $T(\pi)\colon TP \to TB$, and so we have proved that $\frac{T^{\pi}P}{G} \rightarrowtail \frac{TP}{G} \overset{\pi_*}{\twoheadrightarrow} TB$ is an exact sequence of vector bundles over B.

This may be regarded as the Atiyah sequence of P(B,G) but in practice it is generally easier to work with a slight reformulation in which $\frac{T^{\pi}P}{G}$ is replaced by the bundle $\frac{P \times \mathfrak{g}}{G} \to B$ associated to P(B,G) by the adjoint action of G on $\mathfrak{g}$.

Proposition 3.2. The map $j: \frac{P \times \mathfrak{g}}{G} \rightarrowtail \frac{TP}{G}$ induced by

$$P \times \mathfrak{g} \to TP, \qquad (u,X) \mapsto T(m_u)_1(X)$$

(where $m_u: G \to P$, $g \mapsto ug$) is a vector bundle isomorphism over B of $\frac{P \times \mathfrak{g}}{G}$ onto $\frac{T^{\pi}P}{G} \subseteq \frac{TP}{G}$.

Proof: We regard $\frac{P \times \mathfrak{g}}{G} \to B$ as the quotient of the trivial bundle $P \times \mathfrak{g} \to P$ over the action $(u,X)g = (ug, \mathrm{Ad}g^{-1}X)$ (see 2.3(i)). That the map $P \times \mathfrak{g} \to TP$ is smooth can be seen by reformulating $T(m_u)_1(X)$ as $T(m)_{(u,1)}(0_u, X_1)$, where $m: P \times G \to P$ is the action. Thus $P \times \mathfrak{g} \to TP$ is the composite

$$P \times \mathfrak{g} \to TP \times TG \xrightarrow{T(m)} TP$$

where $P \to TP$ is the zero section and $\mathfrak{g} = T(G)_1 \to T(G)$ the inclusion. It is clearly a vector bundle morphism over P.

Now $T(m_{ug})_1(\mathrm{Ad}g^{-1}X) = T(m_{ug} \circ I_{g^{-1}})_1(X)$ and it is easy to check that $m_{ug} \circ I_{g^{-1}} = R_g \circ m_u$. Thus $(ug, \mathrm{Ad}g^{-1}X)$ is mapped to $T(R_g)_u(T(m_u)_1(X))$, which proves that $P \times \mathfrak{g} \to TP$ is G-equivariant and so quotients, by 2.2(ii), to a vector bundle morphism over B

$$\frac{P \times \mathfrak{g}}{G} \to \frac{TP}{G}, \qquad \langle u,X \rangle \mapsto \langle T(m_u)_1(X) \rangle$$

which we denote by j.

That $T(m_u)_1(X) \in T^{\pi}P$, $\forall\, u \in P$, $X \in \mathfrak{g}$ follows from the fact that $\pi \circ m_u$ is constant. On the other hand, $P \times \mathfrak{g} \to T^{\pi}P \subseteq TP$ is clearly injective (because each m_u is so) and since $T^{\pi}P$ and $P \times \mathfrak{g}$ have the same rank, namely $\dim \mathfrak{g}$, it follows that $P \times \mathfrak{g} \to T^{\pi}P$ is a fibrewise isomorphism. Clearly j inherits this property. //

The map $P \times \mathfrak{g} \to TP$ is of course the "fundamental vector field" map $(u,X) \mapsto X^*(u)$ of Kobayashi and Nomizu (1963, p. 51). We shall occasionally use their notation, in which case $j(\langle u,X \rangle) = \langle X^*(u) \rangle$.

Summarizing, we have proved

Proposition 3.3. $\frac{P \times \mathfrak{g}}{G} \overset{j}{\rightarrowtail} \frac{TP}{G} \overset{\pi_*}{\twoheadrightarrow} TB$ is an exact sequence of vector bundles over B. //

The bundle $\frac{P \times \mathfrak{g}}{G} \to B$ is called the adjoint bundle of P(B,G).

Remark 3.4. The reader may like to check that $T^{\pi}P \overset{\subseteq}{\rightarrowtail} TP \overset{\tilde{\pi}}{\twoheadrightarrow} \pi^*TB$ is an exact sequence of vector bundles over P, where π^*TB is the pullback and $\tilde{\pi}$ is the map $X_u \mapsto (u, T(\pi)_u(X_u))$; that π^*TB admits a natural G-action, satisfying the conditions of 2.1, that $\tilde{\pi}$ is equivariant and that $\frac{\pi^*TB}{G} \cong TB$ and that the map $\frac{TP}{G} \twoheadrightarrow TB$ induced by $\tilde{\pi}$ is π_*. //

We proceed now to define a bracket of Lie algebra type on $\Gamma(\frac{TP}{G})$; with this additional structure the exact sequence of 3.3 will be the Atiyah sequence of P(B,G).

From 2.4 we know that $\Gamma(\frac{TP}{G})$ is isomorphic as a C(B)-module to $\Gamma^G TP$. Now $X \in \Gamma TP$ is in $\Gamma^G TP$ precisely if X is R_g-related to itself for all $g \in G$. It therefore follows that $\Gamma^G TP$ is closed under the bracket of vector fields and so we can define a bracket on $\Gamma(\frac{TP}{G})$ by

$$\overline{[X,Y]} = [\bar{X},\bar{Y}], \qquad X,Y \in \Gamma(\tfrac{TP}{G}).$$

The bracket on $\Gamma(\frac{TP}{G})$ inherits the Jacobi identity from the bracket on ΓTP, and also the property of being alternating. For $f \in C(B)$,

$$\begin{aligned}\overline{[X,fY]} &= [\bar{X},(f\circ\pi)\bar{Y}] \\ &= (f\circ\pi)[\bar{X},\bar{Y}] + \bar{X}(f\circ\pi)\bar{Y}.\end{aligned}$$

Recall that a vector field $\mathcal{X}$ on P is called π-projectable if there is a vector field $\mathcal{Y}$ on B such that $\mathcal{X}$ is π-related to $\mathcal{Y}$, that is, such that $T(\pi)_u(\mathcal{X}(u)) = \mathcal{Y}(\pi(u))$, $\forall\, u \in P$, or, equivalently, such that $\mathcal{X}(f\circ\pi) = \mathcal{Y}(f)\circ\pi$, $\forall\, f \in C(B)$. (See, e.g., Greub et al, 1972, 3.13.) It is clear from the definition of π_* that $\bar{X} \in \Gamma^G TP$ is π-related to $\pi_*(X) \in \Gamma TB$ and so

$$\bar{X}(f\circ\pi) = \pi_*(X)(f)\circ\pi.$$

We therefore have,

$$\overline{[X,fY]} = (f\circ\pi)\overline{[X,Y]} + (\pi_*(X)(f)\circ\pi)\bar{Y}$$

so

$$(1) \qquad [X,fY] = f[X,Y] + \pi_*(X)(f)Y, \qquad X,Y \in \Gamma(\tfrac{TP}{G}), \qquad f \in C(B).$$

A bracket on the module of global sections of any vector bundle A over B, which has the property (1) with respect to a morphism $\pi_*: A \to TB$, can be "localized" to sections over any open subset of the base. (See III 2.2.) In the present case the resulting bracket $\Gamma_U(\frac{TP}{G}) \times \Gamma_U(\frac{TP}{G}) \to \Gamma_U(\frac{TP}{G})$ ($U \subseteq B$ open) is easily seen to be equal to that obtained by transporting the bracket on $\Gamma^G_{\pi^{-1}(U)} TP$ to $\Gamma_U(\frac{TP}{G})$ via the $C(U)$-isomorphism $\Gamma_U(\frac{TP}{G}) \to \Gamma^G_{\pi^{-1}(U)} TP$.

From the fact that $\bar{X}$ is π-related to $\pi_*(X)$ it also follows that, for $X,Y \in \Gamma(\frac{TP}{G})$, $[\bar{X},\bar{Y}]$ is π-related to $[\pi_*(X),\pi_*(Y)]$ so since $[\bar{X},\bar{Y}] = \overline{[X,Y]}$ is also π-related to $\pi_*([X,Y])$, and π is onto, it follows that

$$(2) \qquad [\pi_*(X),\pi_*(Y)] = \pi_*([X,Y]), \qquad X,Y \in \Gamma(\frac{TP}{G}).$$

From (2) it follows that the bracket on $\Gamma(\frac{TP}{G})$ restricts to a bracket on $\Gamma(\frac{P\times\mathfrak{g}}{G})$. For, given $V,W \in \Gamma(\frac{P\times\mathfrak{g}}{G})$, $\pi_*([jV,jW]) = [\pi_* jV,\pi_* jW] = [0,0] = 0$ and so there exists a unique section $[V,W]$ of $\frac{P\times\mathfrak{g}}{G}$ such that

$$(3) \qquad [j(V),j(W)] = j([V,W]), \qquad V,W \in \Gamma(\frac{P\times\mathfrak{g}}{G}).$$

This restricted bracket is of course also alternating and satisfies the Jacobi identity. And, for $f \in C(B)$,

$$j([V,fW]) = f[j(V),j(W)] + \pi_*(j(V))(f)j(W)$$

so, since $\pi_*\circ j = 0$, it follows that

$$(4) \qquad [V,fW] = f[V,W], \qquad f \in C(B), \quad V,W \in \Gamma(\frac{P\times\mathfrak{g}}{G}).$$

Thus the bracket on $\Gamma(\frac{P\times\mathfrak{g}}{G})$, unlike that on $\Gamma(\frac{TP}{G})$, is actually a tensor field, and therefore restricts to each fibre. Since each fibre $\frac{P\times\mathfrak{g}}{G}\Big|_x$ is isomorphic to $\mathfrak{g}$, the question arises as to whether the bracket in $\frac{P\times\mathfrak{g}}{G}\Big|_x$ is induced by that in $\mathfrak{g}$.

<u>Proposition 3.5</u>. For $V \in \Gamma(\frac{P\times\mathfrak{g}}{G})$, denote by $\tilde{V} \in C_G(P,\mathfrak{g})$ the corresponding equivariant function $P \to \mathfrak{g}$. (See 1.9.) Then

$$\widetilde{[V,W]}(u) = [\tilde{V}(u),\tilde{W}(u)]_R, \qquad V,W \in \Gamma(\frac{P\times\mathfrak{g}}{G}), \quad u \in P$$

where the bracket on the RHS is the right-hand bracket in $\mathfrak{g}$.

<u>Remark</u>: This result may be expressed as follows: Equation (4) implies that, for $V,W \in \Gamma(\frac{P\times\mathfrak{g}}{G})$ and $x \in B$, $[V,W](x) = [V(x),W(x)]_x$ where $[\ ,\ ]_x$ is the restriction of $[\ ,\ ]$ to $\frac{P\times\mathfrak{g}}{G}\Big|_x$. Proposition 3.5 now states that

$$[\langle u,X\rangle,\langle u,Y\rangle]_x = \langle u,[X,Y]_R\rangle$$

for $u \in \pi^{-1}(x)$, $X,Y \in \mathfrak{g}$. That this bracket is well-defined follows from the fact that Adg is a Lie algebra automorphism for all $g \in G$.

<u>Proof</u>: First note that $\overline{j(V)} \in \Gamma^G TP$ is

$$u \mapsto T(m_u)_1(\tilde{V}(u))$$

and that this vector field has a global flow, namely

$$\phi_t(u) = u \exp t\tilde{V}(u), \qquad t \in \mathbf{R}, \quad u \in P.$$

(Proof: That $\frac{d}{dt}\phi_t(u)\big|_0 = T(m_u)_1(\tilde{V}(u))$ is immediate; for the group law $\phi_t \circ \phi_s = \phi_{t+s}$ we have

$$\begin{aligned}\phi_t(\phi_s(u)) &= u \exp s\tilde{V}(u) \,.\, \exp t\tilde{V}(u \exp s\tilde{V}(u)) \\ &= u \exp s\tilde{V}(u) \,.\, \exp t\mathrm{Ad}(\exp s\tilde{V}(u)^{-1})\tilde{V}(u) \\ &= u \exp t\tilde{V}(u) \,.\, \exp s\tilde{V}(u) \\ &= \phi_{t+s}(u) \,.)\end{aligned}$$

So

$$\begin{aligned}[\overline{jV},\overline{jW}](u) &= -\frac{d}{dt}\left(T(\phi_t)(\overline{jW}(\phi_{-t}(u))\right)\Big|_0 \\ &= -\frac{d}{dt}\left(T(\phi_t \circ m_{\phi_{-t}(u)})_1(\tilde{W}(\phi_{-t}(u))\right)\Big|_0 \,.\end{aligned}$$

Now by using the equivariance of $\tilde{V}$ in a similar manner to the proof of the group law, it can be shown that $\phi_t \circ m_{\phi_{-t}(u)} = m_u$ and so this last expression is actually

$$\begin{aligned}-\frac{d}{dt}\left(T(m_u)_1(\tilde{W}(\phi_{-t}(u))\right)\Big|_0 &= T(m_u)_1\left(-\frac{d}{dt}\tilde{W}(u \exp -t\tilde{V}(u))\Big|_0\right) \\ &= T(m_u)_1\left(-\frac{d}{dt}\mathrm{Ad}(\exp t\tilde{V}(u))\tilde{W}(u)\Big|_0\right) \\ &= T(m_u)_1([\tilde{V}(u),\tilde{W}(u)]_R).\end{aligned}$$

We therefore have

$$\begin{aligned}T(m_u)_1([\tilde{V}(u),\tilde{W}(u)]_R) &= [\overline{jV},\overline{jW}](u) \\ &= \overline{j([V,W])}(u) \\ &= T(m_u)_1(\widetilde{[V,W]}(u))\end{aligned}$$

and the result follows. //

It may seem odd that the bracket on $\Gamma(\frac{P \times \mathfrak{g}}{G})$ should correspond to the right-hand bracket in $\mathfrak{g}$, especially since, for fundamental vector fields, $[A^*,B^*] = ([A,B]_L)^*$ (Kobayashi and Nomizu, 1963, I.4.1). However the fundamental vector field map $\mathfrak{g} \to \Gamma T^{\pi}P$ and $j\colon \Gamma(\frac{P \times \mathfrak{g}}{G}) \to \Gamma^G T^{\pi}P$ are not the same; a fundamental vector field A^* is invariant iff $A \in \mathfrak{g}$ is stable under AdG, that is, iff $\pi(u) \mapsto \langle u,A\rangle$ is a well-defined section of $\frac{P \times \mathfrak{g}}{G}$. 3.5 may seem more reasonable when it is recalled that $\Gamma(\frac{P \times \mathfrak{g}}{G})$ is embedded in ΓTP as the set of right-invariant vertical vector fields; it is only natural then that the bracket on $\Gamma(\frac{P \times \mathfrak{g}}{G})$ should be the right-hand one. This is particularly evident in the case of homogeneous bundles G(G/H,H) - see Example 3.9 below.

In the remainder of this Appendix we will use only this right-hand bracket on the Lie algebra of a Lie group and we now drop the subscript 'R'. For a brief summary of standard formulas, reformulated for this bracket, see B§1.

We can now make the

Definition 3.6. The exact sequence of vector bundles

$$\frac{P \times \mathfrak{g}}{G} \xrightarrow{j} \frac{TP}{G} \xrightarrow{\pi_*} TB,$$

together with the bracket structures on $\Gamma(\frac{TP}{G})$ and $\Gamma(\frac{P \times \mathfrak{g}}{G})$ defined above, is the Atiyah sequence of P(B,G). //

We will use the various properties of the brackets, which have been developed above, without comment in what follows. They reflect, of course, the fact that the Atiyah sequence is a Lie algebroid on B in the sense of III 2.1.

We will often need the following description of the flows of right-invariant vector fields.

Proposition 3.7. (i) Given $\bar{X} \in \Gamma^G TP$ and $u_0 \in P$ there is a local flow $\{\phi_t\}$ for $\bar{X}$ around u_0 defined on a π-saturated open set $\mathcal{U} = \pi^{-1}(U)$, $U \subseteq B$ open, for which $\phi_t(ug) = \phi_t(u)g$, $\forall\, u \in \mathcal{U}$, $g \in G$ and t.

(ii) Given $\bar{X} \in \Gamma^G TP$ with local flow $\{\phi_t\}$ as in (i), the vector field $\pi_*(X)$ on B has local flow ψ_t on U determined by $\psi_t \circ \pi = \pi \circ \phi_t$.

(iii) For $V \in \Gamma(\frac{P \times \mathfrak{g}}{G})$, the vector field $\overline{j(V)} \in \Gamma^G TP$ is complete and has the global flow $\phi_t(u) = u \exp t\tilde{V}(u)$.

Proof: (i) Let $\{\phi_t\}$ be a local flow for $\bar{X}$ defined on an open $\mathcal{O} \subseteq P$ around u_0; write $U = \pi(\mathcal{O})$ and $\mathcal{U} = \pi^{-1}(U)$. It is easy to verify that, for any given $g \in G$,

$\{R_g \circ \phi_t \circ R_{g^{-1}}\}$ is a local flow for $\bar{X}$ on $\mathcal{O}g$. By the uniqueness of local flows it follows that $\{\phi_t\}$ and $\{R_g \circ \phi_t \circ R_{g^{-1}}\}$ must coincide on $\mathcal{O} \cap \mathcal{O}g$. We can thus extend ϕ_t smoothly to the whole of $\mathcal{U} = \bigcup_{g \in G} \mathcal{O}g$ and a repetition of the argument now shows that $R_g \circ \phi_t \circ R_{g^{-1}} = \phi_t$ for all $g \in G$ and t.

(ii) is straightforward and (iii) was proved in the course of 3.5. //

The proof of (i) of course shows that any local flow for $\bar{X}$ which is defined on a π-saturated open set commutes with right-translations. We call such a flow a <u>saturated local flow</u>.

Lastly, we describe the morphism of Atiyah sequences induced by a morphism of principal bundles $F(id,\phi)\colon P(B,G) \to P'(B,G')$ over a fixed base B. It is easily checked that $TF\colon TP \to TP'$ satisfies the conditions of 2.2(ii) and so induces a morphism $F_* = TF^{/G}\colon \frac{TP}{G} \to \frac{TP'}{G'}$ of vector bundles over B. It is also straightforward to check that $F_*^+\colon \frac{P \times \mathfrak{g}}{G} \to \frac{P' \times \mathfrak{g}'}{G'}$, $\langle u,X\rangle \mapsto \langle F(u),\phi_*(X)\rangle$ is a well-defined morphism of vector bundles over B, and that

(5)
$$\begin{array}{ccccc}
\frac{P \times \mathfrak{g}}{G} & \overset{j}{\rightarrowtail} & \frac{TP}{G} & \overset{\pi_*}{\twoheadrightarrow} & TB \\
{\scriptstyle F_*^+}\downarrow & & {\scriptstyle F_*}\downarrow & & \| \\
\frac{P' \times \mathfrak{g}'}{G'} & \overset{j'}{\rightarrowtail} & \frac{TP'}{G'} & \overset{\pi'_*}{\twoheadrightarrow} & TB
\end{array}$$

commutes. Now $TF\colon TP \to TP'$ preserves the Poisson bracket in the sense that if $X \in \Gamma TP$ and $X' \in \Gamma TP'$ are F-related, and Y and Y' likewise, then $[X,Y]$ and $[X',Y']$ are F-related. Since, for $X \in \Gamma(\frac{TP}{G})$, $X' \in \Gamma(\frac{TP'}{G'})$, it is easily verified that $\bar{X}$ and $\bar{X}'$ are F-related iff $X' = F_*(X)$, it follows that

(6)
$$F_*([X,Y]) = [F_*(X),F_*(Y)], \qquad X,Y \in \Gamma(\tfrac{TP}{G}).$$

Since j and j' are injective, the map $\Gamma(\frac{P \times \mathfrak{g}}{G}) \to \Gamma(\frac{P' \times \mathfrak{g}'}{G'})$ also preserves the brackets.

<u>Definition 3.8.</u> $F_*\colon \frac{TP}{G} \to \frac{TP'}{G'}$ in (5) above is called the <u>morphism of Atiyah sequences induced by</u> $F(id,\phi)$. //

Note that $\pi_*\colon \frac{TP}{G} \to TB$ is in fact the morphism of Atiyah sequences induced by $\pi(id,k)\colon P(B,G) \to B(B,\{1\})$ where $k\colon G \to \{1\}$ is the constant morphism onto the trivial group.

Example 3.9. Let H be a closed subgroup of a Lie group G and consider the homogeneous bundle G(G/H,H). Its Atiyah sequence is

(7) $$\frac{G\times\mathfrak{h}}{H} \xrightarrow{j} \frac{TG}{H} \xrightarrow{\pi_*} T(G/H))$$

where $j(\langle g,X\rangle) = \langle T(L_g)_1 X\rangle$ and $\pi_*(\langle X\rangle) = T(\pi)(X)$. There are two alternative formulations of this sequence.

Firstly, the vector bundle isomorphism $G\times\mathfrak{g} \to TG$, $(g,X) \mapsto T(L_g)_1(X)$, respects the right actions of H and so quotients to a vector bundle isomorphism $\mathcal{L}: \frac{G\times\mathfrak{g}}{H} \to \frac{TG}{H}$, where $\frac{G\times\mathfrak{g}}{H}$ is the bundle associated to G(G/H,H) through the adjoint action of H on $\mathfrak{g}$. Likewise there is a vector bundle isomorphism $\mathcal{M}: \frac{G\times(\mathfrak{g}/\mathfrak{h})}{H} \to T(G/H)$ defined by $\langle g, X+\mathfrak{h}\rangle \mapsto T(\pi\circ L_g)_1(X)$, where H acts on the vector space $\mathfrak{g}/\mathfrak{h}$ by $h(X+\mathfrak{h}) = \mathrm{Ad}hX + \mathfrak{h}$. We will show that $\mathcal{M}$ is injective; that it is well-defined and a smooth and surjective vector bundle morphism are easily verified. Suppose $\langle g, X+\mathfrak{h}\rangle$ and $\langle g', X'+\mathfrak{h}\rangle$ have $T(\pi\circ L_g)_1(X) = T(\pi\circ L_{g'})_1(X')$. Then $\pi(g) = \pi(g')$ so $\exists\, h \in H$: $g' = gh$. Now $\pi = \pi\circ R_{h^{-1}}$ so we have

$$T(\pi)_g\{T(R_{h^{-1}})_{gh}\circ T(L_{gh})_1(X') - T(L_g)_1(X)\} = 0.$$

Thus $T(L_g)_1\{\mathrm{Ad}hX' - X\}$ is vertical. But L_g is precisely the map m_g (in the notation of 3.2) so $\mathrm{Ad}hX' - X \in \mathfrak{h}$. This shows that

$$\langle g', X'+\mathfrak{h}\rangle = \langle g, \mathrm{Ad}hX'+\mathfrak{h}\rangle = \langle g, X+\mathfrak{h}\rangle$$

as required. Compare Greub et al, 1973, 5.11.

Thus the sequence of vector bundles (7) can be written as

(8) $$\frac{G\times\mathfrak{h}}{H} \xrightarrow{j_1} \frac{G\times\mathfrak{g}}{H} \xrightarrow{q_1} \frac{G\times(\mathfrak{g}/\mathfrak{h})}{H}$$

where $j_1(\langle g,X\rangle) = \langle g,X\rangle$ and $q_1(\langle g,X\rangle) = \langle g, X+\mathfrak{h}\rangle$.

Secondly, the vector bundle morphism $G\times\mathfrak{g} \to (G/H)\times\mathfrak{g}$, $(g,X) \mapsto (gH, \mathrm{Ad}gX)$ over $\pi: G \to G/H$ respects the action of H on $G\times\mathfrak{g}$ and so induces a vector bundle morphism $\frac{G\times\mathfrak{g}}{H} \to (G/H)\times\mathfrak{g}$, which is easily seen to be an isomorphism. Thus (7) can also be written as

(9) $$\frac{G\times\mathfrak{g}}{H} \xrightarrow{j_2} (G/H)\times\mathfrak{g} \xrightarrow{q_2} T(G/H)$$

where $j_2(\langle g,X\rangle) = (gH, \mathrm{Ad}gX)$ and $q_2(gH,X) = T(\pi\circ R_{g^{-1}})_1(X)$. Again compare Greub et al, 1973, loc. cit.

It would be interesting to have formulae for the bracket on $\Gamma(\frac{TG}{H})$ in terms of (8) or (9). It is certainly not the case that the bracket on $\Gamma((G/H)\times\mathfrak{g})$ transported from $\Gamma(\frac{TG}{H})$ via the composite isomorphism $(gH,X) \mapsto \langle T(R_g),(X)\rangle$ is induced by the pointwise bracket $[(gH,X),(gH,X')] = (gH,[X,X'])$. (If this were so, then whenever G were abelian the bracket on $\Gamma(\frac{TG}{H})$, and therefore the Poisson bracket on $\Gamma T(G/H)$, would be identically zero.)

See also 4.18. //

Many other examples of the Atiyah sequence of a principal bundle may be obtained from Chapter III.

§4. Infinitesimal connections and curvature.

The first advantage of the Atiyah sequence concept is that it allows the standard definitions and basic properties of infinitesimal connections and their curvature forms to be presented quickly and clearly, in an algebraically natural manner. The correspondence between the two standard definitions of a connection is seen to be a particular case of the correspondence between right- and left-split maps in an exact sequence; curvature is seen to measure precisely the extent to which a connection fails to preserve Lie brackets; associated connections, the Bianchi identities and the structural equation appear in a clear and natural algebraic manner. This approach also shows that infinitesimal connection theory should be regarded not so much as a theory about principal bundles as about their first-order approximations - the Atiyah sequence or Lie algebroid.

The account given here is a fairly rapid rehearsal of the Atiyah sequence approach as it applies to the most basic and general concepts of infinitesimal connection theory. At each stage the correspondence of this formulation with the standard one is established. The reader may wish to continue this programme by rewriting further parts of infinitesimal connection theory in terms of Atiyah sequences.

Until 4.18, $P(B,G,\pi)$ is a principal bundle.

The standard account of connection theory begins by defining a connection as a distribution Q on P which is G-invariant (that is, $Q_{ug} = T(R_g)(Q_u)$, $u \in P$, $g \in G$) and horizontal ($\forall\, u \in P$, $T(P)_u = T^{\pi}(P)_u \oplus Q_u$). In terms of the Atiyah sequence, this corresponds to the

Definition 4.1. An infinitesimal connection, or simply connection, in $P(B,G)$ is a vector bundle morphism $\gamma\colon TB \to \frac{TP}{G}$ which is right-inverse to $\pi_*\colon \frac{TP}{G} \to TB$; that is, for which $\pi_* \circ \gamma = \mathrm{id}$. //

For suppose we start with a G-invariant horizontal distribution Q. The G-invariance implies that the action of G on TP restricts to an action of G on Q and it is straightforward to show that Q admits equivariant charts; in the notation of 3.1, the chart

$$\mathcal{U} \times \mathbb{R}^n \to Q_{\mathcal{U}}$$

$$(u,\underset{\sim}{t}) \longmapsto h\psi(u,\underset{\sim}{t},0)$$

(where hX is the horizontal component of $X \in T(P)_u$) is equivariant. So the decomposition $TP = T^{\pi}P \oplus Q$ quotients to the decomposition $\frac{TP}{G} = \frac{T^{\pi}P}{G} \oplus \frac{Q}{G}$. Now $\pi_*\colon \frac{TP}{G} \to TB$ is surjective and its kernel is $\frac{T^{\pi}P}{G}$, so the restriction $\pi_*\colon \frac{Q}{G} \to TB$ is an isomorphism of vector bundles over B. We define $\gamma\colon TB \to \frac{Q}{G} \subseteq \frac{TP}{G}$ to be its inverse.

Conversely, given γ, we define Q to be the preimage under $\natural\colon TP \twoheadrightarrow \frac{TP}{G}$ of $\mathrm{im}\,\gamma \subseteq \frac{TP}{G}$. (Note that $\mathrm{im}\,\gamma$ is a sub vector bundle since γ, being fibrewise injective, is of constant rank.) That Q is a horizontal and G-invariant distribution is easily verified.

Although the "distribution definition" of a connection is usually given pre-eminence by being stated first, practical work is usually done in terms of connection forms:

Definition 4.2. A back-connection in $P(B,G)$ is a vector bundle morphism $\omega\colon \frac{TP}{G} \to \frac{P \times \mathfrak{g}}{G}$ which is left-inverse to $j\colon \frac{P \times \mathfrak{g}}{G} \to \frac{TP}{G}$; that is, for which $\omega \circ j = \mathrm{id}$. //

It would be natural to call ω a connection form, except that this could be confused with the ordinary usage of the term.

A connection form, in the usual sense, is a form $\omega \in A^1(P,\mathfrak{g})$ for which $\omega_{ug}\big(T(R_g)X\big) = \mathrm{Ad}(g^{-1})\circ\omega_u(X)$ $\forall X \in T(P)_u$, $u \in P$, $g \in G$, and $\omega(A^*) = A$, $\forall A \in \mathfrak{g}$. The first of these conditions states that ω, regarded as a map $TP \to P \times \mathfrak{g}$, preserves the actions of G on TP and $P \times \mathfrak{g}$, and so quotients to a map $\omega^{/G}: \frac{TP}{G} \to \frac{P \times \mathfrak{g}}{G}$. The second condition now implies that $\omega^{/G}\circ j = \mathrm{id}$.

Conversely, given a back-connection $\omega: \frac{TP}{G} \to \frac{P \times \mathfrak{g}}{G}$, define $\vec{\omega}: TP \to P \times \mathfrak{g}$ by $\vec{\omega}_u(X) = (\natural_u^{P\times\mathfrak{g}})^{-1}(\omega(\langle X\rangle))$, where $\natural^{P\times\mathfrak{g}}$ is the projection $P \times \mathfrak{g} \to \frac{P \times \mathfrak{g}}{G}$. Now $\vec{\omega}$ is smooth because $\natural^{P\times\mathfrak{g}}\circ\vec{\omega} = \omega\circ\natural^{TP}$, and $\natural^{P\times\mathfrak{g}}$ is a submersion; that it is a connection form is easily verified.

That there is a bijective correspondence between connections and connection forms now follows from the well-known

<u>Proposition 4.3</u>. Let $E' \xrightarrow{\iota} E \xrightarrow{\pi} E''$ be an exact sequence of vector bundles over a common base B. Given a right-inverse $\rho: E'' \to E$ of π there is a unique left-inverse $\lambda: E \to E'$ of ι such that

(1) $$\iota\circ\lambda + \rho\circ\pi = \mathrm{id}_E.$$

Conversely, given a left-inverse λ of ι, there is a unique right-inverse ρ of π such that (1) holds.

In either case $\lambda\circ\rho = 0$ and $E'' \xrightarrow{\rho} E \xrightarrow{\lambda} E'$ is an exact sequence. //

The pair of maps λ, ρ is called a <u>splitting</u> of the exact sequence. λ may be called the <u>left-split map</u> and ρ the <u>right-split map</u>.

Note that the existence of connections in principal bundles now follows from the general result that a fibrewise surjection of vector bundles over a fixed base has a right-inverse.

Before dealing with curvature we need a result concerning associated connections in vector bundles.

Suppose first that $E = \frac{P \times V}{G}$ is the vector bundle associated to P(B,G) via a representation $g \mapsto (v \mapsto gv)$ of G on a vector space V.

<u>Lemma 4.4</u>. If $\mu \in \Gamma E$ and $X \in \Gamma(\frac{TP}{G})$ then $\bar{X}(\tilde{\mu}) \in C(P,V)$ is G-equivariant.

Proof: Let $\{\phi_t\}$ be a saturated local flow for $\bar{X}$ defined in a neighbourhood of $u \in P$ (3.7(i)). Then, for all $g \in G$,

$$\bar{X}(\tilde{\mu})(ug) = \frac{d}{dt}\tilde{\mu}(\phi_t(ug))\Big|_o = \frac{d}{dt}\tilde{\mu}(\phi_t(u)g)\Big|_o = \frac{d}{dt}g^{-1}\tilde{\mu}(\phi_t(u))\Big|_o = g^{-1}\bar{X}(\tilde{\mu})(u),$$

as required. //

We denote the section of E corresponding to $\bar{X}(\tilde{\mu})$ by $X(\mu)$.

Definition 4.5. The <u>action</u> of $\frac{TP}{G}$ on $E = \frac{P \times V}{G}$ is the map

$$\Gamma(\tfrac{TP}{G}) \times \Gamma E \to \Gamma E, \qquad (X,\mu) \mapsto X(\mu). \qquad //$$

Proposition 4.6. The action of $\frac{TP}{G}$ on $E = \frac{P \times V}{G}$ has the following properties. Here $X, X_1, X_2 \in \Gamma(\frac{TP}{G})$, $\mu, \mu_1, \mu_2 \in \Gamma E$, and $f \in C(B)$.

(i) $(X_1 + X_2)(\mu) = X_1(\mu) + X_2(\mu)$

(ii) $(fX)(\mu) = f(X(\mu))$

(iii) $X(\mu_1 + \mu_2) = X(\mu_1) + X(\mu_2)$

(iv) $X(f\mu) = fX(\mu) + \pi_*(X)(f)\mu$

(v) $[X_1, X_2](\mu) = X_1(X_2(\mu)) - X_2(X_1(\mu))$

(vi) The value of $X(\mu)$ at a point $x \in B$ depends only on the value of X at x and the values of μ in a neighbourhood of x.

Proof: (i)-(v) are trivial; we prove (iv) as an example: Recalling that $\widetilde{f\mu} = (f\circ\pi)\tilde{\mu}$, $\overline{fX} = (f\circ\pi)\bar{X}$ and $\bar{X}(f\circ\pi) = \pi_*(X)(f)\circ\pi$, we have $\widetilde{X(f\mu)} = \bar{X}((f\circ\pi)\tilde{\mu})$ $= (f\circ\pi)\bar{X}(\tilde{\mu}) + \bar{X}(f\circ\pi)\tilde{\mu} = \widetilde{fX(\mu)} + \widetilde{\pi_*(X)(f)\mu}$, whence the result.

(vi) follows from the corresponding result for Lie derivatives, or can be proved from (ii) and (iv). //

Using 4.6, the following result is immediate:

Proposition 4.7. If γ is a connection in P(B,G) and E is an associated vector bundle, then

$$\nabla^{\gamma}_X(\mu) = (\gamma X)(\mu), \qquad X \in \Gamma TB, \quad \mu \in \Gamma E$$

defines a linear connection ∇^γ in E, called the <u>connection in</u> E <u>induced by</u> γ. //

That this definition coincides with the usual definition of the induced connection in an associated bundle follows from Kobayashi and Nomizu 1963, III 1.3. In the case of the induced connection in $\frac{P \times \mathfrak{g}}{G}$, there is an alternative formula.

<u>Proposition 4.8</u>. The action of $\frac{TP}{G}$ on $\frac{P \times \mathfrak{g}}{G}$ is given by

$$j(X(V)) = [X, j(V)], \qquad X \in \Gamma(\tfrac{TP}{G}), \quad V \in \Gamma(\tfrac{P \times \mathfrak{g}}{G}).$$

<u>Proof</u>: If $\{\phi_t\}$ is a saturated local flow for $\bar{X}$ (3.7), then $\phi_t \circ m_{\phi_{-t}}(u) = m_u$, $\forall\, u \in P$, and a modification of part of the proof of 3.5 shows that $[\bar{X}, \overline{j(V)}](u) = T(m_{u_1})(\bar{X}(\tilde{V})(u))$, $\forall\, u \in P$. But $T(m_{u_1})(\bar{X}(\tilde{V})(u)) = \overline{j(X(V))}(u)$ and so the result follows. //

<u>Corollary 4.9</u>. If γ is a connection in P(B,G), then the induced connection ∇^γ in $\frac{P \times \mathfrak{g}}{G}$ is given by

$$j(\nabla^\gamma_X(V)) = [\gamma X, jV], \qquad X \in \Gamma TB, \quad V \in \Gamma(\tfrac{P \times \mathfrak{g}}{G}). \qquad //$$

This connection may be called the <u>adjoint connection</u> of γ.

We now proceed to study curvature.

<u>Definition 4.10</u>. Let $\gamma\colon TB \to \frac{TP}{G}$ be a connection in P(B,G). The <u>curvature</u> of γ is the skew-symmetric vector bundle map

$$\bar{R}_\gamma\colon TB \oplus TB \to \frac{P \times \mathfrak{g}}{G}$$

defined by $j(\bar{R}_\gamma(X,Y)) = \gamma[X,Y] - [\gamma X, \gamma Y]$. //

To prove that this is indeed the standard curvature form in disguise requires some preparation. First recall some terminology:

<u>Definition 4.11</u>. (Kobayashi and Nomizu, 1963, §II.5; Greub et al, 1973, §§3.15, 6.6.) Let ρ be a representation of G on a vector space V.

A form $\phi \in A^r(P,V)$ is called <u>equivariant</u> or <u>pseudotensorial of type</u> (ρ, V) if $R^*_g(\phi) = \rho(g^{-1}) \circ \phi$, $\forall\, g \in G$. The set of equivariant r-forms on P with values in V is denoted $A^r(P,V)^G$.

A form $\phi \in A^r(P,V)$ is called <u>horizontal</u> if, at any given point $u \in P$, $\phi(X_1,\ldots,X_r)(u) = 0$ whenever one or more of the $X_i(u)$ is vertical.

A form $\phi \in A^r(P,V)$ is called <u>basic</u> or <u>tensorial of type</u> (ρ,V) if it is both equivariant and horizontal. The set of basic r-forms on P with values in V is denoted $A^r_B(P,V)$. //

Note that the concept of a horizontal form does not depend on the presence of a connection.

<u>Proposition 4.12.</u> (i) There is a bijective correspondence between equivariant r-forms $\phi \in A^r(P,V)^G$ and skew-symmetric vector bundle morphisms $\underline{\phi}\colon \oplus^r(\frac{TP}{G}) \to \frac{P \times V}{G}$. A corresponding pair ϕ, $\underline{\phi}$ are related by the diagram

(2)
$$\begin{array}{ccc}
\oplus^r TP & \xrightarrow{\ \phi\ } & P \times V \\
{\scriptstyle \oplus^r \natural}\big\downarrow & & \big\downarrow{\scriptstyle \natural^{P\times V}} \\
\oplus^r\left(\frac{TP}{G}\right) & \xrightarrow{\ \underline{\phi}\ } & \frac{P \times V}{G}
\end{array}$$

(ii) There is a bijective correspondence between basic r-forms $\phi \in A^r_B(P,V)$ and skew-symmetric vector bundle morphisms $\underline{\phi}\colon \oplus^r TB \to \frac{P \times V}{G}$. A corresponding pair ϕ, $\underline{\phi}$ are related by the diagram

(3)
$$\begin{array}{ccc}
\oplus^r TP & \xrightarrow{\ \phi\ } & P \times V \\
{\scriptstyle \natural \oplus^r}\big\downarrow & & \big\downarrow{\scriptstyle \natural^{P\times V}} \\
\oplus^r\left(\frac{TP}{G}\right) & \xrightarrow{\ \underline{\phi}\ } & \frac{P \times V}{G} \\
{\scriptstyle \oplus^r \pi_*}\big\downarrow & & \big\| \\
\oplus^r TB & \xrightarrow{\ \underline{\phi}\ } & \frac{P \times V}{G}
\end{array}$$

<u>Proof</u>: Let G act on $\oplus^r TP \to P$ by $(X_1 \oplus \cdots \oplus X_r)g = X_1 g \oplus \cdots \oplus X_r g$. It is straightforward to show (using 3.1(i)) that this action satisfies the conditions of 2.1 and that the vector bundle morphism $\oplus^r\colon \oplus^r TP \to \oplus^r(\frac{TP}{G})$ quotients to an isomorphism $\frac{\oplus^r TP}{G} \cong \oplus^r(\frac{TP}{G})$. Given $\phi \in A^r(P,V)^G$, regarded as $\phi\colon \oplus^r TP \to P \times V$, the equivariance of ϕ implies that it quotients (using 2.2(ii)) to a vector bundle morphism $\phi^{/G}\colon \frac{\oplus^r TP}{G} \to \frac{P \times V}{G}$. We let $\underline{\phi}$ be the equivalent morphism $\oplus^r(\frac{TP}{G}) \to \frac{P \times V}{G}$.

Clearly $\underline{\phi}$ inherits skew-symmetry from ϕ and satisfies (2).

Conversely, given a skew-symmetric vector bundle morphism $\underline{\phi}\colon \oplus^r(\frac{TP}{G}) \to \frac{P \times V}{G}$, consider $\underline{\phi}\circ\oplus^r\natural\colon \oplus^r TP \to \frac{P \times V}{G}$. Since $\natural^{P\times V}$ is a pullback over π, there is a unique vector bundle morphism $\phi\colon \oplus^r TP \to P \times V$ such that $\natural^{P\times V}\circ\phi = \underline{\phi}\circ\oplus^r\natural$. Since $\natural^{P\times V}$ is fibrewise an isomorphism, ϕ inherits skew-symmetry from $\underline{\phi}$.

It is straightforward to check that these constructions are mutual inverses.

(ii) Let γ be any connection in $P(B,G)$. If $\phi \in A^r(P,V)^G$ is horizontal, $\phi(X_1,\ldots,X_r)(x)$ vanishes whenever one or more of the $X_i(x)$'s is in $\mathrm{im}(j) = \frac{T^\pi P}{G}$. Therefore the vector bundle morphism $\underline{\phi} = \underline{\phi}\circ\oplus^r\gamma\colon \oplus^r TB \to \frac{P \times V}{G}$ does not depend on the choice of γ. Clearly $\underline{\phi}$ is skew-symmetric, since $\underline{\phi}$ is, and $\underline{\phi}\circ\oplus^r\pi_* = \underline{\phi}$ since each $X_i - \gamma\pi_*(X_i) = j\omega(X_i)$ is in $\frac{T^\pi P}{G}$, where ω is the back-connection corresponding to γ.

Conversely, given a skew-symmetric vector bundle morphism $\underline{\phi}\colon \oplus^r TB \to \frac{P \times V}{G}$, consider $\underline{\phi}\circ\oplus^r\pi_*\colon \oplus^r(\frac{TP}{G}) \to \frac{P \times V}{G}$. This is certainly a skew-symmetric vector bundle morphism and so induces, by (i), an equivariant form $\phi \in A^r(P,V)^G$ which is horizontal since $(\underline{\phi}\circ\oplus^r\pi_*)(X_1,\ldots,X_r)(x)$ vanishes whenever one or more of the $X_i(x)$'s is in $\frac{T^\pi P}{G} = \ker \pi_*$.

Again, it is straightforward to check that these constructions are mutual inverses. //

Of course a special case of (i) was dealt with already in the case of connection forms and back-connections.

Denote by $C^r(\frac{TP}{G}, \frac{P \times V}{G})$ the vector bundle $\mathrm{Alt}^r(\frac{TP}{G}, \frac{P \times V}{G})$ whose fibres are the alternating r-multilinear maps $\oplus^r \frac{TP}{G}\Big|_x \to \frac{P \times V}{G}\Big|_x$, for $x \in B$. Likewise denote by $C^r(TB, \frac{P \times V}{G})$ the vector bundle $\mathrm{Alt}^r(TB, \frac{P \times V}{G})$. Then $\Gamma C^r(\frac{TP}{G}, \frac{P \times V}{G})$ is naturally isomorphic to the $C(B)$-module of alternating bundle morphisms $\oplus^r(\frac{TP}{G}) \to \frac{P \times V}{G}$, and it is trivial to check that the correspondence of 4.12(i) becomes a $C(B)$-module isomorphism of $\Gamma C^r(\frac{TP}{G}, \frac{P \times V}{G})$ with $A^r(P,V)^G$, where the module structure on $A^r(P,V)^G$ is $f\phi = (f\circ\pi)\phi$. Similarly, $\Gamma C^r(TB, \frac{P \times V}{G})$ is isomorphic as a $C(B)$-module to $A^r_B(P,V)$.

It is well-known, and easy to check directly, that the graded module $A^*(P,V)^G$ is closed under the exterior derivative δ. It follows that δ can be transferred to $\Gamma C^*(\frac{TP}{G}, \frac{P \times V}{G})$:

<u>Proposition 4.13.</u>

$$\begin{array}{ccc} A^r(P,V)^G & \xrightarrow{\ \delta\ } & A^{r+1}(P,V)^G \\ \cong\big\downarrow & & \cong\big\downarrow \\ \Gamma C^r\left(\frac{TP}{G}, \frac{P\times V}{G}\right) & \xrightarrow{\ d\ } & \Gamma C^{r+1}\left(\frac{TP}{G}, \frac{P\times V}{G}\right) \end{array}$$

commutes, where the vertical arrows are the isomorphisms of 4.12(i) and, for $\phi \in \Gamma C^r\left(\frac{TP}{G}, \frac{P\times V}{G}\right)$,

$$d\phi(X_1,\ldots,X_{r+1}) = \sum_{i=1}^{r+1}(-1)^{i+1}X_i(\phi(X_1,\ldots\hat{}\ldots,X_{r+1})) + \sum_{i<j}(-1)^{i+j}\phi([X_i,X_j],X_1,\ldots\hat{}\hat{}\ldots,X_{r+1}),$$

where $X_i \in \Gamma(\frac{TP}{G})$.

<u>Proof</u>: First note that, for any $\phi \in A^r(P,V)^G$, and $X_i \in \Gamma(\frac{TP}{G})$, (2) implies that $\phi(\bar{X}_1,\ldots,\bar{X}_r) = \widetilde{\underline{\phi}(X_1,\ldots,X_r)}$ as functions $P \to V$.

Now, for $X_i \in \Gamma(\frac{TP}{G})$, $x \in B$, and any $u \in \pi^{-1}(x)$,

$$\begin{aligned}(\underline{\delta\phi})(X_1,\ldots,X_{r+1})(x) &= \langle u,\ \delta\phi(\bar{X}_1,\ldots,\bar{X}_{r+1})(u)\rangle \\ &= \langle u,\ \sum(-1)^{i+1}\ \bar{X}_i(\phi(\bar{X}_1,\ldots\hat{}\ldots,\bar{X}_{r+1}))(u) \\ &\qquad + \sum(-1)^{i+j}\ \phi([\bar{X}_i,\bar{X}_j],\bar{X}_1,\ldots\hat{}\hat{}\ldots,\bar{X}_{r+1})(u)\rangle \\ &= \langle u,\ \sum(-1)^{i+1}\ \widetilde{X_i(\underline{\phi}(X_1,\ldots\hat{}\ldots,X_{r+1}))}(u) \\ &\qquad + \sum(-1)^{i+j}\ \widetilde{\underline{\phi}([X_i,X_j],X_1,\ldots\hat{}\hat{}\ldots,X_{r+1})}(u)\rangle \\ &= \sum(-1)^{i+1}\ X_i(\underline{\phi}(X_1,\ldots\hat{}\ldots,X_{r+1}))(x) \\ &\qquad + \sum(-1)^{i+j}\ \underline{\phi}([X_i,X_j],X_1,\ldots\hat{}\hat{}\ldots,X_{r+1})(x)\end{aligned}$$

and the result is proved. //

Note that we are using δ as defined by Greub et al, 1972, §§4.3, 4.7, without the factor of $\frac{1}{r+1}$ used by Kobayashi and Nomizu, 1963, I 3.11. Since $\delta^2 = 0$, it follows that $d^2 = 0$. Of course, d is the Lie algebroid coboundary of IV 2.1.

Now suppose that we have a connection γ in P(B,G), and let h: $TP \to Q \subseteq TP$ be the corresponding "horizontal projection" (Kobayashi and Nomizu, 1963, §II.1). Let h^*: $A^r(P,V) \to A^r(P,V)$ be the map dual to h, that is,

$$h^*(\phi)(X_1,\ldots,X_r) = \phi(hX_1,\ldots,hX_r), \qquad X_i \in \Gamma TP.$$

Clearly h^* maps $A^r(P,V)^G$ into $A^r_B(P,V)$.

<u>Lemma 4.14.</u>

$$\begin{array}{ccc} A^r(P,V)^G & \xrightarrow{\;h^*\;} & A^r_B(P,V) \\ \cong \big\downarrow & & \cong \big\downarrow \\ \Gamma C^r\left(\frac{TP}{G}, \frac{P \times V}{G}\right) & \xrightarrow{\;\gamma^*\;} & \Gamma C^r\left(TB, \frac{P \times V}{G}\right) \end{array}$$

commutes, where the vertical arrows are the isomorphisms of 4.12, and γ^* is the map $\phi \mapsto \phi\circ\gamma^r$.

<u>Proof</u>: Take $\phi \in A^r(P,V)^G$; we must prove that $\underline{h^*(\phi)} = \gamma^*(\underline{\phi})$. For $X_i \in \Gamma TB$, $x \in B$, and $u \in \pi^{-1}(x)$,

$$\begin{aligned} \underline{h^*(\phi)}(X_1,\ldots,X_r)(x) &= \underline{h^*(\phi)}(\gamma X_1,\ldots,\gamma X_r)(x) \\ &= \langle u,\ h^*(\phi)(\overline{\gamma X_1},\ldots,\overline{\gamma X_r})(u)\rangle \\ &= \langle u,\ \phi(h(\overline{\gamma X_1}(u)),\ldots,h(\overline{\gamma X_r}(u))\rangle. \end{aligned}$$

Now for any $X \in \Gamma TB$, we have $\gamma X \in \Gamma(\frac{Q}{G})$ (see the discussion following 4.1) so $\overline{\gamma X}(u) \in Q_u$ and so $h(\overline{\gamma X}(u)) = \overline{\gamma X}(u)$. The expression therefore reduces to

$$\langle u,\ \phi(\overline{\gamma X_1},\ldots,\overline{\gamma X_r})(u)\rangle = \underline{\phi}(\gamma X_1,\ldots,\gamma X_r)(x)$$

and the result follows. //

Putting 4.13 and 4.14 together we have

<u>Proposition 4.15.</u> Let $\nabla^Q = h^*\circ\delta$: $A^r(P,V)^G \to A^{r+1}_B(P,V)$ be the covariant exterior derivative induced by the connection Q in P(B,G) (Kobayashi and Nomizu, 1963, §II.5; Greub et al, 1973, §6.12). Then

$$\begin{array}{ccc} A^r(P,V)^G & \xrightarrow{\nabla^Q} & A_B^{r+1}(P,V) \\ \big\downarrow \cong & & \big\downarrow \cong \\ \Gamma C^r(\frac{TP}{G}, \frac{P\times V}{G}) & \xrightarrow{\mathcal{D}^\gamma} & \Gamma C^{r+1}(TB, \frac{P\times V}{G}) \end{array}$$

commutes, where the vertical arrows are the isomorphisms of 4.12, $\mathcal{D}^\gamma$ is the map $\phi \mapsto \gamma^*(d\phi)$, and $\gamma: TB \to \frac{TP}{G}$ corresponds to Q. //

Notice that $\mathcal{D}^\gamma$ is not precisely equal to either of the two exterior covariant derivatives introduced in III§5.

We are at last able to show that the curvature $\bar{R}_\gamma \in \Gamma C^2(\frac{TP}{G}, \frac{P\times \mathfrak{g}}{G})$ defined in 4.10 does indeed correspond to the standard curvature form $\Omega = \nabla^Q(\omega) \in A_B^2(P,\mathfrak{g})$.

Proposition 4.16. Let $\omega \in A^1(P,\mathfrak{g})^G$ be a connection form in P(B,G) and let $\underline{\omega}: \frac{TP}{G} \to \frac{P\times\mathfrak{g}}{G}$ be the corresponding back-connection in the Atiyah sequence of P(B,G). Then

$$\underline{\Omega} = \underline{\nabla^Q(\omega)} = \mathcal{D}^\gamma(\underline{\omega}) = \bar{R}_\gamma,$$

where γ is the connection corresponding to $\underline{\omega}$.

Proof: It has just been proved that $\underline{\nabla^Q(\omega)} = \mathcal{D}^\gamma(\underline{\omega}) = \gamma^*(d\underline{\omega})$, and $\Omega = \nabla^Q(\omega)$ by definition. So it remains to prove that $\bar{R}_\gamma = \gamma^*(d\underline{\omega})$. For $X,Y \in \Gamma TB$,

$$\begin{aligned} d\underline{\omega}(\gamma X,\gamma Y) &= \gamma X(\underline{\omega}(\gamma Y)) - \gamma Y(\underline{\omega}(\gamma X)) - \underline{\omega}([\gamma X,\gamma Y]) \\ &= -\underline{\omega}([\gamma X,\gamma Y]) \qquad (\text{since } \underline{\omega}\circ\gamma = 0) \\ &= \underline{\omega}(\gamma([X,Y]) - [\gamma X,\gamma Y]) \\ &= (\underline{\omega}\circ j)(\bar{R}_\gamma(X,Y)) \\ &= \bar{R}_\gamma(X,Y) \qquad (\text{since } \underline{\omega}\circ j = \text{id}). \qquad // \end{aligned}$$

A "structure equation" for $\bar{R}_\gamma$ can now be easily obtained.

Proposition 4.17. If γ is a connection in P(B,G) and ω is the corresponding back-connection, then

$$(\pi_*)^*\bar{R}_\gamma = d\omega - [\omega,\omega].$$

<u>Proof</u>: For $X, Y \in \Gamma(\frac{TP}{G})$,

$$\begin{aligned} j((d\omega - [\omega,\omega])(X,Y)) &= j(X(\omega Y)) - j(Y(\omega X)) - j\omega[X,Y] - [j\omega X, j\omega Y] \\ &= [X, j\omega Y] + [j\omega X, Y] - j\omega[X,Y] - [j\omega X, j\omega Y] \qquad \text{(using 4.8)} \\ &= [\gamma\pi_* X, j\omega Y] + [X - \gamma\pi_* X, Y] - j\omega[X,Y] \\ &= [\gamma\pi_* X, -\gamma\pi_* Y] + [X,Y] - j\omega[X,Y] \\ &= j(\bar{R}_\gamma(\pi_* X, \pi_* Y)) \qquad \text{(using } j\circ\omega + \gamma\circ\pi_* = \text{id repeatedly)}, \end{aligned}$$

from which the result follows. //

The minus sign in 4.17, compared to the standard equation $\Omega = \delta\omega + [\omega,\omega]$, is due to the use of the right-hand bracket in $\mathfrak{g}$. Notice, on the other hand, that this is a different equation from III 5.13; the Lie algebroid coboundary d used here is not the same as the covariant derivative D^γ of III§5.

4.17 does not possess the importance in the Atiyah sequence formulation of the theory that $\Omega = d\omega + [\omega,\omega]$ does in the standard treatment. The reason for this is straightforward: the standard definition of curvature, $\Omega = h^*(d\omega)$, is difficult to work with in both theoretical and practical calculations and the structure equation is the usual means by which curvature is calculated. The "Lie algebroid curvature" $\bar{R}_\gamma$ is, on the other hand, very easy to work with for almost all theoretical purposes, and it can easily be localized to a family of local 2-forms in $A^2(U_i, \mathfrak{g})$, $U_i \subseteq B$, for computational work. (See III§5.) There is thus no need for an alternative formula. Indeed if the term "structure equation" is to be used at all in this presentation, it should perhaps be applied to the equation $\bar{R}_\gamma = \gamma^*(d\omega)$ proved in 4.16.

We will not develop the general theory any further here since it has already been covered in the abstract Lie algebroid context (Chapters III - V).

We conclude with two examples of how to work with the Atiyah sequence/Lie algebroid formulation in "theoretical" problems.

<u>Example 4.18</u>. Consider a principal bundle $P(B,H,\pi)$ on which a Lie group G acts to the left in the sense of Greub et al, 1973, §6.28: there are actions $G \times P \to P$, $G \times B \to B$ with respect to which π is equivariant, and such that $g(uh) = (gu)h$ $\forall\, g \in G,\ u \in P,\ h \in H$. Denote $u \mapsto gu$, $P \to P$ by L_g and $x \mapsto gx$, $B \to B$ by $\mathcal{L}_g$.

Then G acts on the Atiyah sequence of P(B,H) in the following way. G acts on $\frac{TP}{H}$ by $g\langle X_u\rangle = \langle T(L_g)_u X_u\rangle$, on TB by $gX_x = T(\not{L}_g)_x(X_x)$, and on $\frac{P\times\mathfrak{h}}{H}$ by $g\langle u,X\rangle = \langle gu,X\rangle$. It is easy to check that the projection of each of these vector bundles is equivariant, and that j and π_* are equivariant. For $g \in G$, $(\not{L}_g)_*: \Gamma TB \to \Gamma TB$ denotes the induced map of vector fields, $(\not{L}_g)_*(X)(x) = T(\not{L}_g)_{g^{-1}x}(X_{g^{-1}x}) = g(X(g^{-1}x))$, and $(L_g)_*: \Gamma TP \to \Gamma TP$ denotes the corresponding map of vector fields on P. Also denote by $(L_g)_*$ the maps $\Gamma(\frac{TP}{H}) \to \Gamma(\frac{TP}{H})$ and $\Gamma(\frac{P\times\mathfrak{h}}{H}) \to \Gamma(\frac{P\times\mathfrak{h}}{H})$ defined by $(L_g)_*(\bar{X})(x) = g(X(g^{-1}(x)))$; it then easily follows that $\overline{(L_g)_*X} = (L_g)_*(\bar{X})$ for $X \in (\frac{TP}{H})$. Now

(4) $$(\not{L}_g)_*([X,Y]) = [(\not{L}_g)_*(X),(\not{L}_g)_*(Y)], \quad \forall\ X,Y \in \Gamma TB;$$

from the corresponding result for $(L_g)_*: \Gamma TP \to \Gamma TP$ we get

(5) $$(L_g)_*([X,Y]) = [(L_g)_*(X),(L_g)_*(Y)], \quad \forall\ X,Y \in \Gamma(\tfrac{TP}{H})$$

and thus the corresponding result for $(L_g)_*: \Gamma(\frac{P\times\mathfrak{h}}{H}) \to \Gamma(\frac{P\times\mathfrak{h}}{H})$.

Greub et al, 1973, loc. cit., define a connection (form) $\omega \in A^1(P,\mathfrak{h})$ to be G-invariant if $(L_g)^*\omega = \omega$, $\forall\ g \in G$. We define a connection $\gamma: TB \to \frac{TP}{H}$ to be G-<u>equivariant</u> if it is equivariant with respect to the actions of G on TB and $\frac{TP}{H}$. The reader may check that these two definitions are equivalent. We wish to show that if γ is G-equivariant, then $\bar{R}_\gamma$ is also, that is

(6) $$\bar{R}_\gamma(gX_x,gY_x) = g\bar{R}_\gamma(X_x,Y_x), \quad \forall\ g \in G,\ X_x,Y_x \in T(B)_x,\ x \in B.$$

To prove (6), let X,Y be vector fields on B with the given values at the chosen $x \in B$. Then $\not{L}_{g*}(X)(gx) = gX_x$ and likewise for Y, so

$$j\circ\bar{R}_\gamma(gX_x,gY_x) = j\circ\bar{R}_\gamma((\not{L}_g)_*X,(\not{L}_g)_*Y)(gx)$$

$$= (\gamma[(\not{L}_g)_*X,(\not{L}_g)_*Y] - [\gamma\circ(\not{L}_g)_*X,\gamma\circ(\not{L}_g)_*Y])(gx).$$

Now the G-equivariance of γ implies that $\gamma\circ(\not{L}_g)_* = (L_g)_*\circ\gamma$ and using this and (4), (5), the above becomes

$$(L_g)_*(\gamma[X,Y] - [\gamma X,\gamma Y])(gx) = j((L_g)_*(\bar{R}_\gamma(X,Y))(gx)).$$

So $\bar{R}_\gamma(gX_x,gY_x) = g(\bar{R}_\gamma(X,Y)(x)) = g\,\bar{R}_\gamma(X_x,Y_x)$, as required.

Thus the curvature of a G-equivariant connection in P(B,H) is determined by its values over any one $x \in B$. (Compare Greub et al, 1973, loc. cit.)

The principal example of such an action on a principal bundle is the action of a Lie group G on a homogeneous bundle G(G/H,H). In this case the Atiyah sequence is isomorphic to

(7) $$\frac{G \times \mathfrak{h}}{H} \xrightarrow{j_1} \frac{G \times \mathfrak{g}}{H} \xrightarrow{q_1} \frac{G \times \mathfrak{g}/\mathfrak{h}}{H}$$

where j_1 and q_1 are induced by the corresponding maps in the exact sequence $\mathfrak{h} \rightarrowtail \mathfrak{g} \xrightarrow{T(\pi)_1} \mathfrak{g}/\mathfrak{h}$ (3.9). It is easy to verify that if G acts on each of the bundles in (7) by $g_1\langle g_2,X\rangle = \langle g_1g_2,X\rangle$ ($X \in \mathfrak{h}$, $\mathfrak{g}$ or $\mathfrak{g}/\mathfrak{h}$), then the isomorphism of (7) onto the Atiyah sequence of G(G/H,H) described in 3.9 is G-equivariant. Thus a G-equivariant connection in G(G/H,H) can be identified with a G-equivariant map $\frac{G \times \mathfrak{g}/\mathfrak{h}}{H} \to \frac{G \times \mathfrak{g}}{H}$ which is right-inverse to q_1. We now need the following

<u>Lemma</u>. Let $\frac{G \times V}{H}$ and $\frac{G \times V'}{H}$ be two vector bundles associated to G(G/H,H) via actions ρ,ρ' of H on V,V'. Then every G-equivariant map ϕ: $\frac{G \times V}{H} \to \frac{G \times V'}{H}$ is of the form

(8) $$\phi(\langle g,v\rangle) = \langle g,\phi_1(v)\rangle$$

for some H-equivariant map $\phi_1: V \to V'$, and every H-equivariant map $\phi_1: V \to V'$ defines a G-equivariant map ϕ by (8).

<u>Proof</u>: Let ϕ: $\frac{G \times V}{H} \to \frac{G \times V'}{H}$ be G-equivariant; that is, $\phi(g_1\langle g_2,v\rangle) = g_1\phi(\langle g_2,v\rangle)$ $\forall\, g_1,g_2 \in G$, $v \in V$. Define $\phi_1: V \to V'$ by $\phi(\langle 1,v\rangle) = \langle 1,\phi_1(v)\rangle$. Then $\phi(\langle g,v\rangle) = \phi(g\langle 1,v\rangle) = g\langle 1,\phi_1(v)\rangle = \langle g,\phi_1(v)\rangle$, which establishes (8). That ϕ_1 is H-equivariant follows from $\langle 1,\rho'(h)\phi_1(v)\rangle = \langle h,\phi_1(v)\rangle = \phi(\langle h,v\rangle) = \phi(\langle 1,\rho(h)v\rangle)$ $= \langle 1,\phi_1(\rho(h)v)\rangle$.

The converse is straightforward. //

This lemma is of course a part of the well-known result that the category of G-vector bundles and G-equivariant morphisms over G/H is isomorphic to the category of H-vector spaces and H-equivariant maps.

From the lemma it follows that G-equivariant connections in G(G/H,H) are in bijective correspondence with maps $\gamma_1: \mathfrak{g}/\mathfrak{h} \to \mathfrak{g}$ which are right-inverse to $T(\pi)_1$ and H-equivariant, that is $\gamma_1(\mathrm{Ad}h\, X + \mathfrak{h}) = \mathrm{Ad}h\, \gamma_1(X + \mathfrak{h})$, $\forall\, h \in H$, $X \in \mathfrak{g}$. By chasing around the diagram

$$\begin{array}{ccc}
\dfrac{TG}{H} & \xleftarrow{\ \gamma\ } & T(G/H) \\
\cong \big\uparrow & & = \big\uparrow \\
\dfrac{G \times \mathfrak{g}}{H} & \xleftarrow{\ \frac{\mathrm{id} \times \gamma_1}{H}\ } & \dfrac{G \times \mathfrak{g}/\mathfrak{h}}{H}
\end{array}$$

it can be seen that the connection γ corresponding to γ_1 is given by $\gamma(\langle T(\not{L}_g)_H(X+\mathfrak{h})\rangle) = \langle T(L_g)_1(\gamma_1(X+\mathfrak{h}))\rangle$. It is also easy to check that $\omega_1: \mathfrak{g} \to \mathfrak{h}$, the left split map corresponding to $\gamma_1: \mathfrak{g}/\mathfrak{h} \to \mathfrak{g}$, is the restriction to $\mathfrak{g} \to \mathfrak{h}$ of the connection form $\omega: TG \to G \times \mathfrak{h}$ corresponding to γ.

With these preliminaries established, we can calculate the curvature of a G-equivariant connection γ over the coset $H \in G/H$. Take $X+\mathfrak{h}, Y+\mathfrak{h} \in \mathfrak{g}/\mathfrak{h}$ and write $\xi = \gamma_1(X+\mathfrak{h})$, $\eta = \gamma_1(Y+\mathfrak{h})$. Let $\overleftarrow{\xi}$, $\overleftarrow{\eta}$ denote the left invariant vector fields on G corresponding to ξ, η; then $\langle\overleftarrow{\xi}(1)\rangle = \gamma(\langle X+\mathfrak{h}\rangle)$ and similarly for η and Y. Now

$$\begin{aligned}
\bar{R}_\gamma(X+\mathfrak{h}, Y+\mathfrak{h}) &= d\underline{\omega}(\gamma(X+\mathfrak{h}), \gamma(Y+\mathfrak{h})) && (4.16)\\
&= \delta\omega(\overleftarrow{\xi}(1), \overleftarrow{\eta}(1)) && (4.13)\\
&= \delta\omega(\overleftarrow{\xi}, \overleftarrow{\eta})(1) \\
&= -\omega([\overleftarrow{\xi}, \overleftarrow{\eta}])(1) && (\text{since } \omega(\overleftarrow{\xi}),\ \omega(\overleftarrow{\eta}) \text{ are constant})\\
&= -\omega_1([\xi,\eta]_L) \\
&= \omega_1([\xi,\eta]_R) \\
&= \omega_1([\gamma_1(X+\mathfrak{h}), \gamma_1(Y+\mathfrak{h})]),
\end{aligned}$$

and this, together with (6), completely determines $\bar{R}_\gamma$. (Compare Greub et al, 1973, §§6.30, 6.31.) If $\mathfrak{h}$ is an ideal of $\mathfrak{g}$ the last expression can be expressed as $-\bar{R}_{\gamma_1}(X+\mathfrak{h}, Y+\mathfrak{h})$ as in 4.16, but in general $\gamma_1([X+\mathfrak{h}, Y+\mathfrak{h}])$ has no meaning.

As an explicit example, take the Hopf bundle $SU(2)(\mathbf{S}^2, U(1))$ where $U(1) = \{z \in \mathbf{C} \mid |z| = 1\}$ is embedded in $SU(2)$ by $z \mapsto \begin{pmatrix} z & 0 \\ 0 & \bar{z} \end{pmatrix}$. Identify $\mathfrak{u}(1)$ with $\mathbf{R}$ and $\mathfrak{su}(2)$ with the Lie algebra of all matrices of the form

$$\begin{bmatrix} ix_3 & -x_2 + ix_1 \\ x_2 + ix_1 & -ix_3 \end{bmatrix} \qquad (x_1, x_2, x_3 \in \mathbf{R}).$$

Then $\mathfrak{u}(1) \rightarrowtail \mathfrak{su}(2)$ is $x \mapsto \begin{bmatrix} ix & 0 \\ 0 & -ix \end{bmatrix}$ and $\mathfrak{su}(2)/\mathfrak{u}(1)$ can be identified with $\mathbb{R}^2$. Since $U(1)$ is central in $SU(2)$ the adjoint action of $U(1)$ on $\mathfrak{su}(2)$ is trivial and

$$\gamma_1 : (x_1, x_2) \mapsto \begin{bmatrix} 0 & -x_2 + ix_1 \\ x_2 + ix_1 & 0 \end{bmatrix}$$

is an equivariant right split map of $\mathfrak{u}(1) \rightarrowtail \mathfrak{su}(2) \twoheadrightarrow \mathbb{R}^2$. The corresponding $\omega_1 : \mathfrak{su}(2) \to \mathfrak{u}(1)$ is

$$\begin{bmatrix} ix_3 & -x_2 + ix_1 \\ x_2 + ix_1 & -ix_3 \end{bmatrix} \mapsto x_3.$$

A simple calculation now shows that

$$\bar{R}_\gamma(\underset{\sim}{x}, \underset{\sim}{y}) = 2 \begin{vmatrix} x_1 & x_2 \\ y_1 & y_2 \end{vmatrix}$$

for $\underset{\sim}{x}, \underset{\sim}{y} \in \mathbb{R}^2$. //

Example 4.19. Consider a morphism $\phi(\mathrm{id}_B, f) : P(B,G) \to Q(B,H)$ of principal bundles over a common base, and the induced morphism of their Atiyah sequences (see 3.8)

$$\begin{array}{ccccc} \frac{P \times \mathfrak{g}}{G} & \overset{j}{\rightarrowtail} & \frac{TP}{G} & \overset{*}{\twoheadrightarrow} & TB \\ {\scriptstyle \phi_*^+}\Big| & & {\scriptstyle \phi_*}\Big| & & \Big\| \\ \frac{Q \times \mathfrak{h}}{H} & \overset{j'}{\rightarrowtail} & \frac{TQ}{H} & \overset{\pi'_*}{\twoheadrightarrow} & TB \end{array}$$

Let γ be a connection in $P(B,G)$ with corresponding connection form $\omega \in A^1(P, \mathfrak{g})$, and $\underline{\omega} : \frac{TP}{G} \to \frac{P \times \mathfrak{g}}{G}$. In the abstract Lie algebroid context (III 5.5) we defined the produced connection $\gamma' : TB \to \frac{TQ}{H}$ in $Q(B,H)$ by $\gamma' = \phi_* \circ \gamma$. (That γ' is a connection follows easily from $\pi'_* \circ \phi_* = \pi_*$.) We now show that γ' is the "induced connection" in $Q(B,H)$ in the sense of Kobayashi and Nomizu, 1963, §II.6.

The induced connection form $\omega' \in A^1(Q, \mathfrak{h})$ is characterized by the condition $\phi^*\omega' = f_* \circ \omega$ (op. cit. II.6.1) and in fact only the values of ω' on $\mathrm{im}\, T(\phi) \subseteq TQ$ are given in the standard treatments. Using the Atiyah sequence/Lie algebroid formulation, we can quickly derive the general formula for ω'.

Let ω' and $\underline{\omega}'$ now denote the connection form and back-connection corresponding to γ'. Then $j'\underline{\omega}' = id - \gamma'\pi'_* = id - \phi_*\gamma\pi'_*$. Using this, and the fact that any connection γ in any $P(B,G)$ and its connection form $\omega \in A^1(P,\mathfrak{g})$ are related by

$$\gamma(X_x) = \langle Z_u - \omega_u(Z)^*|_u\rangle \qquad \text{where } Z \text{ is any element of } T(P)_u \text{ with } T(\pi)_u(Z) = X \in T(B)_x,$$

it is straightforward to establish that the connection form $\omega' \in A^1(Q,\mathfrak{h})$ corresponding to γ' is given by

$$\omega'(Y_v) = B + Adh^{-1}f_*(\omega_u(X)), \qquad Y \in T(Q)_v \tag{9}$$

where u is any element of P with $\pi(u) = \pi'(v)$

X is any element of $T(P)_u$ with $T(\pi)_u(X) = T(\pi')_v(Y)$

h is the element of H for which $\phi(u)h = v$

and $B \in \mathfrak{h}$ is determined by $B^*|_v = Y - T(R_h)T(\phi)_u(X)$.

It is straightforward, if tedious, to check directly that this ω' is well-defined and is a connection form in $Q(B,H)$. If $Y_v = T(\phi)_u(X)$ for some $X \in T(P)_u$ then we may use this u and X in (9) and take $h = 1$ so that we get $\omega'(T(\phi)_u X) = f_*(\omega_u(X))$. This confirms that ω' is indeed the induced connection (form) in the ordinary sense.

It may be noted that (9), although it concerns connection forms in the ordinary sense, is most easily derived using the Atiyah sequence/Lie algebroid formulation.

A similar formula to (9) may be derived to express the curvature form Ω' of ω' in terms of the curvature form Ω of ω. However in the Atiyah sequence language we need only note that $\overline{R}_{\gamma'} = \phi^+_*\circ\overline{R}_\gamma$ (see III§5, equation (3)). Here $\phi^+_*: \frac{P\times\mathfrak{g}}{G} \to \frac{Q\times\mathfrak{h}}{H}$ is the map $\langle u,X\rangle \mapsto \langle\phi(u),f_*(X)\rangle$.

Clearly the definition $\gamma' = \phi_*\circ\gamma$ and the resulting equation $\overline{R}_{\gamma'} = \phi^+_*\circ\overline{R}_\gamma$ are considerably simpler than (9) and the corresponding equation for Ω' in terms of Ω, yet $\gamma' = \phi_*\circ\gamma$ and $\overline{R}_{\gamma'} = \phi^+_*\circ\overline{R}_\gamma$ also contain more information than the standard $\phi^*\omega' = f_*\circ\omega$, $\phi^*\Omega' = f_*\circ\Omega$. //

Lastly, the following result is used in II§6.

Proposition 4.20. Let P(B,G) be a principal bundle. Then P admits a Riemannian metric which is invariant under the right action of G.

Proof: The vector bundle $\frac{TP}{G} \to B$ admits a fibre metric (see, for example, Greub et al (1972, p. 67)), and any such metric can be pulled back to a G-invariant Riemannian metric on P via the fibrewise isomorphism $TP \to \frac{TP}{G}$. //

APPENDIX B ON LIE GROUPS AND LIE ALGEBRAS

Throughout these notes, we equip the Lie algebra of a Lie group with the bracket obtained from the right-invariant vector fields. For the Lie theory of differentiable groupoids developed in the main text to be compatible with the standard theory of principal bundles, it is essential to define the Lie algebroid of a differentiable groupoid by right-invariant vector fields; consistency then obliges us to do the same for the Lie algebra of a Lie group.

In §1 we give a brief resumé of the elementary formulas, in terms of this convention. In §2 we list the main properties of the right (Darboux) derivative which to group-valued maps $B \to G$ assigns a Maurer-Cartan form in $A^1(B,\mathfrak{g})$. Our references for Lie groups and Lie algebras are Dieudonné (1972) and Warner (1971).

§1. Definitions and notations.

Let G be a Lie group and $\mathfrak{g}$ the tangent space at the identity. We give $\mathfrak{g}$ the <u>right bracket</u> $[\ ,\ ]$ defined by

$$[X,Y] = [\vec{X},\vec{Y}](1) \qquad X,Y \in \mathfrak{g},$$

where $\vec{X}$ is the right-invariant vector field with $\vec{X}(1) = X$. The <u>left bracket</u> $[\ ,\ ]_L$, which is defined by

$$[X,Y]_L = [\overleftarrow{X},\overleftarrow{Y}](1) \qquad X,Y \in \mathfrak{g}$$

is related to the right bracket by

$$[X,Y]_L = -[X,Y] \qquad X,Y \in \mathfrak{g}.$$

For a morphism of Lie groups $\phi: G \to H$ the induced Lie algebra morphism $T(\phi)_1: \mathfrak{g} \to \mathfrak{h}$ is denoted ϕ_*. The Lie group of Lie group automorphisms $G \to G$ is denoted by $Aut(G)$ (see Hochschild (1952)).

For $g \in G$ the <u>inner automorphism</u> $I_g: G \to G$ is defined by $h \mapsto ghg^{-1}$, and $(I_g)_*: \mathfrak{g} \to \mathfrak{g}$ by $Ad(g)$.

For any Lie algebra $\mathfrak{g}$, the Lie group of Lie algebra automorphisms $\mathfrak{g} \to \mathfrak{g}$ is denoted by $Aut(\mathfrak{g})$, and the Lie algebra of derivations $\mathfrak{g} \to \mathfrak{g}$ by $Der(\mathfrak{g})$. For vector spaces V,W the vector space of linear maps $V \to W$ is denoted by Hom(V,W) and the Lie algebra of endomorphisms $V \to V$ by End(V) or $\mathfrak{gl}(V)$. For any Lie algebra $\mathfrak{g}$, the adjoint representation ad: $\mathfrak{g} \to \mathrm{Der}(\mathfrak{g})$ is defined by $\mathrm{ad}X(Y) = [X,Y]$. The image, ad($\mathfrak{g}$), is the adjoint Lie algebra of $\mathfrak{g}$ and the ideal of inner derivations in Der($\mathfrak{g}$).

For a Lie group G, the exponential map exp: $\mathfrak{g} \to G$ is defined in terms of 1-parameter subgroups and is not affected by the reversal of the bracket on $\mathfrak{g}$. The flow of $\vec{X}$, where $X \in \mathfrak{g}$, is $\phi_t = L_{\exp tX}$.

Rather than change the definition of the bracket on the endomorphism Lie algebra $\mathfrak{gl}(V)$, we retain the standard bracket

$$[X,Y] = X \circ Y - Y \circ X \qquad X,Y \in \mathfrak{gl}(V)$$

and reverse the identification of $T(GL(V))_{id}$ with $\mathfrak{gl}(V)$. The standard identification of $T(GL(V))_{id}$ with $\mathfrak{gl}(V)$ is obtained by regarding GL(V) as an open subset of $\mathfrak{gl}(V)$ and identifying $T(\mathfrak{gl}(V))_{id}$ with $\mathfrak{gl}(V)$ by translation in the vector space $\mathfrak{gl}(V)$ (see, for example, Warner (1971)). We now use the negative of this identification; it can be alternatively expressed by mapping $X \in T(GL(V))_{id}$ to the element

$$\nu \mapsto -\frac{d}{dt} \exp tX(\nu)\Big|_0 \qquad \nu \in V$$

of $\mathfrak{gl}(V)$, where, on the right-hand side, the element of $T(V)_\nu$ is translated to the origin. As a consequence, the representation $\rho_*: \mathfrak{g} \to \mathfrak{gl}(V)$ induced by a representation $\rho: G \to GL(V)$ is now given by

$$\rho_*(X)(\nu) = -\frac{d}{dt} \rho(\exp tX)(\nu)\Big|_0.$$

Consider the formula $\mathrm{Ad}_* = \mathrm{ad}$, which is valid in the standard left-hand theory. Its content is that

$$\frac{d}{dt} \mathrm{Ad}(\exp tX)(Y)\Big|_0 = [X,Y]_L.$$

Multiplying by -1 we obtain

$$-\frac{d}{dt} \mathrm{Ad}(\exp tX)(Y)\Big|_0 = [X,Y]_R$$

which, in the right-hand conventions, also asserts that $\mathrm{Ad}_* = \mathrm{ad}$.

Note however that, for any Lie group G, the representation ad: $\mathfrak{g} \to \mathrm{Der}(\mathfrak{g})$ is the negative of the representation ad in the standard accounts; thus

$$\mathrm{Ad}(\exp X) = e^{-\mathrm{ad}X}$$

for $X \in \mathfrak{g}$.

Lastly, let $\theta \in A^1(G,\mathfrak{g})$ be the right Maurer-Cartan form $\theta(X_g) = T(R_{g^{-1}})(X_g)$. Then θ satisfies the usual Maurer-Cartan equation

$$\delta\theta + [\theta,\theta] = 0$$

with respect to the right bracket.

§2. Formulas for the right derivative

Let G be a Lie group and B a manifold, and let $f\colon B \to G$ be a smooth map. Then the right derivative $\Delta(f)\colon TB \to B \times \mathfrak{g}$ of f is $\mathfrak{g}$-valued 1-form on B defined by

$$\Delta(f)(X_x) = T\left(R_{f(x)^{-1}}\right)(T(f)(X_x)).$$

Alternatively, $\Delta(f)$ is the pullback $f^*\theta$ of the right Maurer-Cartan form θ on G, and so $\Delta(f)$ satisfies the Maurer-Cartan equation

$$\delta(\Delta(f)) + [\Delta(f),\Delta(f)] = 0$$

with respect to the right bracket in $\mathfrak{g}$.

If $G = V$ is a vector space, then $\Delta(f)(X)\colon B \to V$ will be identified with the Lie derivative $X(f)$.

The product rule is

$$(1) \qquad \Delta(f_1 f_2) = \Delta(f_1) + \mathrm{Ad}(f_1)(\Delta(f_2))$$

where $f_1, f_2\colon B \to G$ are two maps. Here and elsewhere the symbol $\mathrm{Ad}(f_1)(\Delta(f_2))$ denotes the map

$$X_x \mapsto \mathrm{Ad}(f_1(x))(\Delta(f_2)(X_x)), \qquad X_x \in T(B)_x.$$

From the product rule it follows that

$$(2) \qquad \Delta(f^{-1}) = -\mathrm{Ad}(f^{-1})(\Delta(f))$$

where f^{-1} denotes the pointwise inverse $x \mapsto f(x)^{-1}$.

If $f\colon B \to G$ and $s\colon B \to G$ are smooth maps, then $I_s(f)$ denotes the map

$x \mapsto s(x)f(x)s(x)^{-1}$. From the product rule it follows that

(3) $$\Delta(I_s(f)) = \mathrm{Ad}(s)\{\mathrm{Ad}(f)(\Delta(s^{-1})) + \Delta(f) - \Delta(s^{-1})\}.$$

In particular,

(4) $$\Delta(R_g \circ f) = \Delta(f), \qquad \Delta(L_g \circ f) = \mathrm{Ad}g(\Delta(f))$$

and

(5) $$\Delta(I_g \circ f) = \mathrm{Ad}g(\Delta(f))$$

where $g \in G$ is fixed, and $f: B \to G$ is a smooth map.

If V and W are vector spaces and $\phi: B \to \mathrm{Hom}(V,W)$ and $f: B \to V$ are smooth maps, then it is easy to see that

(6) $$X(\phi(f)) = X(\phi)(f) + \phi(X(f)), \qquad X \in \Gamma TB.$$

Here, once again, $\phi(f)$ denotes the pointwise evaluation $X \mapsto \phi(x)(f(x))$. If $W = V$ and ϕ takes values in $GL(V)$, then (6) can be rewritten as

(6a) $$X(\phi(f)) = -\Delta(\phi)(X)(\phi(f)) + \phi(X(f)), \qquad X \in \Gamma TB.$$

The reader is urged to check this formula directly. The minus sign and the double appearance of ϕ in the first term on the right-hand side arise from the identification of $T(GL(V))_I$ with $\mathfrak{gl}(V)$ that is adopted in §1.

It is easily verified that if $\phi: G \to H$ is a morphism of Lie groups and $f: B \to G$ is a smooth map, then

(7) $$\Delta(\phi \circ f) = \phi_* \circ \Delta(f).$$

In particular,

(8) $$\Delta(\mathrm{Ad} \circ f) = \mathrm{ad} \circ \Delta(f).$$

This formula may be rewritten as

(8a) $$\Delta(\mathrm{Ad} \circ f)(X)(V) = [\Delta(f)(X), V]$$

where $X \in \Gamma TB$ and $V: B \to \mathfrak{g}$, and the bracket is taken pointwise.

The following result is used in Chapter III.

Proposition 2.1. Let G and H be Lie groups and let B be a manifold. Let $\phi: \mathfrak{g} \to \mathfrak{h}$ be a morphism of Lie algebras and let $f: B \to H$ be a smooth map. Define $\tilde{\phi}: B \to \mathrm{Hom}(\mathfrak{g},\mathfrak{h})$ by $\tilde{\phi}(x) = \mathrm{Ad}(f(x))\circ\phi$. Then $\tilde{\phi}(x)$ is a Lie algebra morphism for $x \in B$, and $X(\tilde{\phi}(V)) = \tilde{\phi}(X(V)) - [\Delta(f)(X),\tilde{\phi}(V)]$ for $X \in \Gamma TB$, $V: B \to \mathfrak{g}$.

Proof: Write $\tilde{\phi}(V)$ as as $(\mathrm{Ad}\circ f)(\phi(V))$ where $\mathrm{Ad}\circ f: B \to \mathrm{Aut}(\mathfrak{h})$ and $\phi(V): B \to \mathfrak{h}$, and apply (6a). This gives $X(\tilde{\phi}(V)) = -\Delta(\mathrm{Ad}\circ f)(X)(\tilde{\phi}(V)) + (\mathrm{Ad}\circ f)(X(\phi(V)))$. Applying (8a) to the first term, and (6) to the expression $X(\phi(V))$, this becomes $X(\tilde{\phi}(V)) = -[\Delta(f)(X),\tilde{\phi}(V)] + (\mathrm{Ad}\circ f)(\phi(X(V)))$. Now $(\mathrm{Ad}\circ f)(\phi(X(V))) = \tilde{\phi}(X(V))$ by definition. //

In the case where $\mathfrak{g} = \mathfrak{h}$ and $\phi \in \mathrm{Aut}(\mathfrak{g})$ this equation can, by (6a), be written more simply as

$$\Delta(\tilde{\phi}) = \mathrm{ad}\circ\Delta(f)$$

where Δ on the left-hand side is with respect to the group $\mathrm{Aut}(\mathfrak{g})$.

APPENDIX C ON VECTOR BUNDLES

In this appendix we assemble some elementary results, definitions and notations which are needed in the text and are not readily accessible elsewhere; we also establish our position on a few matters which are not quite standard.

Our references on vector bundles are Dieudonné (1972) and Greub et al (1972). Throughout the text, vector bundles are real and of finite rank, unless explicitly stated otherwise. Vector bundles are generally denoted (E,p,B), or E for short; the fibre type is generally denoted V and atlases are generally denoted $\{\psi_i\colon U_i \times V \to E_{U_i}\}$.

For a vector bundle E, the C(B)-module of global sections is denoted ΓE; for an open subset U of B, the C(U)-module of local sections defined on U is denoted $\Gamma_U E$.

A morphism of vector bundles from (E,p,B) to (E',p',B') is denoted (ϕ,ϕ_o) where ϕ is the map $E \to E'$ and ϕ_o is the map $B \to B'$. If $B = B'$ and $\phi_o = id_B$ we say that ϕ is a morphism $E \to E'$ <u>over</u> B, or that it is <u>base-preserving</u>.

For vector bundles (E^i,p^i,B^i), i = 1,2, the <u>direct product bundle</u> is $(E^1 \times E^2, p^1 \times p^2, B^1 \times B^2)$; elements are denoted (X,Y) for $X \in E^1$, $Y \in E^2$. For vector bundles (E^i,p^i,B), i = 1,2, over the same base B, the <u>Whitney sum</u> is denoted $(E^1 \oplus E^2,p,B)$ and its elements are written $X \oplus Y$ or X + Y.

For vector bundles E and E' over the same base B we denote by Hom(E,E') the vector bundle over B whose fibres $\mathrm{Hom}(E_x,E'_x)$, $x \in B$, are the vector spaces of linear maps $E_x \to E'_x$, and whose module of global sections gives the C(B)-module of vector bundle morphisms $E \to E'$. For $n \geq 0$, we denote by $\mathrm{Hom}^n(E;E')$ the vector bundle over B constructed in the same way from n-multilinear maps $E^n_x \to E'_x$ and vector bundle morphisms $\oplus^n E \to E'$; the sub-bundles corresponding to alternating and symmetric maps are denoted, respectively, by $\mathrm{Alt}^n(E;E')$ and $\mathrm{Sym}^n(E;E')$.

A morphism $(\phi,\phi_o)\colon (E,p,B) \to (E',p',B')$ is <u>fibrewise-injective</u>, <u>-surjective</u> or <u>-bijective</u> if each $\phi_x\colon E_x \to E'_{\phi_o(x)}$ is, respectively, injective, surjective, or bijective.

Given a vector bundle (E,p,B) and a smooth map $f\colon B' \to B$, the <u>inverse image vector bundle</u> $(f^*E,\bar{p},B')$ is $f^*E = \{(x',u) \in B' \times E \mid f(x') = p(u)\}$ with projection $\bar{p}(x',u) = x'$ and the natural bundle structure. The morphism $(x',u) \mapsto u$, $f^*E \to E$ over $f\colon B' \to B$, is denoted $\bar{f}$.

<u>Definition 1</u>. A morphism of vector bundles $(\phi,f)\colon (E^1,p^1,B^1) \to (E^2,p^2,B^2)$ is a <u>pullback</u> if every morphism of vector bundles $(\psi,f)\colon (E,p,B^1) \to (E^2,p^2,B^2)$ over $f\colon B^1 \to B^2$ can be factored uniquely into

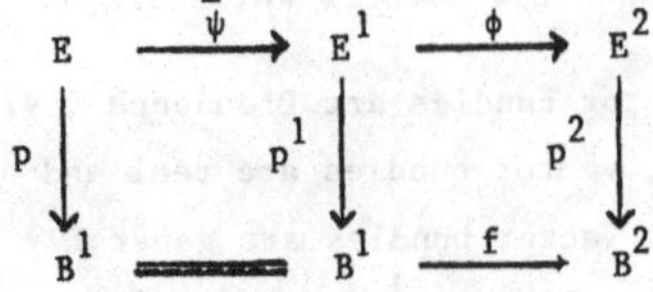

where $\bar{\psi}$ is a vector bundle morphism over B^1. //

<u>Proposition 2</u>. A morphism of vector bundles is a pullback iff it is fibrewise bijective.

<u>Proof</u>: (<=) Suppose $(\phi,f)\colon (E^1,p^1,B^1) \to (E^2,p^2,B^2)$ is fibrewise bijective. Given (ψ,f), as above, define $\bar{\psi}\colon E \to E^1$ fibrewise by $\bar{\psi}_x = (\phi_x)^{-1} \circ \psi_x$, for $x \in B$. It is easy to check, using local charts, that $\bar{\psi}$ is smooth.

(=>) From (<=) it follows that $\bar{f}\colon f^*E^2 \to E^2$ is a pullback. Applying the uniqueness condition in Definition 1 to both ϕ and $\bar{f}$, there is a vector bundle isomorphism $E^1 \to f^*E^2$ such that

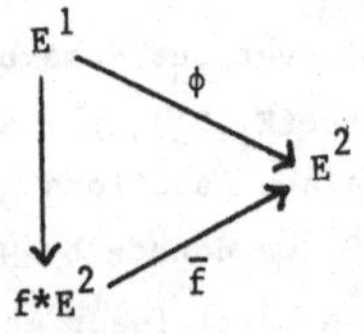

commutes. Since $\bar{f}$ is a fibrewise bijection, it follows that ϕ is also. //

Thus the concepts of pullback, inverse image and fibrewise-bijection are equivalent. Nonetheless, it is useful to distinguish between them.

We now consider the maps induced on modules of sections by maps of vector bundles. If $\phi\colon E \to E'$ is a map of vector bundles over a common base B then the $C(B)$-module morphism $\Gamma E \to \Gamma E'$, $\mu \mapsto \phi\circ\mu$ is simply denoted ϕ. If E and E' are now vector bundles over bases B and B', and $\phi\colon E \to E'$ is a vector bundle map over a diffeomorphism $\phi_o\colon B \to B'$, then the map $\Gamma E \to \Gamma E'$, $\mu \mapsto \phi\circ\mu\circ\phi_o^{-1}$ will usually be denoted $\bar{\phi}$. An alternative formula is

$$\bar{\phi}(\mu)(x') = \phi(\phi_o^{-1}(x'))\big(\mu(\phi_o^{-1}(x'))\big).$$

$\bar{\phi}$: $\Gamma E \to \Gamma E'$ is semi-linear with respect to the map $C(B) \to C(B')$, $f \mapsto f \circ \phi_o^{-1}$, which we will also denote by $\bar{\phi}$; by semi-linearity is meant the equation $\bar{\phi}(f\mu) = \bar{\phi}(f)\bar{\phi}(\mu)$.

In the case of fibrewise-bijections a different construction is appropriate.

<u>Proposition 3</u>. Let ϕ: $E \to E'$ be a fibrewise-bijection over ϕ_o: $B \to B'$. Then for $\mu' \in \Gamma E'$ the map

$$x \mapsto (\phi_x)^{-1}(\mu'(\phi_o(x)))$$

is a smooth section of E, denoted $\phi^{\#}(\mu)$, and $\phi^{\#}$: $\Gamma E' \to \Gamma E$ is semilinear with respect to $f' \mapsto f' \circ \phi_o$, $C(B') \to C(B)$.

<u>Proof</u>: Elementary. //

Interestingly, if ϕ: $E \to E'$ and ϕ_o: $B \to B'$ satisfy all the conditions for being a fibrewise-bijection except that ϕ need not be smooth (or continuous), then smoothness of ϕ follows if $\phi^{\#}$ maps smooth sections to smooth sections.

<u>Theorem 4</u>. Let (ϕ,ϕ_o): $(E,p,B) \to (E',p',B')$ be a fibrewise-bijection. Then the map

$$C(B) \underset{C(B')}{\otimes} \Gamma E' \to \Gamma E, \quad f \otimes \mu' \mapsto f\phi^{\#}(\mu')$$

is an isomorphism of C(B)-modules. Here the tensor product is taken with respect to the C(B')-module structure on C(B) defined by $f'f = (f' \circ \phi_o)f$ and the tensor product is itself a C(B)-module with respect to $f_1(f_2 \otimes \mu') = (f_1 f_2) \otimes \mu'$.

<u>Proof</u>: See, for example, Greub et al (1972, 2.26). //

Lastly, we need the following general construction of inverse image vector bundles.

<u>Proposition 5</u>. Let ϕ^1: $E^1 \to E$ and ϕ^2: $E^2 \to E$ be morphisms of vector bundles over a fixed base B, and suppose that

$$\mathrm{im}(\phi^1_x) + \mathrm{im}(\phi^2_x) = E_x, \qquad \forall x \in B.$$

Then $F = \{u^1 \oplus u^2 \in E^1 \oplus E^2 \mid \phi^1(u^1) = \phi^2(u^2)\}$ is a vector subbundle of $E^1 \oplus E^2$, the maps $\bar{\phi}^1 : F \to E^2,\ u^1 \oplus u^2 \mapsto u^2$ and $\bar{\phi}^2 : F \to E^1,\ u^1 \oplus u^2 \mapsto u^1$ are vector bundle morphisms over B, and

$$\begin{array}{ccc} F & \xrightarrow{\bar{\phi}^1} & E^2 \\ {\scriptstyle \bar{\phi}^2}\downarrow & & \downarrow{\scriptstyle \phi^2} \\ E^1 & \xrightarrow{\phi^1} & E \end{array}$$

is a pullback square in the sense that if E' is another vector bundle on B and $\psi^1 : E' \to E^1$ and $\psi^2 : E' \to E^2$ are morphisms of vector bundles over B such that $\phi^1 \circ \psi^1 = \phi^2 \circ \psi^2$, then there is a unique morphism $\psi: E' \to F$ over B such that $\bar{\phi}^2 \circ \psi = \psi^1$ and $\bar{\phi}^1 \circ \psi = \psi^2$.

<u>Proof</u>: Once it is established that F is a vector subbundle of $E^1 \oplus E^2$, this is merely a formal manipulation. To show that F is a vector subbundle, define $\phi: E^1 \oplus E^2 \to E,\ u^1 \oplus u^2 \mapsto \phi^1(u^1) - \phi^2(u^2)$. Then the condition on the images of ϕ^1, ϕ^2 ensures that ϕ is of maximal rank; hence F, its kernel, is a vector subbundle (for example, Dieudonné (1972, 16.17.5)). //

We denote F above by $E^1 \underset{E}{\oplus} E^2$ and may also refer to it as the pullback bundle.

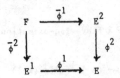

REFERENCES

A.M. Abd-Allah and R. Brown (1980) "A compact-open topology on partial maps with open domain" J. London Math. Soc. (2) 21, 480-486.

R. Almeida (1980) Teoria de Lie para os grupóides diferenciáveis. Tese, Universidade de São Paulo.

R. Almeida and A. Kumpera (1981) "Structure produit dans la catégorie des algèbroids de Lie" An. Acad. Brasil Ciênc. 53, 247-250.

R. Almeida et P. Molino (1985) "Suites d'Atiyah et feuilletages transversalement complets" C.R. Acad. Sc. Paris. t. 300, Serie 1, 13-15.

A. Aragnol (1957) "Espace des tenseurs de type adjoint. Théorèmes d'existence" C.R. Acad. Sc. Paris. 245, 134-136.

R. Arens (1946) "Topologies for homeomorphism groups" Amer. J. Math. 68, 593-610.

M.F. Atiyah (1957) "Complex analytic connections in fibre bundles" Trans. Amer. Math. Soc. 85, 181-207.

M.F. Atiyah, N.J. Hitchin and I.M. Singer (1978) "Self-duality in four-dimensional Riemannian geometry" Proc. R. Soc. London A. 362, 425-461.

R.L. Bishop and R.J. Crittenden (1964) Geometry of Manifolds. Academic Press, New York.

P.I. Booth and R. Brown (1978) "Spaces of partial maps, fibred mapping spaces and the compact-open topology" General Topol. Appl. 8, 181-195.

R.A. Bowshell (1971a) "Abstract velocity functors" Cahiers de Topol. et Geom. Diff. 12, 57-91.

R.A. Bowshell (1971b) "Automorphisms of a connection" contained in Some topics in differential geometry, Ph.D. thesis, Monash University.

W. Browder (1961) "Torsion in H-spaces" Ann. Math. (2) 74, 24-51.

R. Brown (1968) Elements of modern topology. McGraw-Hill, London.

R. Brown (1972) "Groupoids as coefficients" Proc. London Math. Soc. (3)25, 413-426.

R. Brown and G. Danesh-Naruie (1975) "The fundamental groupoid as a topological groupoid" Proc. Edinburgh Math. Soc. 19, 237-244.

R. Brown, G. Danesh-Naruie and J.P.L. Hardy (1976) "Topological Groupoids: II. Covering morphisms and G-spaces" Math. Nachr. 74, 143-156.

R. Brown and J.P.L. Hardy (1976) "Topological groupoids: I. Universal Constructions" Math. Nachr. 74, 143-156.

E. Cartan (1930) From Oeuvres Completes, partie 1, volume 2 (1952): "Le troisième théorème fondamental de Lie" 1143-1145 and 1146-1148; "La théorie de groupes finis et continus et l'analysis situs", 1165-1225.

H. Cartan and S. Eilenberg (1956) Homological Algebra. Princeton University Press, Princeton.

C. Chevalley (1946) Theory of Lie groups. Princeton University Press, Princeton.

A.H.Clifford and G.B.Preston (1961) The Algebraic Theory of Semigroups, volume I. American Mathematical Society, Providence, R.I.

P.M.Cohn (1957) Lie groups. Cambridge University Press.

M.K. Dakin and A.K. Seda (1977) "G-spaces and topological groupoids" Glasnik Matematicki, **12** (32), 191-198.

J. Dieudonné (1972) Treatise on Analysis, volume III. Translator I.G. Macdonald. Academic Press, New York.

A. Douady et M. Lazard (1966) "Espaces fibrés en algèbres de Lie et en groupes" Invent. Math. **1**, 133-151.

J. Dugundji (1966) Topology. Allyn and Bacon, Inc., Boston.

C. Ehresmann (1951) "Les connexions infinitésimales dans un espace fibré différentiable" Colloque de Topologie (Espaces Fibres), Bruxelles, 1950. Georges Thone, Liège; Masson et Cie, Paris. pp. 29-55.

C. Ehresmann (1956) "Sur les connexions d'ordre supérieur" Dagli Atti del V Congresso dell'Unione Matematica Italiana, Pavia-Torino. pp. 344-346.

C. Ehresmann (1959) "Catégories topologiques et catégories differentiables" Colloque de Géometrie Différentielle Globale. Centre Belge de Recherches Mathematiques, Bruxelles, 1958, 137-150.

C. Ehresmann (1961) "Structures feuilletées" Proc. Fifth Canadian Math. Congress. Montreal, 1961. pp. 109-172.

C. Ehresmann (1965) Catégories et structures. Dunod, Paris.

H. Goldschmidt (1978) "The integrability problem for Lie equations" Bull. Amer. Math. Soc. **84**, 531-546.

W.H. Greub (1967) Linear algebra, third edition. Springer-Verlag, New York.

W. Greub, S. Halperin and R. Vanstone (1972) Connections, curvature and cohomology Vol. I. Academic Press, New York.

W. Greub, S. Halperin and R. Vanstone (1973) Connections, curvature and cohomology Vol. II. Academic Press, New York.

W. Greub, S. Halperin and R. Vanstone (1976) Connections, curvature and cohomology Vol. III. Academic Press, New York.

W. Greub and H.R. Petry (1978) "On the lifting of structure groups" Differential Geometrical Methods in Mathematical Physics II, 1977. Ed. K. Bleuler, H.R. Petry and A. Reetz. Lecture Notes in Mathematics #676, Springer-Verlag, Berlin.

A. Haefliger (1956) "Sur l'extension du groupe structural d'une espace fibré" C.R. Acad. Sc. Paris **243**, 558-560.

J.-I. Hano and H. Ozeki (1956) "On the holonomy groups of linear connections" Nagoya Math. J. 10, 97-100.

S. Helgason (1978) Differential geometry, Lie groups, and symmetric spaces. Academic Press, New York.

R. Hermann (1967) "Analytic continuation of group representations, IV" Communications Math. Phys. 5, 131-156.

J. Herz (1953) "Pseudo-algèbres de Lie" C.R. Acad. Sc. Paris t. 236, 1935-1937.

P.J. Higgins (1971) Notes on categories and groupoids. van Nostrand Reinhold, London.

P.J. Higgins (1974) Introduction to topological groups. London Math. Soc. Lecture Note Series, 15. Cambridge University Press.

G. Hochschild (1951) "Group extensions of Lie groups" Ann. Math. 54, 96-109.

G. Hochschild (1952) "The automorphism group of a Lie group" Trans. Amer. Math. Soc. 72, 209-216.

G. Hochschild (1954a) "Lie algebra kernels and cohomology" Amer. J. Math. 76, 698-716.

G. Hochschild (1954b) "Cohomology classes of finite type, and finite-dimensional kernels for Lie algebras" Amer. J. Math. 76, 763-778.

G. Hochschild and G.D. Mostow (1962) "Cohomology of Lie groups" Illinois J. Math. 6, 367-401.

G. Hochschild and J.-P. Serre (1953) "Cohomology of Lie algebras" Ann. Math. 57, 591-603.

S.-T. Hu (1959) Homotopy Theory. Academic Press, New York.

H. Jacobowitz (1978) "The Poincaré lemma for $d\omega = F(x,\omega)$" J. Diff. Geometry 13, 361-371.

F.W. Kamber and Ph. Tondeur (1971) "Invariant differential operators and the cohomology of Lie algebra sheaves" Mem. Amer. Math. Soc., no. 113. American Mathematical Society, Providence, R.I.

S. Kobayashi (1972) Transformation groups in differential geometry. Springer-Verlag, Berlin.

S. Kobayashi and K. Nomizu (1963) Foundations of differential geometry, volume 1. Interscience, New York, London.

S. Kobayashi and K. Nomizu (1969) Foundations of differential geometry, volume 2. Interscience, New York, London.

B. Kostant (1970) "Quantization and unitary representations" Lectures in Modern Analysis and Applications, III, ed. C.T. Taam, Lecture Notes in Mathematics, #170, Springer-Verlag, Berlin, pp. 87-208.

J.-L. Koszul (1960) Lectures on fibre bundles and differential geometry. Notes by S. Ramanan. Tata Institute of Fundamental Research, Bombay.

A. Kumpera (1971) An introduction to Lie groupoids. Núcleo de Estudos e Pesquisas Científicas, Rio de Janeiro.

A. Kumpera (1975) "Invariants différentiels d'un pseudogroupe de Lie. I" J. Diff. Geometry 10, 289-345.

A. Kumpera and D. Spencer (1972) Lie Equations, Vol. I: General Theory. Princeton Univ. Press, Princeton, N.J.

H.B. Lawson, Jr. (1974) "Foliations" Bull. Amer. Math. Soc. 80, 369-418.

H.B. Lawson, Jr. (1977) The quantitative theory of foliations. Conference Board of the Mathematical Sciences, no. 27. American Mathematical Society, Providence, R.I.

K.A. Mackenzie (1978) "Rigid cohomology of topological groupoids" J. Austral. Math. Soc. (Series A) 26, 277-301.

K. Mackenzie (1979) Cohomology of locally trivial groupoids and Lie algebroids. Ph.D. thesis, Monash University, xi + 203 pp.

K. Mackenzie (1980) Infinitesimal theory of principal bundles. Talk given to 50th ANZAAS Conference, Adelaide, 1980. 8 pp.

S. MacLane (1975) Homology, 3rd corrected printing, Springer-Verlag, Berlin.

P. Malliavin (1972) Géométrie différentielle intrinsèque. Hermann, Paris.

W. Miller, Jr. (1972) Symmetry groups and their applications. Academic Press, New York.

J. Milnor (1958) "On the existence of a connection with curvature zero" Comm. Math. Helv. 32, 215-223.

J. Milnor (1963) Morse Theory. Annals of Mathematics Studies, Princeton University Press, Princeton.

J. Milnor (1976) "Curvature of left invariant metrics on Lie groups" Advances in Math. 21, 293-329.

M. Mori (1953) "On the three-dimensional cohomology group of Lie algebras" J. Math. Soc. Japan 5, 171-183.

E. Nelson (1967) Tensor analysis. Princeton University Press, Princeton.

Ngô Van Quê (1967) "Du prolongement des espaces fibrés et des structures infinitésimales" Ann. Inst. Fourier, Grenoble 17, 157-223.

Ngô Van Quê (1968) "Sur l'espace de prolongement différentiable" J. Diff. Geometry 2, 33-40.

Ngô Van Quê (1969) "Nonabelian Spencer cohomology and deformation theory" J. Diff. Geometry 3, 165-211.

Ngô Van Quê and A.A.M. Rodrigues (1975) "Troisième théorème fondamental de réalisation de Cartan" Ann. Inst. Fourier, Grenoble 25, 251-282.

H.K. Nickerson (1961) "On differential operators and connections" Trans. Amer. Math. Soc. 99, 509-539.

R.S. Palais (1957) "A global formulation of the Lie theory of transformation groups" Memoirs Amer. Math. Soc. 22, Amer. Math. Soc., Providence, R.I.

R.S. Palais (1961a) "On the existence of slices for actions of non-compact Lie groups" Ann. Math. 73, 295-323.

R.S. Palais (1961b) "The cohomology of Lie d-rings" Proc. Sympos. Pure Math. Vol. III, pp. 130-137. American Mathematical Society.

R.S. Palais (1965) "Differential operators on vector bundles" Seminar on the Atiyah-Singer Index Theorem, R.S. Palais, Chapter IV. Princeton University Press, Princeton, N.J.

J.P. Pommaret (1977) "Troisième théorème fondamental pour les pseudogroupes de Lie transitifs" C.R. Acad. Sc. Paris 284, 429-432, Série A.

J. Pradines (1966) "Théorie de Lie pour les groupoides différentiables. Relations entre propriétés locales et globales" C.R. Acad. Sc. Paris, t.263, 907-910. Série A.

J. Pradines (1967) "Théorie de Lie pour les groupoides différentiables. Calcul différentiel dans la catégorie des groupoides infinitésimaux" C.R. Acad. Sc. Paris, t.264, p. 245-248. Série A.

J. Pradines (1968a) "Géométrie différentielle au-dessus d'un groupoide" C.R. Acad. Sc. Paris, t.266, p. 1194-1196. Série A.

J. Pradines (1968b) "Troisième théorème de Lie pour les groupoides différentiables" C.R. Acad. Sc. Paris, t.267, p. 21-23. Série A.

J. Pradines (1986) "How to define the graph of a singular foliation" Cahiers de Topol. et Géom. Diff. 26, 339-380.

J. Renault (1980) A groupoid approach to C*-algebras. Lecture Notes in Mathematics, # 793, Springer-Verlag, New York.

G.S. Rinehart (1963) "Differential forms on general commutative algebras" Trans. Amer. Math. Soc. 108, 195-222.

D.J.S. Robinson (1982) A course in the theory of groups. Springer-Verlag, New York.

A.K. Seda (1980) "On measures in fibre spaces" Cahiers de Topol. et Géom. Diff. 21, 247-276.

J.-P. Serre (1965) Lie algebras and Lie groups. W.A. Benjamin, Inc., New York.

U. Shukla (1966) "A cohomology for Lie algebras" J. Math. Soc. Japan 18, 275-289.

I.M. Singer and J.A. Thorpe (1967) Lecture notes on elementary topology and geometry. Scott, Foresman and Company, Glenview.

P.A. Smith (1952) "Some topological notions connected with a set of generators" Proc. Intern. Conf. Math. Cambridge, Mass. 1950. Vol. 2, 436-441.

E.H. Spanier (1966) Algebraic Topology. McGraw-Hill, New York.

M. Spivak (1979) A Comprehensive Introduction to Differential Geometry, Second Edition. Five volumes. Publish or Perish, Berkeley.

J.D. Stasheff (1978) "Continuous cohomology of groups and classifying spaces" Bull. Amer. Math. Soc. **84**, 513-530.

N. Teleman (1972) "A characteristic ring of a Lie algebra extension" Atti. Accad. Naz. Lincei. Rend. Cl. Sci. Fis. Mat. Natur. (8) **52**, 498-506 and 708-711.

S.P. Tsarev (1983) "Which 2-forms are locally curvature forms?" Functional Anal. Appl. **16** (1982), no. 3, 235-237.

W.T. van Est (1953) "Group cohomology and Lie algebra cohomology in Lie groups I,II" Nederl. Akad. Wetensch. Proc. Ser. A. **56**, 484-504.

W.T. van Est (1955a) "On the algebraic cohomology concepts in Lie groups I,II" Nederl. Akad. Wetensch. Proc. Ser. A. **58**, 225-233, 286-294.

W.T. van Est (1955b) "Une application d'une méthode de Cartan-Leray" Nederl. Akad. Wetensch. Proc. Ser. A. **58**, 542-544.

W.T. van Est (1962) "Local and global groups I,II" Nederl. Akad. Wetensch. Proc. Ser. A. **65**, 391-425.

V.S. Varadarajan (1974) Lie groups, Lie algebras, and their representations. Prentice-Hall, Inc. Englewood Cliffs, N.J.

P. ver Eecke (1981) Calculus of jets and higher-order connections. Translated, edited and revised by J.J. Cross and F.R. Smith. Second Edition, Univ. of Melbourne, Mathematics Research Report.

J. Virsik (1969) "A generalized point of view to higher-order connections on fibre bundles" Czech. Math. J. 19(94), 110-142.

J. Virsik (1971) "On the holonomity of higher-order connections" Cahiers de Topol. et Géom. Diff. XII, 197-212.

F.W. Warner (1971) Foundations of differentiable manifolds and Lie groups. Scott, Foresman and Company, Glenview, Illinois.

A. Weil (1958) Variétés Kählériennes. Hermann, Paris.

INDEX

Printed in the United States
By Bookmasters